Lecture Notes in Mathematics

Edited by A. Dold and B. Eckmann

Subseries: USSR
Adviser: L. D. Faddeev, Leningrad

1108

Global Analysis – Studies and Applications I

Edited by Yu. G. Borisovich and Yu. E. Gliklikh

Springer-Verlag
Berlin Heidelberg New York Tokyo 1984

Editors

Yuriĭ G. Borisovich
Yuriĭ E. Gliklikh
Department of Mathematics, Voronezh State University
394693, Voronezh, USSR

Consulting Editor

A. M. Vershik
Department of Mathematics and Mechanics, Leningrad State University
Petrodvorets, 198904, Leningrad, USSR

AMS Subject Classification (1980): 58-02; 58 B, 58 D, 58 C, 58 F, 58 G

ISBN 3-540-13910-9 Springer-Verlag Berlin Heidelberg New York Tokyo
ISBN 0-387-13910-9 Springer-Verlag New York Heidelberg Berlin Tokyo

Printing and binding: Beltz Offsetdruck, Hemsbach / Bergstr.
2146/3140-543210

P R E F A C E

This volume of Lecture Notes in Mathematics presents to English speaking readers the Voronezh University Press series "Novoe v global'nom analize" in which we are involved as editor-in-chief and deputy editor-in-chief respectively (the title may be translated as "New Developments in Global Analysis"). The main purpose of the series is to publish survey (expository) papers, or papers giving a detailed account of important results in Global Analysis and its applications. Among the members of the editorial board there are well-known such as A.T.Fomenko, A.S.Mishchenko, S.P.Novikov, M.M.Postnikov, A.M.Vershik and others.

Each issue of the series has a special title describing its scope and besides the main articles a small number of short communications are included. The present volume comprises the main articles from three issues already published in Russian: Equations on Manifolds (1982), Topological and Geometrical Methods in Mathematical Physics (1983), Geometry and Topology in Global Nonlinear Problems (1984). The articles of V.L.Golo (1983) and A.A.Kirillov (1984) are omitted here. Their material has been published in English in other papers of the authors.

The main articles are selected and commissioned by the editorial board. For this work we are indebted to all the members, especially to academician S.P.Novikov who was editor-in-chief of the second issue (1983).

We hope that this volume will be useful for mathematicians interested in Global Analysis and its applications.

The articles are ordered chronologically. The year of original publication in Russian has been indicated in the Contents.

Yu.G.Borisovich
Yu.E.Gliklikh

Voronezh, May 1984

CONTENTS

1982

1983

1984

TOPOLOGICAL THEORY OF FIXED POINTS ON INFINITE-DIMENSIONAL MANIFOLDS

Yu.G.Borisovich and Yu.E.Gliklikh
Department of Mathematics
Voronezh State University
394693 Voronezh, USSR

The solvability theory of operator equations in infinite-dimensional spaces (i.e. of equations with completely continuous, monotone, condensing, Fredholm, and other operators) has found wide use in the last few decades. Topological characteristics have been constructed which generalize to some extent the concept of Leray-Schauder degree to more general classes of operators and enable one to conclude that solutions exist for a wide class of nonlinear problems (see, for example, [9]).

It must be noted that, besides the theory of Fredholm mappings (where fixed points problems are not solved), equations in vector spaces are considered in the aforementioned studies. As to topological invariants associated with the existence of fixed points, they are constructed using the linearity of space.

In the past years the authors have studied a number of problems where the existence of a solution is related to that of the fixed points of mappings acting in nonlinear spaces. At the time of undertaking these studies we had known only about the work of F.E. Browder [17] who by homological methods has constructed the Lefschetz number for compact mappings of Banach manifolds which can be imbedded in a Banach space as a neighbourhood retract. However, the requirement that the image of the whole manifold be compact is very restricted and is generally not fulfilled in applications. This inspired the authors to undertake further studies of the problem.

With this in view we worked out a general method of using an auxiliary vector field in imbodying space, and succeeded in constructing Lefschetz's number and a certain homotopic characteristic in analogy with index which enable fixed points to be found in the manifold for much wider classes of mappings: locally compact, condensing, and others.

A survey of the studies on fixed points theorem on Banach manifolds, and several applications of the theorem are available in [14] . A part of this is covered in [9] .

The present work is the continuation of the aforementioned studies. Since the publication of [14] , we have succeeded in extending the theory to new classes of mappings and in finding its other applications. Besides, the presentation will make this work understandable even to non-specialists.

It has six sections. In Section 1 the fixed points theorem of mappings in nonlinear spaces is discussed in the general form. The differences between the methods of constructing topological invariants in finite- and infinite-dimensional situations are also considered.

The theory of rotation of vector field with various classes of operators in infinite-dimensional linear spaces is briefly reviewed in Section 2. It is difficult to immediately generalize this theory to the mappings of nonlinear spaces, but a study of the latter is finally reduced to the theory of vector fields.

The detailed construction of Lefschetz's number is described in Section 3 by going over to an auxiliary vector field in imbodying space. This is illustrated by an example of locally compact mappings. It is shown that the construction of the invariant is independent of the choice of the space where a manifold is imbedded, of inclusion, retraction, etc. [11] .

Using the same technique of going over to an auxiliary vector field, one succeeds in defining a certain analog of rotation of vector field, i.e. a homotopic characteristic equal to the algebraic number of fixed points inside the domain of the manifold. This is described in Section 4 [12] .

The Lefschetz number and the homotopic characteristic are constructed for condensing and locally condensing mappings in Section 5 [20, 21]. Here it must be noted that at present it is difficult to construct a general theory of condensing mappings in nonlinear spaces, because the definition of the abstract measure of non-compactness is based on the concept of convex closure [34] . We have used the metric determination of the measure of non-compactness, and considered the condensing mappings of Finsler manifolds which can be isometrically imbedded into some Banach space as a neighbourhood retract.

Weakly compact and close to them mappings are considered in Section 6. In view of the specific character of the intake of weak topology on

Banach manifolds [24, 31] , use has here been made of rotation of weakly compact vector fields, which is constructed in [1, 2] .

Some of the applications are briefly described at the end of every section.

1. Discussion of the Problem

Let us discuss the specific character of the fixed point theorem on a linear space and some methods of solving this theorem in nonlinear finite- and infinite-dimensional cases.

Let a continuous mapping $F:R^n \longrightarrow R^n$ be given. Obviously, point $X \in R^n$ is a fixed point of F if and only if $FX = X$, where X is a zero of mapping (of the vector field) $\Phi = I - F$, i.e. $\Phi X = 0$.

The zeros of the vector field Φ are found by using a topological invariant, i.e. rotation of the vector field (see, for example, [26, 28]) which, in particular, can be determined by the theory of obstructions.

Suppose that S^{n-1} is the boundary of ball $B^n \subset R^n$ and $\Phi X \neq 0$ on S^{n-1}. Thus, $\Phi : S^{n-1} \longrightarrow R^n \setminus 0$. The homotopic class of this mapping - an element of the homotopic group $\pi_{n-1}(R^n \setminus 0) = \mathbb{Z}$ ($\mathbb{Z}$ is a group of integers) - is an integer valued characteristic which is known as rotation $\gamma(\Phi, S^{n-1})$ of the vector field Φ on S^{n-1}.

Obviously, when $\gamma(\Phi, S^{n-1}) \neq 0$, in B^n there exists a zero of the field Φ (fixed point of mapping F); for the vector fields Φ_0 and Φ_1 , $\gamma(\Phi_0, S^{n-1}) = \gamma(\Phi_1, S^{n-1})$ if and only if Φ_0 and Φ_1 belong to one homotopic class of the mapping S^{n-1} in $R^n \setminus 0$, i.e. there exists a homotopy that joins Φ_0 with Φ_1 and has no zeros on S^{n-1}.

If X_0 is an isolated zero of the field Φ , then the rotation $\gamma(\Phi, S^{n-1})$ on the boundary of S^{n-1} of a sufficiently small ball containing X_0 is independent of S^{n-1}, and is known as index $\gamma(X_0)$ of the point X_0.

Note that rotation additively depends on the domain; it can be defined on the boundary of an arbitrary restricted open domain R^n.

The construction described can be readily generalized to the case of sections of vector bundles. Let $E \longrightarrow M^n$ be a vector bundle. Let S^{n-1}

be the boundary of the domain $B^n \subset M^n$, homotopic to a ball, and let $\Phi : S^{n-1} \longrightarrow E$ be the section of bundle E, and $\Phi \chi \neq 0$. Since B^n is contractible, the bundle E over B^n is trivial [36] , i.e. Φ can be taken as a mapping from S^{n-1} in $F^k \setminus 0$, where F^k stands for the vector space - standard fibre of bundle E. The homotopic class $\Phi : S^{n-1} \longrightarrow F^k \setminus 0$ (an element of group $\pi_{n-1}(F^k \setminus 0)$ is an obstruction to extending Φ from S^{n-1} onto B^n without zeros and possesses the properties of rotation of vector field.

If E is a tangent bundle of TM^n, then the sections are vector fields on M^n, and the obstruction constructed is a direct generalization of rotation of the vector field. In this case, the rotation can be defined in a much simple manner [16] .

It is clear that on a nonlinear manifold M^n the problems concerning fixed point mapping of $f: M^n \longrightarrow M^n$ and the zero of vector field are not equivalent. However, with the theory of obstructions one succeeds in constructing a homotopic characteristic equal to the algebraic number (to the sum of indices) of fixed points inside the domain of the manifold [16] ; (also see [37]).

Let M^n be a finite-dimensional manifold and $\bar{B}^n$ be a closed domain in M^n, homomorphic to a ball of space R^n. Let S^{n-1} be the boundary of $\bar{B}^n$. Suppose that $f : \bar{B}^n \longrightarrow M^n$ is a continuous mapping that does not have fixed points on S^{n-1}. We shall now consider a bundle

$$(\bar{B}^n \times M^n) \setminus \Delta \longrightarrow \bar{B}^n$$

where Δ is a diagonal in $\bar{B}^n \times M^n$. This bundle is locally trivial [35, p 377] and is therefore trivial because B^n is contractible [36] . In other words, there exists a homeomorphism

$$\varphi : (\bar{B}^n \times M^n) \setminus \Delta \longrightarrow \bar{B}^n \times (M^n \setminus pt)$$

(pt is a point of M^n) which is obviously extended up to the homeomorphism on the entire $\bar{B}^n \times M^n$.

As per the condition, f gives a mapping of the pair $(\bar{B}^n, S^{n-1})$ into the pair $(\bar{B}^n \times M^n, (\bar{B}^n \times M^n) \setminus \Delta)$. Denote by p the natural projection of the pair $(\bar{B}^n \times M^n, (\bar{B}^n \times M^n) \setminus \Delta)$ on the pair $(M^n, M^n \setminus pt)$. Superposition of $p \circ \varphi \circ f : (\bar{B}^n, S^{n-1}) \longrightarrow (M^n, M^n \setminus pt)$ determines the element $\delta(f, \bar{B}^n)$ of the relative homotopic group $\pi_n(M^n, M^n \setminus pt) = \mathbb{Z}$.

<u>Definition 1</u>. The number $\delta(f, \bar{B}^n)$ is called the homotopic character-

istic of mapping f on B^n.

Note that, unlike the rotation of vector field, the homotropic characteristic is determined from mapping f on the closure of $\bar{B}^n$, but not on the boundary only.

From the definition it immediately follows that, for $\delta(f, \bar{B}^n) \neq 0$, there is a fixed point within B^n, and that $\delta(f_o, \bar{B}^n) = \delta(f_1, \bar{B}^n)$ if and only if f_o is homotopically equivalent to f_1 without fixed points on S^{n-1}.

A similar construction can be given also for more complex domains B^n [16] if the bundle $(\bar{B}^n \times M^n)\setminus\Delta \longrightarrow \bar{B}^n$ is trivial. This requirement is always fulfilled if, for example, M^n is a Lie's group.

If M is a compact manifold, then the Lefschetz number Λ_f is correctly defined as alternated sum of traces of mappings induced by continuous mapping $f: M \longrightarrow M$ in homology groups of the manifold. Here we shall not describe the construction of the Lefschetz number, but shall refer the reader to textbooks or monographs on algebraic topology (see, for example, [10]). We shall list only its properties which may be required in further discussion: if $\Lambda_f \neq 0$, then in M there exists a fixed point of mapping f ; Λ_f equal to the algebraic number of fixed points of f on M; if f_o is homotopically equivalent to f_1, then $\Lambda_{f_o} = \Lambda_{f_1}$.

Note that the Lefschetz number can be constructed for spaces which are more general than a manifold. For example, the Lefschetz number is correctly determined for continuous mappings of compact absolute neighbourhood retracts (ANR).

We shall now discuss the possibility of constructing topological invariants for finding fixed points on infinite-dimensional manifolds.

It is quite natural that for certain classes of mappings of infinite-dimensional manifolds the homotopic characteristic or the Lefschetz number can be determined by constructing finite-dimensional approximations, using finite-dimensional invariants, and by subsequent limit transition. In linear spaces this technique leads, for example, to the concept of Leray-Schauder degree. However, in considering infinite-dimensional nonlinear manifolds one encounters difficulties in constructing finite-dimensional approximations and as well as in proving that a homotopic characteristic is independent of the choice of approximations. This appreciably narrows the number of considered classes of mappings and manifolds.

F.E. Browder [17] has used direct methods of algebraic topology for determining Lefschetz's number in an infinite-dimensional case. Also he has shown that for a compact (i.e. the image of the whole manifold is compact) mapping of a topological space, which can be imbedded into some Banach space as a neighbourhood retract, the Lefschetz number is correctly determined by a general formula; of course, the summation is taken over an infinite number of indices. The invariant so constructed has all common properties of the Lefschetz number : it is retained when there is a homotropy in the class of compact mappings; from its difference from zero follows the existence of a fixed point, etc.

Also there are direct constructions of the Lefschetz number for some classes of local-compact mappings [38, 39] .

In a linear space for widening a class of operators that admit topological invariants it is advantageous to make use of contraction on an invariant subset having "good" properties. Let us now formulate some general principles of existence of an invariant subset [15] .

Let X be a topological space. Denote by 2^X the totality of all the closed subsets of X with exponential topology. If X is a metric space, then the Hausdorff metric $\rho(A,B)$ can be considered in 2^X.

Let E be a linear real topological space with an isolated cone K of nonnegative elements [27] .

A collection $\{\Phi_\lambda\}$ of closed sets $\Phi_\lambda \in 2^X$ is called the Φ - system, provided the following conditions are satisfied:

(a) non-empty intersection of any collection of sets Φ_λ belongs to the Φ -system;

(b) for any $A \in 2^X$ one finds $\Phi_\lambda \supset A$.

The mapping $\chi : 2^X \longrightarrow K$ is called the distinguishing mapping, if it satisfies the following axioms:

1. $\chi(A) \leqslant \chi(B)$, provided $A \subset B$;
2. $\chi(A \cup R) \leqslant \chi(A)$ if $\chi(R) = 0$;
3. for any $A \in 2^X$, $\chi\left(\bigcap_{\Phi_\lambda \supset A} \Phi_\lambda\right) = \chi(A)$;
4. for every point $x \in X$ $\chi(x) = 0$.

From (1) and (2) it follows that

2'. $\chi(A \cup R) = \chi(A)$ if $\chi(R) = 0$.

Let us denote by ker χ the collection of sets $A \in 2^X$ such that $\chi(A) = 0$ (a kernel of the distinguishing mapping χ).

Suppose that $\Omega \subset X$ is an open set. We call the mapping $F: \bar{\Omega} \longrightarrow X$ to be compatible with the distinguishing mapping χ, if

1. a family of ker χ is invariant over F, i.e. $\chi(\overline{FA}) = 0$, provided $\chi(A) = 0$ for the set $A \subset \Omega$;

2. for any $A \subset \Omega$ $\chi(\overline{FA}) \not\geq \chi(A)$, if $\chi(A) \neq 0$.

<u>Theorem 1</u>. [15] Every compatible with χ mapping $F: \bar{\Omega} \longrightarrow X$ has a set $\Phi \in \{\Phi_\lambda\}$ such that

1. $F(\Phi \cap \Omega) \subset \Phi$.
2. $\Phi \in$ ker χ.
3. $\Phi \supset R$; here, R is the given set from ker χ.

A set of fixed points of mapping F belong to ker χ. Thus, according to Theorem 1, a set Φ can be constructed that would contain all fixed points of F.

In a nonlinear infinite-dimensional case, theoretically it is possible to determine, using Theorem 1, topological invariants for finding the fixed points. If one succeeds, say, in constructing a Φ-system and a distinguishing mapping χ on a nonlinear manifold X such that ker χ consists of compact ANR, then for a compatible with χ mapping $f: X \longrightarrow X$ the Lefschetz number can be defined as Lefschetz number of restriction on an invariant compact ANR containing all the fixed points of f.

However, the construction of suitable Φ-systems in nonlinear spaces is an important problem that has yet to see its solution.

2. Rotation of Vector Field in Infinite-Dimensional Linear Spaces

Here we shall briefly review the theory of rotation of compact and close to them vector fields in infinite-dimensional linear spaces. The methods of this theory are not directly extended to mappings in non-linear spaces, but their study will finally lead to the theory of vector fields.

Let E be a Banach space, $\Omega \subset E$ be an open set, and let $F:\bar{\Omega} \longrightarrow E$ be a continuous mapping with compact image. Such mappings and corresponding vector fields $\Phi = I - F$ will be called completely continuous mappings.

Consider a finite ε-net of the set $F(\bar{\Omega})$ and its linear closure $\tilde{E}$. There exists a natural ε-shift of compact set $F(\bar{\Omega})$ into E, which is known as Schauder's projector. Let us denote it by P. The finite-dimensional approximation $PF:\bar{\Omega} \longrightarrow \tilde{E}$ of operator F determines the finite-dimensional vector field $\tilde{\Phi} = I - PF : \bar{\Omega} \cap \tilde{E} \longrightarrow \tilde{E}$. If $\Phi : \dot{\Omega} \longrightarrow E \setminus 0$, then $\tilde{\Phi}: (\Omega \cap \tilde{E})^{\bullet} \longrightarrow \tilde{E} \setminus 0$ for a sufficiently small ε. Rotation $\gamma(\tilde{\Phi}, (\Omega \cap \tilde{E})^{\bullet})$ (see Section 1) is called (according to M.A. Krasnosel'sky) the rotation of a completely continuous vector field Φ, and is denoted by $\gamma(\Phi, \dot{\Omega})$. The reader interested in detailed description of the construction and proof of the validity of definition is referred, for example, to [26, 28].

The rotation $\gamma(\Phi, \dot{\Omega})$ is the homotopic characteristic of vector field Φ, because for a completely continuous homotopy without zeros on $\dot{\Omega}$ the rotation is conserved. If $\gamma(\Phi, \dot{\Omega}) \neq 0$, then in Ω there exists a fixed point of the operator F.

It can be shown that from the topological viewpoint the rotation of M.A. Krasnosel'sky's vector field is equivalent to Leray-Schauder degree.

This construction applies also to locally convex spaces [30].

The concept of relative rotation of a completely continuous vector field has also proved to be of much use in modern topological theory of nonlinear mappings [3, 7, 8]. Let $L \subset E$ be a closed convex set, $L \cap \Omega \neq \emptyset$, and let $F : \Omega \cap L \longrightarrow L$ be a completely continuous mapping such that $\Phi = I - F : (\Omega \cap L)^{\bullet} \longrightarrow E \setminus 0$, where $(\Omega \cap L)^{\bullet}$ is the relative boundary in L. The relative rotation $\gamma(\Phi, (\Omega \cap L)^{\bullet}, L)$ of vector field Φ with respect to subspace L can be determined by the method of finite-dimensional approximations or with complete index of solutions in the sense of Leray [29]. It has the same properties as the "absolute" rotation.

Relative rotation has proved to be very convenient, in particular, in the theory of weakly compact vector fields [1, 2].

As it is known, a Banach space with weak topology, defined by a total set of functionals from an adjoint space, is a locally convex space, i.e. in this case use can be made of the theory of rotation of vector field in a locally convex space. However, the closures of weakly open sets in a Banach space are very "large", and the requirement that the image of such a set be weakly compact is restrictive.

Let E be a Banach space and let B be a convex closed in weak topology set. B is called to be almost bounded if for an open set V the intersection of $B \cap V$ is bounded by norm in E.

Let Ω be a bounded open in weak topology set in B (for instance, $B \cap V$ can be considered as Ω). Suppose that on a weak closure of Ω in B is defined a weakly continuous operator F with values in B such that weak closure of image $F\Omega$ is weakly compact, and there are no fixed points on $(B \cap \Omega)^{\bullet}$. In this case, we can consider relative rotation of a weakly compact vector field $I - F$ on $(B \cap \Omega)^{\bullet}$; the so constructed theory of rotation of weakly compact fields proved useful in studying a number of problems related to the theory of differential equations, and others.

If F maps weakly convergent sequences into strongly convergent sequences, then the rotation of the field $I-F$ can be considered to be completely continuous and weakly compact. It has been shown that these rotations coincide. Perturbations of such a field caused by a weakly compact field have also been studied [4, 5] .

If as B we take a cone or a wedge in E, then it is possible to obtain a statement for the existence of positive and semi-positive solutions [2, 4, 6] . Consideration of operator equations in direct sum of spaces with a cone (wedge) obtained as direct sum of corresponding cones (wedges) proved to be of significance for various applications. Moreover, we can consider its weak or strong topology in each of the direct factors.

Transition to non-compact maps F can be effected in two ways : by constructing a "good" spectrum of finite-dimensional approximations or by finding a "good" invariant subspace. Both these techniques are used in modern studies (monotone and condensing mappings).

For condensing mappings, preference is given to the second technique. We shall describe in brief how Theorem 1 of Section 1 enables us to introduce the concept of rotation of vector field with a condensing operator F.

Let X be a Banach space, Φ_λ be a convex closed set, and $\ker \chi$ be compact sets. Then χ is a measure of noncompactness according to the definition given in [34]. The mapping compatible with the measure of noncompactness are called the condensing mappings.

Suppose that $\bar{\Omega}$ is a closure of an open set in X. For a condensing mapping $F : \bar{\Omega} \longrightarrow X$ there exists a compact convex set Φ such that $\bar{\Omega} \cap \Phi \neq \emptyset$, $\Omega \cap \Phi$ contains all fixed points of the operator F and $F(\Omega \cap \Phi) \subset \Phi$. Now the rotation of the vector field I-F on $\dot{\Omega}$ [34] can be determined as relative rotation. Its value is independent of the choice of the set Φ, provided Φ is sufficiently "large" [33].

Yu.I. Sapronov has proved the existence of a fundamental set Φ(def.in [25]). As, according to Dugunji's theorem, a convex set is a retract of Banach space, the mapping F on Ω can be changed to a homotopic mapping with values in Φ. This is a condensing homotopy and, as Φ is a fundamental set, has no fixed points on the boundary of $\dot{\Omega}$. Thus, every homotopic class of condensing mappings contains a completely continuous mapping. This construction can be regarded as a simple method of introducing rotation. Details of the construction are available in [32].

The rotation of condensing vector fields in a locally convex space has now been constructed [34].

3. Locally Compact Mappings. The Lefschetz Number

In this section we shall describe in detail the construction of Lefschetz's number, based on the use of an auxiliary vector field. We shall illustrate this construction by an example of locally compact mappings having a compact iteration [11] (other locally compact mappings are also considered).

Let X be a topological space that can be imbedded into a Banach space as a neighbourhood retract. Let $f:X \longrightarrow X$ be a continuous locally compact mapping, $f^n(X)$ being a compact mapping for certain $n \geqslant 1$. Imbed X into a Banach space E and consider $r:U \longrightarrow X$ a retraction of an open neighbourhood U of the set X in E onto X. Now by formula $F(u) = f \circ r\,(u)$ we shall determine the mapping $F:U \longrightarrow U$. Hence, F is continuous and locally compact and $F^n(U) \subset X$ is a compact set. Note that all the fixed points of F lie in $F^n(U)$.

From the compactness of $F^n(U)$ it follows that a finite system of open sets Ω_α exists in U such that $\Omega = \bigcup_\alpha \Omega_\alpha \supset F^n(U)$ and $F(\Omega_\alpha)$

is a compact set for every α . Therefore, $F(\overline{\Omega})$ is also a compact set. Note that there are no fixed points of the mapping F on the boundary $\dot{\Omega}$.

Let us now consider rotation $\gamma(I-F, \dot{\Omega})$ of a completely continuous vector field $I-F$ on the boundary $\dot{\Omega}$.

Definition 2. $\gamma(I-F, \dot{\Omega})$ is called the Lefschetz number of f on X.

We shall prove the validity of this definition.

Lemma 1. Λ_f is independent of the choice of U and r.

Proof. Suppose that U_1 exists in E and U_2 is a neighbourhood of X, and $r_1: U_1 \longrightarrow X$ and $r_2: U_2 \longrightarrow X$ are retractions. Let $U = U_1 \cap U_2$ and Ω_ε be an ε-neighbourhood of $f^n(X)$ in E, $U \supset \Omega_\varepsilon$ such that $F_1(\Omega_\varepsilon) = f \circ r_1(\Omega_\varepsilon)$ and $F_2(\Omega_\varepsilon) = f \circ r_2(\Omega_\varepsilon)$ are compact sets. There exists a number $\delta > 0$ such that for $u \in \Omega_\delta$, $\|F_1(u) - F_2(u)\| < \varepsilon$, i.e. a straight segment γ that joins $F_1(u)$ with $F_2(u)$ belongs to Ω_ε. Retracting γ on X we obtain a path in X which joins $F_1(u)$ with $F_2(u)$. It is easy to see that the homotopy obtained is compact on $\Omega_\delta \times [0,1]$ and has no fixed points on the boundary $\dot{\Omega}_\delta$. Therefore, the rotation of fields $I-F_1$ and $I-F_2$ on $\dot{\Omega}$ are equal, and this proves the lemma.

Lemma 2. Λ_f is independent of the choice of E and the imbedding.

Proof. Let E_1, E_2 be Banach spaces, $i_1: X \longrightarrow E_1$, $i_2: X \longrightarrow E_2$ be the imbeddings, and $r_1: U_1 \longrightarrow i_1X$ and $r_2: U_2 \longrightarrow i_2X$ be retractions, on X, of corresponding neighbourhoods. Let $F_1 = f \circ r_1$ and $F_2 = f \circ r_2$. Let us take a Banach space $E = E_1 \oplus E_2$ with $\|e\| = \|p_1 e\|_1 + \|p_2 e\|_2$ (p_i is a natural projection of E on E_i) and natural injection of X into E $i = (i_1, i_2)$. We shall consider a set $U = U_1 \oplus U_2$. Obviously, iX is an r-retract of U, where $r = (r_1, r_2)$. On U is defined the mapping $F: U \longrightarrow iX$ induced by retraction r. Let Ω_1 , Ω_2 be those neighbourhoods of the set $f^n(X)$ in U_1 and U_2 which determine the Lefschetz numbers Λ_{1f} and Λ_{2f} as rotations $\gamma(I - F_1, \dot{\Omega}_1)$ and $\gamma(I-F_2, \dot{\Omega})$. It is sufficient to prove that these numbers coincide. Further we shall show that each of them coincides with $\gamma(I-F, \Omega)$, where $\Omega = \Omega_1 \oplus \Omega_2$. With this in view, we go over to a homeomorphic to U set $\varphi(U)$, where $\varphi(u) = u - i_2 \circ i^{-1} \circ r(u)$. The image $\varphi(U)$ is retracted onto $i_1(X)$ with the same fibres as U onto iX. Note that $\varphi(U) \cap E_1 = U_1$ and $\varphi(\Omega) \cap E_1 = \Omega_1$. Mapping φ induces a mapping $F^*: \varphi(U) \longrightarrow \varphi(U)$ by the rule

$F^*(u - i_2 \circ i^{-1} \circ r(u)) = F(u) - i_2 \circ i^{-1} \circ F(u)$. We have $\gamma(I-F, \dot{\Omega}) =$
$= \gamma(I-F^*, \varphi(\dot{\Omega}))$. As the image F^* lies in i_1X, $\gamma(I-F, \varphi(\dot{\Omega}))=$
$= \gamma(I-F^*, \dot{\Omega}_1)$ according to rotations product theorem [26, p 134] . In view of the fact that F^* on $\bar{\Omega}_1$ coincides with F_1, $(I-F^*, \varphi(\dot{\Omega}))=$
$= \gamma(I-F_1, \dot{\Omega}_1)$. Hence $\gamma(I-F, \dot{\Omega}) = (I-F_1, \dot{\Omega}_1)$. A similar result is obtained for $\gamma(I-F_2, \dot{\Omega}_2)$, which proves the lemma.

Assume f has only isolated fixed points on X. By virtue of compactness of f^n , the fixed points are finite in number. From the general properties of rotation follows the following statement.

Theorem 2. Suppose that f has only isolated fixed points on X. Let j_i be the index of mapping F in a neighbourhood of a fixed point x_i. Then $\Lambda_f = \sum_i j_i$.

If X is a compact manifold, then the classical Lefschetz number Λ^*_f is defined.

Theorem 3. Let X be a compact smooth manifold. Then $\Lambda_f = \Lambda^*_f$.

Proof. The mapping $f: X \longrightarrow X$ can be approximated by a homotopic to it smooth mapping having isolated fixed points. At this homotopy, Λ and Λ^* are obviously conserved, and it may be assumed, without any restriction of generality, that f is a mapping with isolated fixed points. In this case, Λ^*_f coincides with the sum of indices of fixed points x_i with respect to X. On the other hand, for computing Λ_f which is equal to the sum of indices j_i of mapping F at points x_i, we make use of the rotations product theorem and restrict ourselves to the maps of the manifold X that contain points x_i. We have $\Lambda_f = \Lambda^*_f$.

We shall now study the properties of the Lefschetz number.

Definition 3. We shall say that f_0 is homotopic to $f_1 (f_0 \sim f_1)$, if there exists a continuous and locally compact mapping $\varphi : X \times [0,1] \to X$ having a compact iteration φ^l, $l \geqslant \max(n_0, n_1)$ such that

$$\varphi|_{X \times \{0\}} \equiv f_0 \qquad \varphi|_{X \times \{1\}} \equiv f_1 .$$

Theorem 4. If $f_0 \sim f_1$, then $\Lambda_{f_0} = \Lambda_{f_1}$.

Proof. Let us consider a compact set $\varphi^l(X \times [0,1])$ and cover it with a finite family of neighbourhoods Ω_α such that the image $\varphi(\Omega_\alpha \times [0,1])$ is compact for any α . Let Ω be a union of these neighbourhoods. Then $\Phi = \varphi \circ r: r^{-1}(\Omega \times [0,1]) \longrightarrow r^{-1} X$ is a homotopy of F_0 and F_1 without fixed points on the boundary $\dot{\Omega}$.

Now from the rotation property it follows that $\Lambda_{f_0} = \Lambda_{f_1}$.

Theorem 5. If $\Lambda_f \neq 0$, then f has a fixed point in X.

Proof. As $\gamma(I-F, \dot{\Omega}) \neq 0$, there exists a fixed point x^* of mapping F in Ω. Then, according to construction, $x^* \in X$ and $f(x^*) = x^*$.

Theorem 6. If X is contractible, then f has a fixed point in X.

Proof. Let φ_t be a contracting homotopy, i.e. $\varphi_t(X)$ is continuous, $\varphi_0 = id$, and $\varphi_1(X) = x^* \in X$. The image of the product $X \times [0,1]$, under superposition of $\varphi_t \circ f^n$, is compact. Let Ω be a neighbourhood of this image such that $f(\Omega)$ is compact. It is understood that on $\dot{\Omega}$ the homotopy $\varphi_t \circ f$ has no fixed points. It is easy to see that the index of the only fixed point x^* of mapping $\varphi_1 \circ f$ equals 1. Therefore, the rotation of the vector field $\gamma(I-F, \dot{\Omega})$ is also equal to 1. By the definition of Λ_f and by Theorem 5, f has a fixed point.

Theorem 7. Assume in addition that X is a smooth Banach manifold. Suppose that the image $f: X \longrightarrow X$ belongs to a closed submanifold $M \subset X$. Then $\Lambda_f = \Lambda_{f|M}$ ($f|M$ is the restriction of mapping f onto M).

Proof. Let $\hat{U}$ be a tubular neighbourhood of M in X. It is known that a homotopy $r_t: \hat{U} \times [0,1] \rightarrow \hat{U}$ exists such that $r_0 = id$ on $\hat{U}$, and $r_1: \hat{U} \longrightarrow M$ is a retraction. Let $r_X: U_X \longrightarrow X$ be a retraction on X of a neighbourhood in E. We shall consider a neighbourhood $U_M = r_X^{-1} \hat{U}$ which is retractable on M by retraction $r_M = r_1 \circ r_X$. It is easy to see that the mappings $F_X = f \circ r_X$ and $F_M = f \circ r_M$ are homotopic on a certain Ω, $f^n(X) \subset \Omega \subset U_M$ with a compact homotopy $f \circ r_t \circ r_X$, without fixed points, on $\dot{\Omega}$.

Note that the construction of this section is trivially generalized to the following case: X may be imbedded as a neighbourhood retract into a normed space or a locally convex space; $f: X \longrightarrow X$ is continuous, locally compact and has a compact set of fixed points (for example, f has a compact attractor or is asymtotically compact, cf. [38, 39]). We restricted ourselves to a detailed analysis of the situation which is most often encountered in applications.

Here we shall discuss an example of the applications [22] .

Let M be a compact Riemannian manifold, V(t,m) and W(t,m) be continuous, nonautonomous vector fields on M, periodic in t with a period $\omega > 0$. We shall now consider a differential equation of the type

$$\dot{\gamma}(t) = V(t, \gamma(t)) + \| W(t-h, \gamma(t-h)) \qquad (*)$$

where $0 < h \leqslant \omega$, symbol $\|$ means Riemannian parallelism along the solution of the equation. Such equations appear in geometric description of some complex mechanical systems.

Since parallel translation has been determined only along C^1 smooth curves, the right-hand side of (*) has a meaning only if γ is C^1 smooth curve.

If V(t,m) is smooth in m, then in a Banach manifold $C^1([0,\omega], M)$ of C^1-smooth mappings of interval $[0,\omega]$ into M the action of the shift operator $u_1(\omega)$ along trajectories of (*) is correctly defined. It has been shown that $u_1(\omega)$ is continuous and locally compact, $(u_1(\omega))^2 C^1([0,\omega], M)$ is compact in $C^1([0,\omega], M)$, and the manifold $C^1([0,\omega], M)$ itself can be imbedded into a Banach space as a neighbourhood retract. Thus, $\Lambda_{u_1(\omega)}$ is correctly defined. Using certain homotopies and a restriction onto a finite-dimensional manifold homeomorphic to M, one succeeds in showing that $\Lambda_{u_1(\omega)} = \chi(M)$, i.e. is equal to Euler's characteristic of M. Hence, if $\chi(M) \neq 0$, then there exists a fixed point $u_1(\omega)$, being the periodic solution of equation (*).

For other applications, we refer the reader to [11, 14].

4. Locally Compact Mappings. Homotopic Characteristic

The technique developed in Section 3 makes it possible also to construct an analog of vector field rotation [12]. The construction, as in Section 3, is valid also for much wider classes of mappings, cf [38, 39]. Here, we shall again discuss in detail the most useful case for applications.

Let X be a topological space that can be imbedded into Banach space E as a neighbourhood retract, and $\Theta \subset X$ be an open domain. Let $f: \Theta \longrightarrow X$ be a continuous, locally compact mapping such that f^n is compact for a certain $n \geqslant 1$, and there are no fixed points of f on the boundary of Θ.

Let U be a neighbourhood of X in E and $r: U \longrightarrow X$ be a retraction. Using the equality $F = f \circ r$, we shall define the mapping F, and shall consider a set $f^n(\Theta) \cap \Theta$. This set is compact and, as f is locally compact, there exists a neighbourhood $\Omega \supset f^n(\Theta) \cap \Theta$

in U such that $F(\overline{\Omega})$ is compact. Naturally, there are no fixed points of F on the boundary of Ω. Without any loss of generality, it may be assumed that $r\overline{\Omega} \subset \overline{\Theta}$.

<u>Definition 4</u>. The rotation $\gamma(I-F, \dot{\Omega})$ of vector field I-F on the boundary $\dot{\Omega}$ is called the homotopic characteristic $\delta(f, \overline{\Theta})$ of mapping f on $\overline{\Theta}$. If $f^n(\overline{\Theta}) \cap \overline{\Theta} = \emptyset$, then, by definition, we set $\delta(f, \overline{\Theta}) = 0$.

It is proved, as in the case of Lefschetz number, that $\delta(f, \overline{\Theta})$ is independent of the choice of U and retraction, E, and imbedding.

Of course, if X is a Banach space and f is a completely continuous mapping, then the homotopic characteristic of f on $\overline{\Theta}$ coincides with classical rotation of the vector field $\gamma(I-f, \dot{\Theta})$. This immediately follows from the rotations product theorem [26]. Similarly, if Θ contains only isolated fixed points, then $\delta(f, \overline{\Theta})$ is equal to the sum of their indices. Using this fact, it is not hard to prove that if X is a finite-dimensional manifold, then the characteristic constructed coincides with the homotopic characteristic in the sense of Definition 1 (Section 1).

The natural outcome of the considered case is the following definition of homotopy.

<u>Definition 5</u>. The mappings $f_0 : \overline{\Theta} \longrightarrow X$ and $f_1 : \overline{\Theta} \longrightarrow X$ are homotopic, if they are homotopic in the sense of Definition 3, and for any $t \in [0,1]$, $x \in \dot{\Theta}$, the condition $\varphi(x,t) \neq x$ is fulfilled.

It is not hard to show that from $f_0 \sim f_1$ follows $\delta(f_1, \overline{\Theta}) =$
$= \delta(f_2, \overline{\Theta})$.

Similarly to Theorem 7 is proved

<u>Theorem 8</u>. Additionally suppose that X is a smooth manifold and an image of f belongs to a closed submanifold M. Then $\delta(f, \overline{\Theta}) =$
$= \delta(f|_{\overline{\Theta} \cap M}, \overline{\Theta} \cap M)$.

From the definition it immediately follows that for $\delta(f, \overline{\Theta}) \neq 0$ in Θ there exists a fixed point of f.

As an example we shall consider an operator of integral type S^{ω} whose fixed points are ω-periodic integral curves of the ω-periodic vector field $\xi(t, g)$ on Lie group G [23]. It has been constructed on the basis of integral type operators [18, 19], in whose construction use is made of Riemannian parallel translation.

Operator S^{ω} acts in a Banach manifold $C^1([0,\omega], G)$ of C^1 mappings of interval $[0,\omega]$ into G, and is continuous and locally compact. If V is a bounded open set in G, then on the closure $C^1([0,\omega], V)$ in $C^1([0,\omega], G)$ the second iteration of S^{ω} is compact.

Of course, we can consider a homotopic characteristic of S^{ω} on C^1 neighbourhoods in $C^1([0,\omega], G)$, whose image is compact. However, the use of cl $C^1([0,\omega], V)$ is preferred, because in this case one succeeds in computing the homotopic characteristic. It is shown in [23] that, when certain natural conditions are fulfilled, $\delta(S^{\omega}, \mathrm{cl}\, C^1([0,\omega],V)) = \delta(\xi(0, g), \overline{V})$, i.e. it is an obstruction to the extending of $\xi(0,g)$ from the boundary V onto V without zeros (see Section 1).

Other examples illustrating the use of homotopic characteristic are available, for example, in [12-14] .

5. Condensing and Locally Condensing Mappings

This section is devoted to the construction of Lefschetz's number and the homotopic characteristic for condensing and locally condensing mappings [20, 21]. Such mappings of linear spaces have been studied in detail (see, for instance, the bibliography of [34]). At present it is difficult to extend the general theory of condensing operators to non-linear spaces, since in the definition of the abstract measure of non-compactness use is made of the concept of convex closure of a set. Possibly, we may succeed in defining the analogues of the measures of noncompactness in arbitrary spaces, using the concept of Φ-system and the distinguishing mapping, introduced in [15](see Section 1). We shall make use of the measures of noncompactness defined in metric terms.

It is said that a Finsler metric is given on Banach manifold X, if in every tangent space T_xX a norm is given that turns T_xX into a Banach space. Finsler's manifold X is isometrically imbedded by a smooth imbedding i in a Banach space E, if in every tangent space the norm is a restriction on this space of the norm in E.

Note that the Finsler metric gives on X a metric ρ as infimum of the lengths of curves that join the points, i.e. turns X into a metric space.

Let X be the Finsler manifold that permits isometric imbedding as a neighbourhood retract in a Banach space E.

Definition 6. $\chi(\Omega) = \inf\{\varepsilon;$ set Ω has a finite ε-set in $X\}$ is called the Hausdorff measure of noncompactness of the set $\Omega \subset X$.

Definition 7. $\alpha(\Omega) = \inf\{d:$ set Ω permits its partition into a finite number of subsets with diameters less than $d\}$ is called the Kuratowski measure of noncompactness of the set $\Omega \subset X$.

χ and α have important properties: they equal to zero on compact sets and only on them; from $\Omega_1 \subset \Omega_2$ it follows that $\chi(\Omega_1) \leqslant \chi(\Omega_2)$ and $\alpha(\Omega_1) \leqslant \alpha(\Omega_2)$.

Further in this section, by the measure of noncompactness ψ we shall mean χ or α.

Definition 8. We shall say that a continuous operator $f: X \longrightarrow X$ condenses relative to ψ with a constant $q < 1$, if for any bounded set $\Omega \subset X$ the inequality $\psi(f\Omega) < q\,\psi(\Omega)$ is realized.

Definition 9. A continuous operator $f: X \text{---} X$ is called a locally condensing operator relative to ψ, if for every point $x \in X$ there exists a neighbourhood U such that the inequality $\psi(f\Omega) < \psi(\Omega)$ is fulfilled for any set $\Omega \subset U$.

Suppose that f condenses relative to ψ with a constant $q < 1$, and X has a finite diameter. We shall consider a set $f^{\infty}X = \bigcap_{k=1}^{\infty} f^k X$, where f^k is the k-th iteration of f.

Lemma 3. $f^{\infty}X$ is compact.

Proof. Assume that it is not so. Let $\psi(f^{\infty}X) = a > 0$. Pick up k so that $q^k\psi(X) < a$. We have $\psi(f^kX) < \psi(f^{\infty}X)$, this contradicts the imbedding $f^{\infty}X \subset f^kX$. The lemma is proved.

Note that $f^{\infty}X$ contains all fixed points of f.

We shall imbed X isometrically into a Banach space E. In E, using the natural metric of normed space ρ_E, we can determine the noncompactness measure on E, which corresponds to ψ. Operator f on X may, in general, be not condensing relative to ψ_E. Nonetheless, the following statement holds.

Lemma 4. For every point $x \in X$ there exists a neighbourhood $V_x \subset X$ such that for any set $\Omega \subset V_x$ we have $\psi_E(f\Omega) < \psi_E(\Omega)$, i.e. f locally condenses relative to ψ_E.

This statement follows from the fact that in a small neighbourhood of every point the metrics ρ and ρ_E are close to each other, the difference being infinitely small (of the order higher than unity).

Let $r: U \longrightarrow X$ be a retraction of open neighbourhood U of manifold X in E on X. Using the equality $F = f \circ r$, we determine the operator $F: U \rightarrow U$. Clearly, F is continuous and locally condensing relative to Ψ_E, and the set $f^{\infty}X$ contains all fixed points of F. Arguing in the same manner as in Section 3, we may consider a neighbourhood Ω of the compact set $f^{\infty}X$, where F condenses and, on whose boundary there are no fixed points of F.

Definition 10. The rotation $\gamma(I-F, \dot{\Omega})$ of the condensing vector field I-F on the boundary of Ω is called the Lefschetz number Λ_f of mapping f on manifold X.

Similarly to the proofs of Lemmas 1 and 2, it may be shown that Λ_f is independent of the choice of U and retraction, E and isometric imbedding. Necessary modification of the proof is left to the reader.

From the properties of rotation of condensing vector fields, it follows that, for $\Lambda_f \neq 0$, operator f has a fixed point on X; in addition, if f_0 and f_1 are homotopic in the class of operators that condense with a constant $q < 1$ on X, then $\Lambda_{f_0} = \Lambda_{f_1}$.

Clearly, if f is a compact mapping, then Λ_f coincides with the Lefschetz number constructed in Section 3.

The construction described above enables the Lefschetz number to be defined for locally condensing mappings of manifold X, if their iteration is compact. Indeed, it is necessary to consider the compact set f^nX instead of $f^{\infty}X$. Since f locally condenses, F will be a locally condensing mapping relative to Ψ_E and a condensing mapping on a neighbourhood in E.

It is not difficult, as in Section 4, to define homotopic characteristic for Finsler manifolds mappings that condense on a closure of open domain Θ with a constant $q < 1$, or are locally condensing and have compact iteration on $\overline{\Theta}$. As a simple excercise, we leave this construction to the reader.

The example of a condensing operator is the shift operator along the trajectories of a functional differential equation (FDE) on a smooth manifold M [20]. It acts on a manifold of continuous curves, is continuous and condenses with $q < 1$ relative to the Hausdorff non-compactness measure of the natural Finsler metric. It has been shown that the Lefschetz number equals the Euler characteristic of the manifold M.

An example of a locally condensing operator is described in [21]. Other examples of condensing and locally condensing mappings can be found in [14].

6. Weakly Compact Mappings

In the previous section we have considered the condensing mappings of Finsler manifolds. By introducing other special classes of Banach manifolds one can extend the general method of constructing the Lefschetz number and the homotopic characteristic to new classes of mappings.

Here we shall study the weakly continuous mappings. For this, we shall first describe the concept of bimanifold, i.e. of the Banach manifold on which it is possible to correctly define a weak topology. This theory is discussed in [24, 31].

The main difficulty in introducing a weak topology on an arbitrary Banach manifold is that the map on a manifold, which can be regarded as domain in model Banach space, is not an open set in the weak topology of model space. This difficulty can be overcome by considering the topology of weak convergence, instead of weak topology on a model space, where closed sets are defined as sets containing limits of all weakly convergent sequences of their points.

Let E be a Banach space. On E we shall determine a topology ω in the following manner. The open sets in ω are such subsets of E which intersect with any bounded set $B \subset E$ along open sets in a topology on B induced by weak topology E. Obviously, ω is stronger than the weak topology on E.

<u>Statement</u>. If E is a reflexive space, then the topology ω coincides with the topology of weak convergence.

Other cases, when ω coincides with the topology of weak convergence, are described in [24, 31].

Let E be a reflexive Banach space and let τ be the strong topology (topology of norm) on E.

Definition 11. The bimanifold with a model on E is called the manifold whose atlas satisfies the following properties :

(a) the maps of the atlas are open sets in the topology ω (and, hence, τ) in E;

(b) functions permitting transition from one map to another (transformation of coordinates) are homeomorphic both in the topology τ and ω .

Examples of bimanifolds are open sets in E in the topology ω , Hilbert space spheres, and several manifolds of mappings.

Consider a bimanifold X such that

(a) X may be imbedded in a reflexive Banach space E_1, and imbedding is continuous in the topology ω on X and on E_1;

(b) there is a neighbourhood U in the topology ω of manifold X in E_1, which is retracted onto X by retraction r continuous in the topology ω .

Call the mapping $f:X \longrightarrow X$ the weakly compact mapping if it is continuous in the topology ω and the image $fX \subset X$ is compact in ω .

As before, we shall define the mapping $F:U \longrightarrow U$ using formula $F = for$. Obviously, F is continuous in the topology ω and the image F(U) is compact in ω . By the Eberlein-Shmulyan theorem, F(U) is weakly compact in E_1 and is, therefore, weakly closed and bounded by the norm set in E_1, because E_1 is reflexive.

Let $\overline{B}$ be a ball of space E_1, which contains a bounded set F(U). By the definition of topology ω , we have $\overline{B} \cap U = \overline{B} \cap V$, where V is the weakly open set. It is not hard to see that there exists a weakly open set $V_1 \supset F(U)$ such that a weak closure V_1 is contained in V.

Consider a set $\overline{B} \cap \overline{V}_1 \subset \overline{B} \cap U$. The operator F is defined on $\overline{B} \cap \overline{V}_1$ and, obviously, there are no fixed points F on the relative boundary $(\overline{B} \cap \overline{V}_1)^{\bullet}$. Thus, the relative rotation $\gamma(I-F, (\overline{B} \cap \overline{V}_1)^{\bullet}, \overline{B}))$ can be defined, i.e. use can be made of the construction of rotation of weakly compact vector fields (see [1, 2] and Section 2 of this paper).

Definition 12. The rotation $\gamma(I-F, (\overline{B} \cap \overline{V}_1)^{\bullet}, \overline{B})$ is called the Lefschetz number Λ_f of a weakly compact mapping f on X.

Lemma 5. Λ_f is independent of the choice of ball B, neighbourhood U and retraction r, space E_1 and imbedding.

Proof. The fact that Λ_f is independent of the choice of B follows from the following. If B_1 is another ball containing F(U), then $B_2 = B \cap B_1$ is a convex set and $\gamma(I-F, (\bar{B} \cap \bar{V}_1)^{\bullet}, \bar{B}) = \gamma(I-F, (\bar{B}_2 \cap \bar{V}_1)^{\bullet}, \bar{B}_2)$, $\gamma(I-F, (\bar{B}_1 \cap \bar{V}_1)^{\bullet}, \bar{B}_1) = \gamma(I-F, (\bar{B}_2 \cap \bar{V}_1)^{\bullet}, \bar{B}_2)$, i.e.

$$\gamma(I-F, (\bar{B} \cap \bar{V}_1)^{\bullet}, \bar{B}) = \gamma(I-F, (\bar{B}_1 \cap \bar{V}_1)^{\bullet}, \bar{B}_1).$$

Arguing in the same manner as before, it can be proved that Λ_f does not depend on the choice of U, r, E, and imbedding.

The constructed Lefschetz number has, obviously, usual properties. The fact that it is not equal to zero leads to the existence of a fixed point in X. This number is conserved under a homotopy in the class of weakly compact mappings, etc. The proofs are similar to those given earlier. Modifications involving the use of weak topologies and relative rotation can be easily carried out by the reader.

The construction described is generalized to the case of locally compact weak mappings that have a weakly compact iteration f^n, $n \geq 1$. In fact, it is necessary to consider, as in Section 3, an open set U_1 of a weakly compact set $f^n X$ in U such that $f(U_1)$ is weakly compact. Further, the above construction is given after replacing U by U_1.

Also one can construct a homotopic characteristic of a weakly continuous (i.e. of a continuous in topology ω) mapping f on the closure of an ω-open set $\circledH \subset X$ such that $f(\bar{\circledH})$ is compact in ω and there are no fixed points on $\dot{\circledH}$.

REFERENCES

(Items I-I3,I5,I6,I8-23,25-28,32-34,36 are in Russian)

1. Borisovich, Yu.G. Rotation of Weakly Continuous Vector Fields, Dokl. AN SSSR, 1960, vol. 131, N 2.
2. Borisovich, Yu.G. Rotation of Weakly Continuous Vector Fields, Trudy Tbil. mat. in-ta im. Razmadze AN GSSR, 1961, vol. 27.
3. Borisovich, Yu.G. About an Application of the Concept of Rotation of Vector Field, Dokl. AN SSSR, 1963, vol. 153, N 1.
4. Borisovich, Yu.G. To the Browder Fixed Point Theorem, Trudy seminara po funkts. analizu, Voronezh, 1963, issue 7.
5. Borisovich, Yu.G. About Perturbation of a Completely Continuous Vector Field by a Weakly Continuous Operator, Trudy seminara po funkts. analizu, Voronezh, 1963, issue 7.
6. Borisovich, Yu.G. On Bifurcation Points of Positive and Semipositive Operators, In: Functional Analysis and the Theory of Functions,

Kazan, 1967, issue 4.
7. Borisovich, Yu.G. On Relative Rotation of Compact Vector Fields in Linear Spaces, Trudy seminara po funkts. analizu, Voronezh, 1969, issue 12.
8. Borisovich, Yu.G. Relative Rotation of Compact Vector Fields and the Lefschetz Number, Trudy seminara po funkts. analizu, Voronezh, 1969, issue 12.
9. Borisovich, Yu.G. Topology and Nonlinear Functional Analysis (Report presented at an International Congress on Topology), Uspehi Mat. Nauk, 1979, vol. 34, N 6.
10. Borisovich, Yu.G., Bliznyakov, N.M., Izrailevich, Ya.A., and Fomenko, T.N. An Introduction to Topology, Vysshaya shkola, Moscow, 1980.
11. Borisovich, Yu.G. and Gliklikh, Yu.E. The Lefschetz Number of Mappings of Banach Manifolds and the Relatedness Theorem, Trudy mat. fac. VGU, Voronezh, 1973, issue 11.
12. Borisovich, Yu.G. and Gliklikh, Yu.E. On Generation of a New Type of Closed Curves from Periodic Orbit of Vector Field Upon Varying the Metric of Riemannian Manifold, Trudy NIIM, Voronezh State University, Voronezh, 1975, issue 20.
13. Borisovich, Yu.G. and Gliklikh, Yu.E. On a Class of Closed Curves, Which Appears in the Theory of Vector Fields on Riemannian Manifolds, Deposited in VINITI on January 29, 1978, No. 2164-78.
14. Borisovič, Ju.G. and Gliklih, Ju.E. Fixed Points of Mappings of Banach Manifolds and some Applications. Nonlinear Analysis: Theory, Methods and Applications, 1980, vol. 4, N 1.
15. Borisovich, Yu.G. and Sapronov, Yu.I. To the Topological Theory of Compact Contractible Mappings, Trudy seminara po funkts. analizu, Voronezh, 1969, issue 12.
16. Borisovich, O.Yu. and Khaikin, A.L. On the Concept of Rotation of Vector Field on Manifold, Trudy NIIM VGU, Voronezh, 1975, issue 20.
17. Browder, F.E. Fixed Point Theorems on Infinite-Dimensional Manifolds, Trans. AMS, 1965, vol. 119, N 2.
18. Gliklikh, Yu.E. Integral Operators on Manifolds, Trudy mat. fac. VGU, Voronezh, 1971, issue 4.
19. Gliklikh, Yu.E. About a Generalization of the Hopf-Rinow Theorem on Geodesics, Uspehi Mat. Nauk, 1974, vol. 29, N 6.
20. Gliklikh, Yu.E. About a Shift Operator Along the Trajectories of Functional Differential Equations on Smooth Manifolds, Trudy NIIM VGU, Voronezh, 1975, issue 17.
21. Gliklikh, Yu.E. On Pseudo-Integral Curves of Functional Differential Equations on Riemannian Manifolds, In: Methods of Solving

Operator Equations, Voronezh, 1978.

22. Gliklikh, Yu.E. On an Analog of Differential Equations with Discrete Delay on Riemannian Manifolds, Deposited in VINITI on March 20, 1980, No. 1089-80 Dep.
23. Gliklikh, Yu.E. and Khaikin, A.L. On Periodic Integral Curves of Vector Fields on Lie Groups, In: Applied Analysis, Voronezh, 1979.
24. Graff, R.A. Elements of Nonlinear Functional Analysis, Memoirs of AMS, 1978, vol. 206.
25. Zabreiko, P.P., Krasnosel'sky, M.A., and Strygin, V.V. On Invariance Principles of Rotation, Izv. vuzov ser. mat., 1972, vol. 5.
26. Krasnosel'sky, M.A. Topological Methods in the Theory of Nonlinear Integral Equations, Gostekhizdat, Moscow, 1956.
27. Krasnosel'sky, M.A. Positive Solutions of Operator Equations, Gostekhizdat, Moscow, 1962.
28. Krasnosel'sky, M.A. and Zabreiko, P.P. Geometrical Methods of Nonlinear Analysis, Nauka, Moscow, 1975.
29. Leray, J. Théorie des points fixes: indices total et nomber de Lefschetz, Bull. Soc. Math. France, 1959, vol. 87.
30. Nagumo, M. Degree of Mapping in Linear Local Convex Topological Spaces, Amer. J. of Math., 1951, vol. 73, N 3.
31. Penot, J.-P. Weak Topology on Functional Manifolds. Global Analysis and its Applications, Vienna, IAEA, 1974, vol. 3.
32. Sapronov, Yu.I. To Homotopic Classification of Condensing Mappings, Trudy mat. fac. VGU, Voronezh, 1972, N 6.
33. Sadovsky, B.N. On Noncompactness Measures and Condensing Operators, In: Problems of Mathematical Analysis of Complex Systems, Voronezh, 1968, issue 2.
34. Sadovsky, B.N. Limiting Compact and Condensing Operators, Uspehi Mat. Nauk, 1972, vol. 27, N 1.
35. Spanier, E. Algebraic Topology, New York, 1966.
36. Fuks, L.B, Fomenko, A.T., and Gutenmakher, V.L. Homotopic Topology, Moscow State University, Moscow, 1969.
37. Fuller, F.B. The Homotopy Theory of Coincidences, Ann. of Math., 1954, vol. 59, N 2.
38. Fournier, G. Généralisations du théorème de Lefschetz pour des espases non-compacts, I, II, III.- Bull. Acad. Polon. Sci., Ser. Math. Astron. Phys., 1975, vol. 23, N 6, pp 693-699, 701-706, 707-711.
39. Fenske, C.C. and Peitgen, H.-O. On Fixed Points of Zero Index in Asymptotic Fixed Point Theory, Pacific J. Math., 1976, vol. 66, N 2, pp 391-410.

Translated from the Russian by P.K. Dang

THE STRUCTURE OF EXTENSION ORBITS OF LIE ALGEBRAS

A.S.Mishchenko
Department of Mechanics and Mathematics
Moscow University
117234 Moscow, USSR

1. Statement of the Problem

In Hamiltonian mechanics, for integrating a dynamic system with n degrees of freedom, it is sufficient in most cases to know only the first n integrals. This situation is known as Liouville complete integrability of a Hamiltonian system (Journ. d. Math., 1855, vol. 20). Now we shall give exact formulation of the statement, following [1, p 235]. Suppose that on a symplectic $2n$-dimensional manifold M there are n functions $F_1, F_2, \dots, F_n$ whose Poisson brackets are equal to zero: $\{F_i, F_j\} = 0$. Assume that on a compatible manifold of the level of functions F_i

$$P_\xi = \{x: F_i(x) = \xi_i, \ i = 1, \dots, n\}$$

these are functionally independent functions, i.e. their differentials are linearly independent at every point of the manifold P_ξ. Then P is a smooth manifold which is invariant with respect to an Hamiltonian flow with Hamilton function $H = F_i$. If the manifold is compact and connected, then it is diffeomorphic to an n-dimensional torus, and the Hamiltonian flow is linear in angular coordinates of the torus. There are more assertions of integrability of an Hamiltonian flow in quadratures, which will not be considered here.

From the algebraic point of view, the conditions imposed on the system of first n integrals $F_1, F_2, \dots, F_n$ can be formulated as follows. Let G be a linear space in the space of all functions on a sympletic manifold M generated by functions $F_1, F_2, \dots, F_n$. Then G is the finite-dimensional commutative Lie algebra with respect to Poisson bracket. And if by dG we denote the space of differentials of functions from G, then dim G = dim dG and on the compatible manifold of level $P_\xi = \{x: F(x) = \langle \xi, F \rangle, F \in G\}$ the equality 2dim G = dim M holds.

In [2] the author has proposed a generalization of the Liouville complete integrability condition. It resides in rejecting the demand that the finite-dimensional algebra C of first integrals should be commutative. A corresponding statement sounds as follows [2, Theorem 4.1]. Suppose that G is the finite-dimensional Lie algebra of functions on sympletic manifold M. Let further dim G = dim dG on a submanifold of level P_ξ and dim G + ind G = dim M. Then, if P_ξ is a compact manifold, then every connection component of P_ξ is diffeomorphic to a torus of dimension k = ind G, and the Hamiltonian flow is linear in angular coordinates of the torus. Recall that indG denotes the index (rank) of algebra G, equal to codimensionality of the orbit of co-adjoint representation, the orbit being in its general position (i.e. equal to minimum co-dimensionality of the orbit).

In particular, if G is the commutative Lie algebra, then dim G = ind G and we arrive at a classical condition of complete integrability : 2 dim G = dim M. In the case of non-commutative complete integrability the dimension of the manifolds of the level of first integrals P_ξ is less than one half of that of the symplectic manifold M. This corresponds to degeneration of the Hamiltonian system.

In a number of cases it has been noted that together with the non-commutative algebra of first integrals the Hamiltonian systems permit also some commutative algebras of integrals, which still satisfy the classical conditions of complete integrability. Geometrically this implies that invariant toruses P_ξ of small dimension are grouped into some (also invariant) toruses of half dimension, on which the Hamiltonian flow is conditionally periodic (degenerating periodically in some directions). In this case a natural hypothesis appear [2] : the Liouville complete integrability condition in the commutative and classical sense should follow from the complete integrability condition in the non-commutative sense. In other words, if G is the finite-dimensional Lie algebra satisfying the complete integrability conditions, then there exists another commutative Lie algebra G' which also satisfies the complete integrability conditions. The proof of this hypothesis may be reduced to a problem on algebraic structure of finite-dimensional Lie algebras by assuming the functions from unknown algebra G' to be functionally dependent on the functions of algebra G.

The aforementioned reduction of the problem consists in the following. We denote by G^* a linear space dual to the algebra G, and by $\mathcal{G}$ the Lie group associated with the algebra G. Then there exists a canonical mapping $\varphi : M \longrightarrow G^*$ which at every point $x \in M$ matches the linear functional on the algebra G: $\langle \varphi(x), f \rangle = f(x)$. The canonical mapping

φ is equivariant with respect to the symplectic action of the group $\mathcal{G}$ on the manifold M and the co-adjoint action in the space G^*. For functions on G^*, one can determine the Poisson brackets by making use of the standard symplectic structure on every orbit of the co-adjoint representation. The preimages of these functions on the manifold M appears to have a compatible Poisson bracket. This means that for constructing the commutative Lie algebra of functions on the manifold M it is sufficient to construct a commutative algebra of functions on the space G^*. The latter is now a universal problem: its solution is independent of a concrete manifold M, but depends only on the algebraic structure of the algebra G, exactly on the structure of orbits of its co-adjoint representation.

It must be noted that the complete integrability condition can be generalized in the non-commutative sense. A.V. Strel'tsov (see [3]) has proved that in this condition one may disregard dim G = dim dG, and represent the remaining requirements as dim dG + dim dH_ξ = dim H, where H_ξ is an annihilator of vector $\xi \in G^*$. This means that the canonical mapping $\varphi : M \longrightarrow G^*$ transforms the mapping M not on the general position orbits, but on special orbits. The reduction of the problem pertaining to the determination of a sufficiently large number of first integrals in involution leads to the study of not only general position orbits, but to special orbits also. Geodesic flows on symmetric spaces provide us with examples of finite-dimensional algebras of first integrals whose dimension exceeds that of the manifold itself [4].

2. Semisimple Algebras and Borel Subalgebras

A series of semisimple (complex and some real forms) Lie algebras is the first example of the series of algebras for which a sufficiently large number of functions in involution have been found on the co-adjoint representation [5]. In the case of semisimple Lie algebra G, the dual space G^* is identified, using the Cartan-Killing (non-degenerate) form, with the algebra G and the co-adjoint representation with the adjoint representation. Let f be a polynomial function on algebra G, which is invariant with respect to the adjoint representation, i.e. constant on the orbits. Then a family of functions $V = \{f(x + \lambda a)\}$ parametrized with the given function f and parameter λ forms the finite-dimensional commutative algebra. If vector a is in the general position (is a regular element of algebra G), then the dimension of dim dV equals 1/2(dim G + ind G). In particular, if O_ξ

is an orbit of adjoint representation, then

$$\dim(dV/O_{\xi}) = 1/2\ (\dim G - \mathrm{ind}\ G) = 1/2 \dim O_{\xi}$$

Thus, for semisimple Lie algebras the problem of reducing non-commutative complete integrability to commutative integrability is solved positively. Exactly, if G is the semisimple Lie algebra on the symplectic manifold M and $g_1, \ldots, g_n$ is the linear basis in algebra G, then one can find functions f_j polynomially depending on g_k such that f_j are pairwise present in involution and yield a complete set of first integrals on the manifold M.

Note that the indicated reduction of non-commutative conditions of complete integrability to commutative ones in a class of infinitely smooth functions can be accomplished in a trivial manner for all finite-dimensional algebras. Indeed, let $\xi \in G^*$ be a covector in general position. Then there exists a neighbourhood $U \ni \xi$ and coordinates $t_1, \ldots, t_r, p_1, \ldots, p_s, q_1, \ldots, q_s$ in this neighbourhood such that

(a) every orbit of co-adjoint representation is given by equations $t_i = \mathrm{const}$;

(b) the symplectic form on orbits has the shape : $\sum_{i=1}^{s} dp_i \wedge dq_i$. Let $\varphi(x)$ be a finite function of one variable with a sufficiently small support, which is identically equal to unity in some neighbourhood of zero. Then we assume

$$F_i(t_1, \ldots, t_r, p_1, \ldots, p_s, q_1, \ldots, q_s) =$$

$$= (p_i^2 + q_i^2) \prod_{j=1}^{s} \varphi(p_j^2 + q_j^2)\ \varphi(t_1^2 + \ldots + t_r^2)$$

Functions F_i are finite, functionally independent in a neighbourhood of the point ξ , and pairwise found in involution. Continuing in a trivial manner the functions F_i on the entire space G^*, we obtain the desired reduction. Therefore in the earlier obtained result attention should be paid to the possibility of performing reduction in a class of polynomial functions.

A number of papers are devoted to constructing a quite large number of functions in involution on different solvable and nilpotent algebras. A.A. Arkhangel'sky [6] (for the algebra of triangular matrices) and V.V. Trofimov [7, 8] (for an arbitrary Borel subalgebra of semisimple Lie algebra) propose to use not only the invariants of the algebra, but also some of its subinvariants for constructing functions.

Of instrest in this direction is the observation that goes back to [9]. If $G_0 \supset G_1 \supset \ldots \supset G_i \supset \ldots$ is a chain of imbedded subalgebras, p_i: $G^* \longrightarrow G_i^*$ are natural projections, and $f_{i,\alpha}$ are the invariants of algebra G, then the functions $f_{i,\alpha} \cdot p_i$ are pairwise found in involution. The indicated method of constructing functions on a dual space to algebra G may prove convenient for arbitrary algebras. T.A. Pevtsova [10] has proved that for semisimple Lie algebras there exist standard chains of subalgebras whose invariants give a complete set of functions in involution. It is not clear whether these functions depend on the functions that are constructed in [3].

3. Extensions of Semisimple Algebras

An affine algebra is the extension of linear algebra, obtained with the use of commutative algebra of parallel transfers. Examples related to affine algebra are available in rigid-body dynamics. The equations of motion of a heavy rigid body can be interpreted as Euler equations of geodesic flow of a left-invariant metric on an affine group [11]. The equations of motion of a rigid body in viscous liquid also admit interpretation of Euler equations on an affine group, as it is shown in the report of H. Weber (Am. Math. Pure Appl., 1878, ser. 2, vol. 9).

The task of constructing functions on dual spaces of the extension of semisimple algebras rests on the study of the structure of co-adjoint representation orbits. Of interest in this direction are the reports of M. Rais [12, 13] where the index (rank) of the algebra which is a semidirect product of the semisimple and commutative algebras is computed by representation ρ. The formula used for computing the index of algebra $V \times_\rho G$ is of the form

$$\operatorname{ind}(V \times_\rho G) = \operatorname{ind}\rho + \operatorname{ind} G_0 . \qquad (1)$$

Here, G_0 is an annihilator of covector in general position in the representing space V^* and $\operatorname{ind}\rho$ is the codimension of the representation orbit ρ. To describe the invariants of algebra $V \times_\rho G$ we need a more exact description of the co-adjoint representation annihilator of algebra $V \times_\rho G$. Call $P_0 \subset V \times_\rho G$ the annihilator of covector $(x_0^*, \xi_0^*) \subset V^* \times G^*$. Denote by G_0 the annihilator of the covector x_0^*, and by $H_0 \subset G_0$ the annihilator of the covector $\xi_0^* \mid G_0 \subset G_0^*$. Then the following holds

<u>Theorem 1.</u> (a) $P_0 \cap V = \{\rho_G^* x_0^*\}^\perp$;

(b) $\pi(P_0) = H_0$, where $\pi : V \times_\rho G \longrightarrow G$ is the natural projection.

On the strength of Theorem 1, T.A. Pevtsova has assertained the following statement.

Theorem 2. Let G be the simple complex Lie algebra and ρ be an irreducible representation. Then on the space $(V \times_\rho G)^*$ there are N functionally independent rational functions, which are pairwise found in involution, and

$$N = 1/2\,(\dim (V \times_\rho G) + \operatorname{ind} (V \times_\rho G))$$

The algebra $(V \times_\rho G)$ admits a chain of subalgebras such that all the indicated functions are pre-images of invariants of subalgebras.

More complex than the above considered is the case of semidirect product of the algebra G and the commutative algebra V. As before, we denote by P the algebra $V \times_\rho G$, where ρ is the representation of algebra G in the algebra of endomorphisms of algebra V. Let P_0 be an annihilator of the covector $p_0^* = (x_0^*, \xi_0^*) \subset P^* = V \times G^*$. The algebra P acts on spaces V and V^*. These actions are denoted by B and B^*, respectively.

Theorem 3. $V \cap P_0 = \{B_p^* (x_0^*)\}^{\perp}$.

The description of the projection $\pi(P_0) \subset G$ is more complex. Let $P_1 \subset P$ be an annihilator of the covector x_0^* in the presentation of B^*. Restrict the covector p_0 on P_1. Denote by P_2 the annihilator, in P_1, of the covector p_0^*/P_1 and by V_0 the annihilator, in V, of the covector x_0^*.

Theorem 4. $\pi(P_0) = \pi(P_2)$, $P_2 \cap V = V_0$.

Corollary.

$$\operatorname{ind} P = \operatorname{ind}(ad_V + \rho) + \dim P_2 - \operatorname{ind} V. \qquad (2)$$

In the case of the commutative algebra V we have $\operatorname{ind}(ad_V + \rho) = \operatorname{ind} \rho$, $P_1 = V \oplus G_0$, $\dim P_2 = \dim V + \operatorname{ind} G_0$, $\operatorname{ind} V = \dim V$ we arrive at the Rais formula (1). This is a particular case of formula (2).

Note that in Theorems 3 and 4 it is sufficient to represent the algebra P as an average term of exact sequence.

$$0 \longrightarrow V \longrightarrow P \xrightarrow{\pi} G \longrightarrow 0$$

Theorems 3 and 4 give new information on the structure of co-adjoint representation orbits of already arbitrary algebras in the decomposition theorems of Lie algebra on a nonpotent radical and the semi-simple part.

References

1. Arnol'd, V.I. Mathematical Methods of Classical Mechanics, Moscow, 1974.
2. Mishchenko, A.S. and Fomenko, A.T. A Generalized Method of Liouville Integration of Hamiltonian Systems, Func. Anal., 1978, vol. 12.
3. Mishchenko, A.S. and Fomenko, A.T. Integration of Hamiltonian Systems with Non-commutative Symmetries, Trudy seminara po vektornomu i tenzornomu analizu, 1980, issue 20.
4. Mishchenko, A.S. Integration of Geodesic Flows on Symmetric Spaces, Trudy seminara po vektornomu i tenzornomu analizu, 1981, issue 21.
5. Mishchenko, A.S. and Fomenko, A.T. Euler Equations on Finite-Dimensional Lie Groups, Izv. AN SSSR, ser. mat., 1978, vol. 42.
6. Arkhangel'sky, A.A. On Integration of Euler Equation on Triangular Matrices Algebra, Mat. sb., 1979, vol. 108, issue 1.
7. Trofimov, V.V. Euler Equations on Borel Subalgebras of Semisimple Lie Algebras, Izv. AN SSSR, ser. mat., 1979, vol. 43, issue 3.
8. Trofimov, V.V. Finite-Dimensional Representation of Lie Algebras of Completely Integrable Systems, Mat. sb., 1980, vol. 111, issue 4.
9. Vergene, M. La structure de Poisson sur l' algebre symmétrique d'une algèbre de Lie nilpotente, Bull. Soc. Math. France, 1972, vol. 100.
10. Pevtsova, T.A. One Method of Constructing the Commutative Algebra of Integrals on Lie Algebras, Tez. dokl. 7-i Vsesoyuz. konf. po sovremennym problemam geometrii, Minsk, 1979.
11. Bocharov, A.V. and Vinogradov, A.M. An Addendum to the Report "Structure of Hamiltonian Mechanics", Uspehi Math. Nauk, 1977, vol. 32, issue 4.
12. Rais, M. La représentation coadjonte du group affine, Ann. Inst. Fourier, 1978, vol. 28, N 1.
13. Rais, M. L'indice des produits semi-product $E \times_{\rho} g$, C.R. Acad. Sci., 1978, vol. 237, N 4A.

BRANCHING OF SOLUTIONS OF SMOOTH FREDHOLM EQUATIONS

Yu.I.Sapronov
Department of Mathematics
Voronezh State University
394693 Voronezh, USSR

In a framework of the analysis of Fredholm equations on Banach manifolds we shall study some problems related to branching of nonlinear operator equations.

Despite the large number of publications on this topic (references to a majority of them, which are adjacent to the classical trend, can be found in [1, 2, 6, 7, 12, and 15]), the flow of papers dedicated to branching is at present not declining, but, on the contrary, increasing. This is, first of all, associated with the application of new and effective methods developed in the theory of singularities of smooth mappings and, particularly, the theory of catastrophes (see [8, 14, 16, 18, 20, 33-37, 43]). Penetration of some important and perspective methods of smooth topology into the branching theory may also be noted (see [8, 9, 14, 16, 18, 25, 29, 33, 37, and 40]).

Among the problems which have been touched upon in this report, we have studied the most simple topological structures of germs of sets of solutions in isolated singular solutions, the equations finite definiteness conditions, and the normal forms of branching equations. In passing we have briefly reviewed the use of some of the aforementioned new methods in the branching theory.

1. Simple Branchings of Solutions

Let M, N be Banach C^∞-manifolds which can be modelled by real Banach spaces E, F. Let $f: M \longrightarrow N$ be the smooth (i.e. of class C^∞) mapping. Then the equation

$$f(x) = b, \quad b \in N, \quad x \in M \tag{1}$$

is called the smooth Fredholm equation. Further it is assumed that ind $f > 0$. By $\Sigma(f)$ we denote the set of singular points : $\Sigma(f) = \{x \in M : \dim \operatorname{Coker} df(x) > 0\}$. The solution $x = a$ to equation (1) is called simple if $a \bar{\in} \Sigma(f)$. Otherwise, it is called singular. The singular solution a is called a point of simple branching if a is an isolated point of the set $\Sigma(f) \cap f^{-1}(b)$. The germ of the set $f^{-1}(b)$ at point a we denote by $N(f,a)$. Call a the point of correct branching if a is a point of simple branching, $N(f, a)$ does not coincide with the germ of one-point set and a smooth functional $\varphi(x)$ is defined in some neighbourhood U of the point a such that

$$\ker d\varphi(x) \cap \ker df(x) \neq \ker df(x), \quad \forall x \in u \cap f^{-1}(b)$$

The latter condition is equivalent to the following : there exists such a representative N' of the germ $N(f, a)$ (which for "good" selection of U and in the case of simple branching is a compact manifold with boundary and one and only one singular point a) on which the functional φ is defined and does not have critical points in the domain N' a and a is a point of its global minimum or maximum on N'. An equivalence is readily established if we take into consideration the property of local properness of Fredholm mappings [22]. For definiteness we shall assume further that a is a point of minimum and $\varphi(a) = 0$. Call φ the carpeting functional.

By $C(W)$ we denote a standard cone on a compact manifold W without boundary (i.e. $C(W)$ is a cylinder $W \times [0,1]$ in which the lower base is contracted to a point). Clearly $C(W)$ is a manifold of dimension $\dim W + 1$ with a unique singular point. We call the germ $N(f, a)$ to be a conoid if some of its representative N' is a submanifold with a unique singular point u such that there exists a homeomorphism $\hat{\tau} : C(W) \longrightarrow N$ which transforms into a diffeomorphism when the singular points are discarded [29, 32] . In lieu of $\hat{\tau}$ we may consider $\tau = \hat{\tau} \circ p$, $\tau : W \times [0,1] \longrightarrow N'$, where $p: W \times [0,1] \longrightarrow C(W)$ is the factor mapping (call τ the parametrization of conoid $N(f, a)$).

<u>Theorem 1</u>. Let a be a point of correct branching for equation $f(x)=b$. Then $N(f, a)$ is a conoid.

<u>Proof</u>. As the Fredholm mapping is locally proper, we may choose such a representative N' of the germ $N(f, a)$ which is a compact manifold (with a unique singular point). Let $\varphi(x)$ be the carpeting functional. The values of φ and $\partial N'$ are bounded above by a positive number d. Let $0 < c < d$. Then the diffeomorphic type of manifold $W_c = \{ x \in N' : \varphi(x) = c \}$ is independent of C, and the mapping $\hat{\tau}: C(W) \longrightarrow \{x \in N' : \varphi(x) \leqslant c \}$ given by a shift operator along the trajectories of the $\varphi|_{N'}$ realizes the necessary homeomorphism.

Remark 1. Using 9 , it is not hard to prove that in the case of analyticity of manifolds M, N and the mapping f, every point of simple branching is simultaneously a point of correct branching.

That in a smooth manifold it is not so is evidenced by the following example of scalar equation on a plane

$$\exp(-1/x^2 + y^2) \sin(1/x^2 + y^2) = 0$$

The set of solutions to this equation is a collection of concentric circles of radii $r_n = (\pi n)^{-1/2}$, $n = 1, 2, \ldots$ All the solutions are regular excepting zero. Hence, we have an example of simple branching. However, this is not the case of correct branching as any smooth functional on the circle has no less than two critical points.

It is clear that every example of a scalar equation with improper branching of solutions is at the same time an example of an infinite order critical point (in virtue of the known Samoilenko-Tougeron theorem [3, 17]).

Remark 2. For correct branchings we can determine the branching index. Let H be any homotopy invariant defined on a category of compact C^∞-manifolds. Then the value of invariant H on the manifold $W = N' \cap \varphi^{-1}(c)$ (see proof of Theorem 1) is called the H-index of branching at a point a, and is denoted by $\chi_H(f, a)$. It is not difficult to prove that $\chi_H(f, a)$ is independent of the choice of the carpeting functional φ.

2. Finitely Defined Equations

In the theory of singularities of smooth mappings the finite definiteness with respect to several kinds of mappings' equivalence relations are studied [3, 4, 21]. The most simple of them - the contact equivalence is of particular interest for studying the structure of the set of solutions to equation (1). This is described in detail in the following section. Here, we shall consider a weak equivalence relation of equations at a point, which has been derived and studied in [30, 31].

Let $f_1, f_2: M \longrightarrow N$ be a pair of Fredholm mappings. Say that equations $f_1(x) = b_1$ and $f_2(x) = b_2$ are similar at a point $(a_1, a_2) \in M \times M$ if the germs $N(f_1, a_1)$ and $N(f_2, u_2)$ are equisingular, i.e. they are either simultaneously empty or, on the contrary, some of their representatives are homeomorphic and diffeomorphic when the singular points are discarded. When $a_1 = a_2$ we write $a_1 \in M$ instead of $(a_1, a_2) \in M \times M$.

Call the equation $f(x) = b$ to be r-definite (where r is a positive integer) at point a if for every Fredholm mapping $g: M \longrightarrow N$, which in a has a contact of the r-th order with f,

$$j_a^r(f) = j_a^r(g)$$

the equations $f(x) = b$ and $g(x) = b$ are similar in a.

Theorem 2. If a is a solution to equation $f(x) = b$ and this equation is r-definite in a, then a is the correct branching point of this equation.

Proof. First we shall show that branching is simple. To do this, we shall make use of local representation (in maps) of Fredholm mapping [22] of the type

$$\tilde{f}(u, v) = (\psi \circ f \circ \varphi^{-1})(u, v) = (u, h(u, v)), \; dh(0,0) = 0$$

where

$$u \in \operatorname{Im} d\tilde{f}(0,0), \; v \in \ker d\tilde{f}(0,0), \quad h(u, v) \in R$$

is linear complement to $\operatorname{Im} d\tilde{f}(0, 0)$ in F, and

$$\varphi : U(a) \longrightarrow \operatorname{Im} d\tilde{f}(0,0) \times \ker d\tilde{f}(0,0), \qquad \psi : V(b) \longrightarrow F$$

are suitably chosen local diffeomorphisms. Thus, the Fredholm equation $f(x) = b$ in a neighbourhood of the point a is smoothly equivalent to the finite-dimensional equation $h(0, V) = 0, V \in \ker df(0,0)$. Now the proof of simplicity reduces to the following statement.

Lemma 2.1. Let $h: (\mathbb{R}^{n+p}, 0) \longrightarrow (\mathbb{R}^n, 0)$ be a smooth mapping such that zero is not an isolated point of the set $h^{-1}(0) \cap \sum(h)$. Then for any integer $r \geqslant 1$ we can find a mapping $\omega: (\mathbb{R}^{n+p}, 0) \longrightarrow (\mathbb{R}^n, 0)$ with zero r-jet at zero ($j_0^r(w) = 0$) such that for any neighbourhood of zero U in $\mathbb{R}^{n+p}$ the set $(f^{-1}(0) \cap U) \setminus 0$, $f = h + w$ is not a manifold.

Proof. Let $\{x_k \in \mathbb{R}^{n+p}\}_{k=1}^{\infty}$ be a sequence such that $x_k \longrightarrow 0$, $x_k \in h^{-1}(0) \cap \sum(h)$. Let $\{\varphi_k(x)\}$ be a family of smooth functions satisfying the following requirements:

1. $\varphi_k(x) = 1$ in some neighbourhood x_k, $k = 1, 2, \ldots$;
2. $(\operatorname{supp} \varphi_k) \cap (\operatorname{supp} \varphi_s) = \emptyset$ for $\forall \{k, s \mid k \neq s\}$.

Let $\sigma_k : (U, 0) \longrightarrow (\mathbb{R}^n, 0)$ be diffeomorphisms such that the first row of all matrices are zero rows. Assume

$$\omega = |x|^{2r} \left(\sum_{k=1}^{\infty} \lambda_k \varphi_k(x) \, g_k(x) \right)$$

where $g_k(x) = \sigma_k(|x - x_k|^2, \ \xi - \ \xi_2^k, \ldots, \xi_n - \xi_n^k)$, $x = (\xi_1, \ldots, \xi_{n+p})$ and λ_k is so chosen that

1. the given row converges in the class C^∞;

2. $rk(dh(x_k) + \lambda_k |x_k|^{2r} \ dg_k(x_k)) = n - 1.$

From these constructions it follows that the mapping h + has at points x_k singularities of the type $\sum p+1, o$ [3, 4] . In this case, every germ $N(h + \omega, x_k)$ is equisingular to the germ at the origin of coordinates of zero set of non-degenerate quadratic form of (p+1) variable. Thus, $N(h + \omega, x_k)$ is either a germ of hyperboloid in its singular point, or a germ of the single-point set (if the corresponding quadratic form is elliptic). In all these cases, $N(h + \omega, x_k)$ is not a germ of the smooth p-dimensional manifold. This proves the lemma.

We shall now prove the correctness of branching. Arguing in the same manner as before, we can readily ascertain that testing of correctness of branching in a for equation f(x) = b reduces to establishing the correctness of branching the equivalent finite-dimensional equation h(v) = 0 at zero.

<u>Lemma 2.2</u>. Let $\hat{\tau} : (C(W), *) \longrightarrow (\mathbb{R}^{n+p}, a)$ be a continuous mapping of the cone, which is a smooth imbedding, if the vertex of the cone is discarded. Then in some neighbourhood $\mathcal{O}$ of the image $K = \hat{\tau}(C(W))$ the functional φ is defined such that its restriction $\varphi|_{K \setminus a}$ has no critical points.

<u>Proof</u>. Let $a = \hat{\tau}(*)$, $\{U_n(a)\}$ be a system of contracting neighbourhoods that converges to a: diam $U_n(a) \longrightarrow 0$. By φ_n we denote the C^∞-function defined on $\mathcal{O}$ and satisfying the following conditions:

1. $a \bar{\in} \operatorname{supp} \varphi_n$;

2. $\frac{\partial}{\partial t}(\varphi_n \circ \tau)(t, w) > 0$ if $\tau(t, w) \in K \setminus U_n(a)$;

3. $\|\varphi_n\| \, C^n(\mathcal{O}) \leq 2^{-n}$.

The existence of such a function is a standard fact of smooth topology [5].

We shall now consider a function $\varphi = \sum_{n=1}^{\infty} \varphi_n$. From conditions (1)-(3)

it follows that $\varphi \in C^{\infty}(\mathcal{O})$, $\varphi(a) = 0$ and $\varphi|_{K \setminus a}$ has no critical points. The proof of the theorem is complete (see Remark 1).

Homotopy or, exactly, smooth deformation is a standard way of ascertaining the equivalence of equations.

Consider a smooth family of Fredholm equations

$$f_{\lambda}(x) = b, \quad \lambda \in [0, 1] \tag{1^0}$$

Let here a be a point of correct branching (1^0), and $\varphi_{\lambda}(x)$ be a smooth family of the corresponding carpeting functionals given on a unique neighbourhood $U(a)$ (for all $\lambda \in [0,1]$). Then the family of equations (1^0) is called the correct deformation.

<u>Theorem 3</u>. For correct deformation (1^0) the topological type of the germ $N(f_{\lambda}, a)$ is independent of λ.

The proof of the theorem is reduced to the verification of invariance of the topological and differential type preimage of moving regular value of the Fredholm mapping $\Phi : (U(a) \times [0,1]) \longrightarrow N \times \mathbb{R} \times \mathbb{R}$ given by the correspondence $(x,t) \longmapsto (f_{\lambda}(x), \varphi_{\lambda}(x), t)$. If we note that Φ is proper, as a mapping onto its image, then this testing becomes a standard argument in smooth topology [8].

For investigating the behaviour of solutions of certain nonlinear equations, it is necessary to know the tests of finite definiteness which are convenient from the computational point of view and are sufficiently general at the same time. With such tests may be grouped the following

<u>Theorem 4</u>. Suppose the mapping $p: (\mathbb{R}^n, 0) \longrightarrow (\mathbb{R}^m, 0)$ is polynomial, $\deg p \leqslant r$. Let on an open set $\mathcal{O}$, $0 \in \mathcal{O}$, for the mapping $e(x,q) = p(x) + \sum_{|\alpha|=r+1} q_{\alpha} x^{\alpha}$; $q = (q_{\alpha})_{|\alpha| = r+1}$; $q_{\alpha} \in \mathbb{R}^m$ holds, for any q_0, l, the estimate

$$\left| \left(\frac{\partial e}{\partial x}(x, q)\right)^{*} h \right| \geqslant C(q_0, l)\, |x|^r\, |h| \tag{2}$$

with some constant $C(q_0, l)$ if $(x, q) \in e^{-1}(0)$ and

$$|q - q_0| \leqslant l|x| \tag{2.1}$$

Then the equation $p(x) = 0$ is r-definite in zero.

<u>Proof</u>. Every smooth mapping $f : (\mathcal{O}, 0) \longrightarrow (\mathbb{R}^m, 0)$, for which $j_0^r(f) = j_0^r(P)$, can be written as $f(x) = e(x, q(x))$, where $q(x)$ is a smooth

mapping $\mathcal{O} \longrightarrow Q = \{q = (q_\alpha)_{|\alpha| = r+1}\}$. We shall first show that there is a neighbourhood of zero $U \in \mathcal{O}$ in which zero is a unique singular solution to every equation $f(x) = 0$, $f(x) = e(x, q(x))$, $q(0) = q_o$, $|\partial q / \partial x| \leqslant 1$. This follows from the following inequalities :

$$|(\frac{\partial f}{\partial x})^* h| = |(\frac{\partial e}{\partial x})^* h + (\frac{\partial e}{\partial q} \quad \frac{\partial q}{\partial x})^* h| \geqslant$$

$$\geqslant |(\frac{\partial e}{\partial x})^* h| - |(\frac{\partial q}{\partial x})^* (\frac{\partial e}{\partial q})^* h| \geqslant$$

$$\geqslant C(q_o, 1)|X|^r |h| - 1d|x|^{r+1} |h| = \tag{2.2}$$

$$= (C(q_o, 1) - 1d|x|)|x|^r |h|$$

Here

$$d = \max_{x \neq 0} \{ \|(\frac{\partial e}{\partial q}(x, q))^*\| \ |x|^{-(r+1)} \}$$

Now we shall establish that the functional $\varphi(x) = |x|^2$ is a carpeting for the considered equation on a neighbourhood U, $0 \in U \subset \mathcal{O}$, and for all q(x) satisfying the relations

$$|\frac{\partial q}{\partial x}| \leqslant 1, \quad x \in U, \quad q(0) = q_o$$

First note that the points $x \in f^{-1}(0) \cap U$, where the local carpeting condition is not satisfied, are solutions to the equation

$$x = (\frac{\partial e}{\partial x})^* h + (\frac{\partial q}{\partial x})^* (\frac{\partial e}{\partial q})^* h, \quad h \in \mathbb{R}^m \tag{3}$$

From (3) follow the estimate

$$|x - (\frac{\partial e}{\partial x})^* h| \leqslant 1d|X|^{r+1} |h| \tag{4}$$

and the estimate

$$|X| \geqslant |(\frac{\partial e}{\partial x})^* h| - 1d|X|^{r+1} |h| \geqslant \delta |x|^r |h| \tag{5}$$

(see inequality (2.2)) if $C(q_o, 1) - 1d|x| \geqslant \delta$.

From (5) we obtain the condition for h

$$h \leqslant 1/\delta |X|^{r-1} \tag{6}$$

Denying that $\varphi(x) = |x|^2$ is carpeting in any neighbourhood will involve, according to lemma on the selection of curves [9], the existence of an analytical curve $(x(t), q(t), h(t)) \in \mathbb{R}^n \times Q \times \mathbb{R}^m$, $t \in (0,1]$

(and for $x(t)$ and $q(t)$: $x(0) = 0$, $q(0) = q_o$, $t\in[0,1]$) satisfying inequality (4) together with the imbedding $(x(t), q(t)) \in G = e^{-1}(0)$, $\forall t$, and relations (2.1), (6). Let here $x(t) = \alpha t^p + \ldots$, $p>0$. From (2.1) and (4) it follows, in particular, that

$$|q(t) - q_o| = o(t^{p-1}) \quad , \quad |x - (\frac{\partial e}{\partial x})^* h| = o(t^{2p-1})$$

Considering that $(x(t), q(t)) \in G$, i.e. $\frac{\partial e}{\partial x}\dot{x} + \frac{\partial e}{\partial q}\dot{q} = 0$

we may write the following relations :

$$\frac{d}{dt}(\frac{|x|^2}{2}) = (\dot{x}, x) = (\dot{x}, (\frac{\partial e}{\partial q})^* h) + o(t^{3p-2}) =$$

$$= ((\frac{\partial e}{\partial x})\dot{x}, h) + o(t^{3p-2}) = -((\frac{\partial e}{\partial q})\dot{q}, h) + o(t^{3p-2})$$

As

$$|(\frac{\partial e}{\partial q})\dot{q}, h)| \leqslant \|\frac{\partial e}{\partial q}\| \, |\dot{q}| \, |h| \leqslant d\,\delta^{-1}|x|^2 (t^{p-2}) = o(t^{3p-2})$$

then

$$\frac{d}{dt}(\frac{|x|^2}{2}) = o(t^{3p-2}) \quad \text{and} \quad x^2 = o(t^{3p-1}).$$

But, on the other hand, $|x|^2 = (\alpha t^p + \ldots, \alpha t^p + \ldots) =$
$= \alpha^2 t^{2p} + \ldots \sim t^{2p}$.

We have obtained a contradiction which signifies the falsity of the assumption that functional $\varphi(x) = |x|^2$ is not a carpeting functional in a neighbourhood of zero. For final proof of the theorem it remains to consider deformation $f_\varepsilon(x) = e(x, \varepsilon q(x))$ and refer to Theorem 3 on correct deformation.

From the theorem proved we can readily obtain the following

Theorem 5. Let the mapping $p : (\mathbb{R}^n, 0) \longrightarrow (\mathbb{R}^m, 0)$ be polynomial, $\deg p \leqslant r$. Then for r-definiteness of the equation it is sufficient that the estimate holds

$$|(\frac{\partial p}{\partial x})^* h| \geqslant c|x|^{r-1}|h| \qquad (7)$$

if $|p(x)| \leqslant b|x|^{r+1}$, $c = c(b)$.

Proof. Let us consider $e(x, q) = p(x) + \sum q_\alpha x^\alpha$. Denote the second term by $\Omega(x)$. Then we have

$$|(\frac{\partial e}{\partial x})^* h| = |(\frac{\partial P}{\partial x})^* h + (\frac{\partial \Omega}{\partial x})^* h| \geq c|x|^{r-1} - \|(\frac{\partial \Omega}{\partial x})^*\| |h| \geq$$

$$\geq c|x|^{r-1}|h| - d|x|^r|h| \geq \delta|x|^r|h|$$

if $p(x) \leq b|x|^{r+1}$, $\delta \leq c - dx$, $\|(\frac{\partial \Omega}{\partial x})^*\| < d|x|^r$.

Since at small $|x|$: $|x|^{r-1} > |x|^r$, the estimate is fulfilled from the condition of the previous theorem. This proves the theorem.

Theorem 6. Let $V: \mathbb{R}^n \longrightarrow \mathbb{R}^m$ be a polynomial homogeneous mapping of degree r: $V(\lambda x) = \lambda^r V(x)$, and zero be a unique singular solution to equation $V(x) = 0$. Then this equation is r-definite.

Proof. It is required to establish estimate (7). To do this, it is sufficient to know the lower bound of $|dV(x)^* h|$ at the intersection of $V^{-1}(0)$ and unit sphere, and then to make use of the homogeneity of V(x).

The statements of Theorems 5 and 6 can be readily applied to the Fredholm equation. For Fredholm equations we shall give a variant of certain theorem [37] which is a direct consequence of Theorem 6. We consider a quadratic differential [3, 17] of the smooth Fredholm mapping f: (M, a)⟶(N, b) which in non-invariant form is defined as mapping

$$\delta(f) : \ker d\tilde{f}(0) \longrightarrow \operatorname{Coker} d\tilde{f}(0)$$

given by the equality

$$\delta(f)(h) = \lim_{t \to 0} t^{-2} f(th) \quad (\operatorname{mod} \operatorname{Im} d\tilde{f}(0))$$

where

$$\tilde{f} = \tau \circ f \circ \sigma^{-1}, \quad \tau : (V,b) \longrightarrow (F,0), \quad \sigma : (U, a) \to (E,0)$$

are local diffeomorphisms (maps).

A quadratic differential is a polynomial homogeneous mapping of degree two.

Theorem 7. Suppose that equation $\delta(f)(h) = 0$, $h \in \ker d\tilde{f}(0)$ has a unique singular solution (an automatically zero solution). Then the Fredholm equation $f(x) = b$ is 2-definite at point a, and the germ N(f, a) is equisingular to the germ $\delta(f)^{-1}(0)$ at zero.

In studying stratified mappings, R. Thom [41] formulated some theorems that are directly related to the considered problem of finite definiteness. In proving one of them he actually formulated certain conditions of finite definiteness of equations. These conditions were later revised by V.R. Zachepa [31] who obtained some results concerning finite definiteness of Fredholm equations. Necessary and sufficient conditions for r-definiteness are also ascertained in [31] . The theorems of [31] were used in [30] for studying "small" solutions to the nonlinear Karman equation from the theory of thin elastic shells. For certain natural conditions the Karman equation proved to be 3-definite, and by its polynomial representative of degree 3 was written out the first asymptotics of "small" solutions.

3. Contact Equivalent Fredholm Germs

In this section we shall use the following notations:

$\mathcal{E}(E, F)$ is a space of smooth germs : $(E, 0) \longrightarrow (F, .)$;

$\mathfrak{m}(E, F) = \{ \tilde{f} \in \mathcal{E}(E, F),\ f(0) = 0 \}$;

$\mathfrak{m}^k(E, F) = \{ \tilde{f} \in \mathfrak{m}(E, F),\ d^i f(0) = 0,\ i \leq k \}$;

$D(E) = \{ f \in \mathcal{E}(E, E),\ df(0) \in GL(E) \}$;

$\mathrm{Diff}_o(E) = D(E) \cap \mathfrak{m}(E, E)$;

$\Phi_p \mathcal{E}(E, F) = \{ f \in \mathcal{E}(E, F)$, f is the Fredholm, ind f = p ;

$\Phi_p \mathfrak{m}(E, F) = \Phi_p \mathcal{E}(E, F) \cap \mathfrak{m}(E, F)$;

$\mathcal{E}(E) = \mathcal{E}(E, \mathbb{R})$;

$\mathfrak{m}(E) = \mathfrak{m}(E, \mathbb{R})$.

The germs $\tilde{f}, \tilde{g} \in \Phi_p \mathfrak{m}(E, F)$ are called contact equivalent germs if there exist $\tilde{\varphi} \in \mathrm{Diff}_o(E)$, $\tilde{A} \in \mathcal{E}(E, GL(F))$ such that $A(x)g(x) = f(\varphi(x))$. The germ f is called the contact r-definite if every germ $\tilde{g} \in \Phi_p \mathfrak{m}(E, F)$ having such an r-jet : $j_o^r(f) = j_o^r(g)$ is contact equivalent to $\tilde{f}$.

Obviously, from the contact equivalence of $\tilde{f}$ and $\tilde{g}$ follows the similarity of Fredholm equations $f(x) = 0$ and $g(x) = 0$, and from the contact r-definiteness follows the r-definiteness of the equation $f(x) = 0$. The inverse is not true; this is proved by the following example of the scalar equation on a plane : $x(x^2 + y^2)^2 = 0$. It is 5-definite, but the germ of the function $f(x) = x(x^2 + y^2)^2$ at zero is not contact finitely defined (i.e. it is not r-definite for any $r < \infty$). The latter is explained by the presence of complex singular solutions : $x = \pm iy$. It appears that if the Fredholm equation $f(x) = 0$ is analytic, then for contact finite definiteness of f at zero it is necessary and sufficient that the zero singular solution be isolated from all other (including complex) solutions. This statement follows from the Hilbert theorem on zeros and from the below given finite-dimensional version of the Mather theorem [42].

Theorem 8. Let $f \in \Phi_p \mathfrak{M}(E, F)$ and let for any $\tilde{\alpha} \in \mathfrak{M}^{r+1}(E, F)$ the equation

$$X(x)\, f(x) + df(x)U(x) = \alpha(x) \tag{8}$$

$$\tilde{X} \in \mathcal{E}(E, L(E, F)),\ \tilde{U} \in \mathfrak{M}(E, F)$$

be solvable. Then the germ $\tilde{f}$ is contact r-definite.

The proof is given as per the known scheme [42, 17]. First we consider a family $g(x, t) = f(x) + th(x)$, $0 \leq t \leq 1$, with arbitrary $h \in \mathfrak{M}^{r+1}(E, F)$. Then we find a pair $\{A(x, t), \varphi(x, t)\}$ as solution to the equation

$$A(x,t)\, g(\varphi(x,t), t) = f(x),\quad \varphi(x, 0) = x,\quad A(x, 0) = id_F .$$

The last identity is equivalent to the following :

$$\frac{\partial A}{\partial t} g(\varphi(x, t), t) + A \frac{\partial g}{\partial x} \cdot \frac{\partial \varphi}{\partial t} + A \frac{\partial g}{\partial t} = 0$$

Considering $\frac{\partial g}{\partial t} = h$ this identity may be replaced by the system of equations

$$X(x, t)\, g(x,t) + \frac{\partial g}{\partial x}(x, t)\, U(x, t) + h(x) = 0$$

$$\frac{\partial \varphi}{\partial t} = U(\varphi, t),\quad \varphi(x, 0) = x ,$$

$$X(\varphi(x, t), t) = A^{-1}(x, t) \frac{\partial A}{\partial t}(x, t),\quad A(x, 0) = id_F$$

The solvability of this system of equations follows from (8) and the corresponding theorem on solvability of initial value problem of an ordinary differential equation in a Banach space with smooth right-hand side [4, 18].

A study of deformations of contact r-definite germs is made in [39] .

In particular, the existence of versal deformation with a finite basis is ascertained and the normal forms of deformations are given.

Here we shall consider analogous problems, but for a more special case of deformations of zero index germs. In this connection it may be noted that the classical task of studying the behaviour of small solutions of an equation with parameter

$$f(x, \lambda) = 0, \quad \operatorname{ind} \frac{\partial f}{\partial x} = 0 \tag{9}$$

demand, from the general point of view, that the family $\tilde{f}_\lambda$, $f_\lambda(x) = f(x, \lambda)$ (i.e. the deformations) be analysed as particular deformation of the germ f_0, $f_0(x) = f(x, 0)$. For this, it is necessary to study all possible deformations with the aim of constructing their normal forms; this usually comes to constructing a canonical versal deformation [11, 25]. Further we shall make use of the concept of subordinate deformation : deformation $f(x, \lambda)$, $\lambda \in \Lambda$ is subordinate to the deformation $g(x, \mu)$, $\mu \in M$, if

$$f(x, \lambda) = A(x, \lambda)\, g(\varphi(x, \lambda), \mu(\lambda)), \ \tilde{\varphi}_\lambda \in D(E), \mu(\lambda) \in C^\infty,$$

$$\mu(0) = 0$$

Let the germ $\tilde{f}_0$ be finite* at zero [3, 4, 10] . This means a finite dimension of the ring of germs at the zero of C^∞-functionals in module of the ideal generated by functionals of the form $\alpha(f_0(x))$, $\tilde{\alpha} \in \mathfrak{M}(F)$. The mentioned factor-ring is called the local ring particularly at the zero of the mapping f_0. Denote it by $Q(f_0)$. Let the rank $Q(f_0)$ be equal to q (this means that $q = \dim \ker df_0(0)$). Then in virtue of finiteness of f_0, a linearly independent set of linear functionals

$$l = \{l_1, l_2, \dots, l_q\}, \quad l_i \in E^*, \quad i = 1, \dots, q$$

may be found such that $Q(f_0)$ is generated by cosets of functionals of the form

$$l^r(x), \quad r \in R \subset \mathbb{Z}_+^q, \text{ card } R = \dim Q \tag{10}$$

Here $\mathbb{Z}_+^q$ is a set of integer-valued multi-indices of dimension q with non-negative components $r = (r_1, r_2, \dots, r_q)$ and $l^r(x) = l_1^{r_1}(x) \dots l_q^{r_q}(x)$.

---

* The finiteness condition conforms to the so-called coarse case in [7] and to a quasi-regular case in [6].

Using the Malgrange-Weierstrass preparatory theorem 5 it may be proved that the deformation $f(x, \lambda)$ in the case examined by us is equivalent to the deformation $g(x, \lambda)$, where

$$g(x, \lambda) = f_o^k(x) + \sum_{r \in R} \alpha_r(\lambda)\, l^r(x) \tag{11}$$

Here, $k = \dim Q$, $f_o^{(k)}(x)$ is a segment of Taylor's series of order k for f_o, and $\alpha_r(\lambda)$ is a function of the class C^∞ with values in F, and equals zero when $\lambda = 0$.

Expression (11) could be further simplified by "discarding" the unnecessary monomial terms (of type (10)). Equation $g(x, \lambda) = 0$, where g is of the form (11), can be studied in a complex domain by "going out" in variables x. In this connection it may be noted that the following statement follows from the results of [10, 19] : the multiplicity of f_o at zero (i.e. $\dim Q(f_o)$) agrees with the topological index of zero for the complex extension $f_o^k(x)$.

Call the equations with g of the type (11) the equations in the Malgrange-Weierstrass form.

As illustrating examples, we shall consider cases of the most simple singularities of mapping f_o.

(a) Let $\dim \ker df_o(0) = 1$, $\dim Q = k$. Then f_o has a Morin singularity at zero: $Q(\tilde{f}_o) \cong \mathbb{R}[t]/\langle t^k \rangle$ [21]. Equation (9) in this case is equivalent to equation

$$Ax + \Big(l^k(x) + \sum_{j=0}^{k-1} \alpha_j(\lambda)\, l^j(x)\Big)e = 0, \quad \alpha_j(0) = 0 \tag{12}$$

where A is a linear Fredholm operator of zero index; l is a linear functional; e is a vector in space F; $\dim \ker A = 1$, $\ker A \,\bar{\in}\, \ker l$, $e \,\bar{\in}\, \operatorname{Im} A$. Let vector g be the generator in ker A and $l(g) = 1$. Then vector x is written as $x = tg + v$, $v \in \ker l$, $t \in \mathbb{R}$. Now we may go over from equation (12) to equation

$$t^k + \sum_{j=0}^{k-1} \alpha_j(\lambda) t^j = 0 \tag{13}$$

which is the Lyapunov-Schmidt branching equation [6, 7] . Clearly, equation (9) has a simple correct branching at zero if and only if the generated by it equation (13) has a simple correct branching at zero, and the zero germ of the set of solutions to (13) is equi-singular to

$N(f, (0,0))$.

If the left-hand side of equation (13) is a function with a finite-to-one critical point, then equation (9) has correct branching 9 .

(b) Let $\dim \ker df_o(0) = 2$, $\dim Q(f_o) = 4$ ($\dim Q < 4$ only for Morin singularities). Then (see [21]) $Q(f_o) = \mathbb{R}[t_1,t_2] / \langle t_1^2 \pm t_2^2, t_1t_2 \rangle$ and equation (9) is equivalent to equation

$$\begin{aligned} Ax &+ (l_1^2(x) \pm l_2^2(x))e_1 + l_1(x)l_2(x)e_2 + l_1^2(x)(\alpha_{20}(\lambda)e_1 + \\ &+ \beta_{20}(\lambda)e_2) + l_1(x)(\alpha_{10}(\lambda)e_1 + \beta_{10}(\lambda)e_2) + \\ &+ l_2(x)(\alpha_{01}(\lambda)e_1 + \beta_{01}(\lambda)e_2) + \alpha_{00}(\lambda)e_1 + \\ &+ \beta_{00}(\lambda)e_2 = 0 \end{aligned} \tag{14}$$

where $\dim \ker A = 2$, $e_1, e_2 \in F$ is a pair of linearly independent vectors in module of the subspace ImA, and the linear functionals l_1 and l_2 remain to be linearly independent when restricted on ker A. Let $g_1, g_2 \in \ker A$, $l_i(g_l) = \delta_{ij}$. Writing x as $x = t_1g_1 + t_2g_2 + v$, $v \in \ker l_1 \cap \ker l_2$, equation (14) can be replaced by a system of two scalar equations

$$\begin{aligned} &t_1^2 \pm t_2^2 + \alpha_{20}(\lambda)\, t_1^2 + \alpha_{10}(\lambda)t_1 + \alpha_{01}(\lambda)t_2 + \alpha_{00}(\lambda) = 0 \\ &t_1t_2 + \beta_{20}(\lambda)t_1^2 + \beta_{10}(\lambda)t_1 + \beta_{01}(\lambda)t_2 + \beta_{00}(\lambda) = 0 \end{aligned} \tag{15}$$

The system of equations (15) can be reduced to one scalar equation by eliminating one of the variables t_1 or t_2.

The conclusions are similar to those in (a).

Here it must be noted that for equations in the Malgrange-Weierstrass form the correctness does not follow from the simplicity of branching. In fact, we consider the following example of equation (13):

$$t^2 - \exp(-1/\lambda^2)\sin 1/\lambda = 0, \quad \lambda \in \mathbb{R}$$

Here small solutions are determined when $\lambda \in [1/\pi(2n + 1), 1/\pi 2n]$; there graphs are a sequence of ovals converging to the zero point. In this case the simplicity condition is fulfilled. However, the branching cannot be correct because every smooth functional on a closed oval has no less than two critical points.

Instead of the last example we may consider a somewhat more complicate example of equation (13) (with simple and incorrect branching)

$$t^2 + [\mu^2 - \exp(-1/\lambda^2)(1 + \sin 1/\lambda)]^2 - \exp(-2/\lambda^2) = 0$$

which differs from the previous one in that here the set of solutions is connected and locally connected.

The fact that every deformation of the finite-to-one germ $\tilde{f}_0$ is equivalent to the deformation of type (9) is indicative of the existence of versal deformation [11, 25] with finite basis. For example, the deformation

$$f_0^{(k)}(x) + \sum_{r \in R} \alpha_r l^r(x), \quad \alpha_r \in K, \quad K \oplus \operatorname{Im} df_0(0) = F$$

will be versal. We can point to other versal deformations that have a more convenient analytical form in one or the other sense. Ambiguity of the analytical form leads to the task of chosing a normal (canonical) form.

Applied to branching equations, this problem is discussed in detail in the following section.

4. The Lyapunov-Schmidt Method

In studying concrete Fredholm equations with a parameter use is generally made of the method of finite reduction or, more exactly, the Lyapunov-Schmidt method [2,6, 7] . Many publications dedicated to various aspects of using this method have appeared in the last few years [18, 22, 34, 38]. We shall discuss the question of chosing a normal form for the branching equation, and certain problems related to computation of its solutions.

Before giving necessary definition of a branching equation, we note that in the further discussion the deformation $\tilde{f}_\lambda$ is assumed to belong to the set $\mathfrak{M}(E, F)$, i.e. $f(0, \lambda) = 0$ when $\forall \lambda \in \Lambda^p$. The set Λ^p is an open or closed p-dimensional domain containing zero. The non-zero function $x = x(\lambda)$ will be called the small solution to equation $f(x, \lambda) = 0$ such that $f(x(\lambda), \lambda) = 0$. The function $x(\lambda)$ is assumed to be defined not on the whole domain Λ^p , but only on a part of it.

In virtue of the theorem on local representation of Fredholm's mapping [22] it may be asserted that a smooth family exists such that

$$\{ \tilde{\varphi}_\lambda \in \mathrm{Diff}_0(E) \}_{\lambda \in U^p},\ 0 \in U^p \subset \Lambda^p$$

where U^p is open in Λ^p, that is in a neighbourhood $\mathcal{O}$ of zero in E :

$$f(\varphi(x, \lambda), \lambda) = (j(u), g(u, v, \lambda))$$

where j(u) is a linear isomorphism $L \to R$, L is a direct complement to

$$N = \ker \frac{\partial f}{\partial x}(0,0), \quad R = \mathrm{Im} \frac{\partial f}{\partial x}(0,0), \quad u \in L, \quad v \in N,$$

$$g(u, v, \lambda) \in K;$$

K is a direct complement to R, $\frac{\partial g}{\partial x}(0, 0, \lambda) = 0$.

From differential calculus in a Banach space [18] follows the possibility of representing

$$g(u, v, \lambda) = \sigma(v, \lambda) + B(u, v, \lambda)u$$

where $\sigma(v, \lambda) = g(0, v, \lambda)$, $B(u, v, \lambda)$ is a smooth mapping $\mathcal{O} \times U^p \to L(L, K)$. Here, the equality

$$(j(u), g(u,v, \lambda)) = A(u, v, \lambda)(j(u), \sigma(v, \lambda))$$

holds in which $A(u, v, \lambda)$ is a smooth mapping $\mathcal{O} \times U^p \to GL(F)$ given by the correspondence

$$(u, v, \lambda) \longmapsto \begin{pmatrix} id_R & 0 \\ B(u,v, \lambda)j^{-1} & id_K \end{pmatrix}$$

(recall that $F = R \oplus K$).

Thus, we have ascertained that the deformation $f(x, \lambda)$ is equivalent to the deformation $(j(u), \sigma(v, \lambda))$. The equation

$$\sigma(v, \lambda) = 0 \tag{16}$$

is called the branching equation. The essence of the Lyapunov-Schmidt method consists in replacing (9) by (16). There are other ways of deriving the branching equation [6, 7, 38] . To disgress from the method of deducing the branching equation, we may define it as follows : any equation (16) is called the branching equation conforming to the deformation f(x,) if the following condition holds : f(x,) is equivalent to deformation of the form $(j(u), \sigma(v, \lambda))$, $j: L \to R$ is a linear morphism, $v \in N$, $\sigma(v, \lambda) \in K$. Here, L and R are as before. Clearly, for this definition $\frac{\partial\sigma}{\partial v}(0, \lambda) \equiv 0$. Further, for any two branching equations $\sigma(v, \lambda) = 0$ and $\tau(v, \lambda) = 0$ the deformations $\sigma(v, \lambda)$ and $\tau(v, \lambda)$ are equivalent.

From the definition follows also the isomorphism of local rings :

$$Q(\tilde{f}_0) = Q(\tilde{\sigma}_0), \qquad Q(\tilde{f}) = Q(\tilde{\sigma}) \tag{17}$$

Here $\tilde{\sigma}_0$ is the germ of mapping $\sigma_0(v) = \sigma(v, 0)$.

Relation (17) may be used for invariant determination of the branching equation.

Further we shall consider only equation (16). In so doing, we assume $N = K = \mathbb{R}^n$, $\sigma = (\sigma_1, \ldots, \sigma_n)$, $v = (v_1, \ldots, v_n)$, $\sigma_0 = (\sigma_{01}, \ldots, \sigma_{0n})$.

Let $(p_1, \ldots, p_n)$ be a set of polynomials such that their germs are generators in the ideal $\langle \tilde{\sigma}_{01}, \ldots, \tilde{\sigma}_{0n} \rangle$ of the ring $\mathcal{E}(n) = \mathcal{E}(\mathbb{R}^n)$ and these polynomials do not contain terms that appear in $\mathfrak{m}\langle \tilde{\sigma}_{01}, \ldots, \tilde{\sigma}_{0n} \rangle$, where $\mathfrak{m} \subset \mathcal{E}(n)$ is the maximal ideal. Let further $w_1, \ldots, w_k$ be the monomial basis of local ring $Q(\tilde{\sigma}_0) = \mathcal{E}(n)/\langle \tilde{\sigma}_{01}, \ldots, \tilde{\sigma}_{0n} \rangle$ (i.e. the co-sets of these monomials form a basis in factor Q). Finally, we consider the following deformation:

$$\mathscr{X}(v, \xi_1, \ldots, \xi_k) = p(v) + \sum_{j=1}^{k} \xi_j w_j(v) \tag{18}$$

Call it the principal deformation. From the above given arguments follows the following

<u>Theorem 9</u>. The principal deformation is versal for a class of smooth deformations of the germ $\tilde{\sigma}_0$.

From this theorem it follows that the branching equation conforming to deformation $f(x, \lambda)$ may be represented as

$$\tau(v, \lambda) = p(v) + \sum_{j=1}^{k} \xi_j(\lambda) w_j(v) = 0 \tag{19}$$

Here, $\xi_j(\lambda)$ are smooth functions. This form will be called the normal form.

<u>Remark</u>. From [13, 19, 26] it follows that a set of polynomials

$q_1, \ldots, q_k$ can be chosen for which the following equality (applied to any deformation of the form (19) holds:

$$\xi_j = \frac{1}{(2\pi i)^n} \int\limits_{|p_m(z)|=\varepsilon,\ \forall m} \frac{\tau(z,\lambda)\, q_j(z)}{p_1(z)\cdot\ldots\cdot p_n(z)}\, dz_1 \wedge \ldots \wedge dz_n$$

where $z = (z_1, \ldots, z_n) \in \mathbb{C}^n$.

These polynomials are found from the conditions

$$k\, \tilde{q}_i\, \tilde{w}_j = \delta_{ij}\, \tilde{J} \pmod{\langle \tilde{p}_1, \ldots, \tilde{p}_n \rangle}$$

$$J = \det \frac{\partial p}{\partial z}$$

The examples given in the previous section show that equations in the form (18), (19) (as equations of collections of variables (v, λ)) may not have correct branching although they have the same simplicity of branching. All examples of this kind are necessarily followed by non-analticity of the functions $\xi_1(\lambda), \ldots, \xi_k(\lambda)$.

Computation of branching equations in concrete examples is a difficult task, as a rule. The main difficulty resides in the large number of one-type, but cumbersome calculations [6, 7, 15] . Usually one succeeds in calculating only finite Taylor approximations of the functions $\xi_j(\lambda)$. That is why the problem of finite definiteness of equation (19) arises. The following statement is a direct corollary of Theorem 3.

Theorem 10. Let the coefficients $\xi_j(\lambda)$ of deformation (19) be polynomials. Assume that the following conditions are fulfilled.

1. $\deg p \leq r$, $\deg \xi_j \leq r - \deg w_j$

2. In a neighbourhood of zero of the space $\mathbb{R}^n \times \mathbb{R}^p$ holds the estimate

$$\left\|\left(\frac{\partial \tau}{\partial v}\right)^* h\right\| + \left\|\left(\frac{\partial \tau}{\partial \lambda}\right)^* h\right\| \geqslant c(b)\left(|v|^{r-1} + |\lambda|^{r-1}\right)|h|$$

as soon as $|\tau(v,\lambda)| < b\left(|v|^{r+1} + |\lambda|^{r+1}\right)$, where the component $C(b)$ depends on b. Then equation (19) is r-definite.

This statement enables us to judge about the topological structure of the collection of the graphs of small solutions to an r-definite equation of the type (19) upon substituting the functions $\xi_j(\lambda)$ by their Taylor approximations.

The next step in the study of (19) is to estimate the perturbations of small solutions depending on the perturbations of the equation. In [7] is described a general approach of asymptotic approximations to simple small solutions, i.e. in the given to such small solutions $V(\lambda)$ of equation (19) for which $\det \frac{\partial \tau}{\partial V}(V(\lambda), \lambda) \neq 0, \ \forall \lambda \in \Lambda^P$. We shall now formulate one of the results 7 , as applied to equation (19).

Theorem 11. Suppose the continuous function $V(\lambda)$ is defined on a closed set $X \subset \Lambda^P$ for which zero is the limiting point, $V(0) = 0$. Assume the following conditions hold:

$$|\tau(V(\lambda), \lambda)| \leq \alpha |\lambda|^{r+1} \tag{20}$$

$$\left\| \frac{\partial \tau}{\partial V}(V(\lambda), \lambda) h \right\| \geq \beta |\lambda|^s |h|, \quad s \leq r \tag{21}$$

$$\left\| \frac{\partial^2 \tau}{\partial V^2}(V(\lambda) + h, \lambda) \right\| \leq \gamma |\lambda|^1 (1 + |h|), \quad r + 1 \geq 2s \tag{22}$$

Then for some $\nu > 0$ and a neighbourhood of zero U^P on $U^P \cap X$ a simple small solution $u(\lambda)$ to equation (19), for which $|u(\lambda) - V(\lambda)| \leq \nu |\lambda|^{r-s+1}$, is defined.

Proof. [7]. A small solution to equation (19) may be found as a fixed point of operator

$$Tu = u, \quad u \in L_\nu(U^P) \tag{23}$$

where

$$L_\nu(U^P) = \{u \in C(U^P \cap X, E), \ |u(\lambda) - V(\lambda)| \leq \nu |\lambda|^{r-s+1}\}$$

$$(Tu)(\lambda) = u(\lambda) - \left[\frac{\partial \tau}{\partial \lambda}(V(\lambda), \lambda)\right]^{-1} \tau(u(\lambda), \lambda)$$

From differential calculus we get formula [18]:

$$\begin{aligned} \tau(u(\lambda), \lambda) &= \tau(V(\lambda), \lambda) + \frac{\partial \tau}{\partial V}(V(\lambda), \lambda)(u(\lambda) - v(\lambda)) + \\ &+ \left[\int_0^1 (1-s) \frac{\partial^2 \tau}{\partial V^2}(V(\lambda) + s(u(\lambda) - v(\lambda)))ds\right] \times \\ &\times (u(\lambda) - v(\lambda))^{(2)}. \end{aligned}$$

Here, in virtue of estimates (20), (21), and (22), the following inequalities hold :

$$|V(\lambda) - (Tu)(\lambda)| \leq \nu|\lambda|^{r-s+1} \text{ for}$$
$$\beta^{-1}(\alpha + \gamma\nu)^r|\lambda|^{r+l-2s+1}(1 + \nu|\lambda|^{r-s+1})) \leq \nu$$

$$|(Tu_1)(\lambda) - (Tu_2)(\lambda)| \leq q\,|u_1(\lambda) - u_2(\lambda)| \text{ for}$$
$$2\nu\beta^{-1}\gamma(1 + 2\nu|\lambda|^{r-s+1})|\lambda|^{l+r-2s+1} \leq q$$

These inequalities imply that for a sufficiently small $|\lambda|$ the closed set $L_\nu(U^p)$ is transformed into itself by operator T. And T acts in L_ν as contracting operator if the neighbourhood U^p is small enough. Hence, in virtue of the Banach theorem, there exists a fixed point.

In conclusion we must mention about further possible ways of studying equation (19).

Not long ago the Kronecker method of eliminating an unknown with subsequent application of Newton's polygonal method was the main technique of studying branching equations [6, 7]. In the last few years some reports have appeared where use is made of Newton's general polyhedrons [14, 16, 24, 25, 27, 28]. The application of Newton's polyhedrons eliminates the need of using the Kronecker method.

Using the Newton polygonal method we can construct the Taylor approximations of small solutions of any order. This construction is a sequence of steps, each of which is a generalized σ-process [16, 25, 28]. In [36] the σ-process is used for analysing the branching equation generated by Carman equation from thin elastic shell theory [6, 12].

For brevity we assume the mapping $\tau(v, \lambda)$ to be a polynomial mapping: $\tau(v, \lambda) = \sum a_{pq} V^p \lambda^q$, $\lambda \in R^1$. Then by the generalized σ-process (or resolution of a singularity) we mean the following construction : in equation (19) we substitute

$$V_k = u_k \mu^{\beta_k}, \quad \lambda = \mu^{\beta_0}, \quad k = 1, \ldots, n$$

The mapping $\tau(v, \lambda)$ now takes the form

$$\tau(V, \lambda) = \sum_{p,q} a_{pq} v^p \lambda^q = \sum_{p,q} a_{pq} u^p \mu^{\langle p,\beta\rangle + q\beta_0}$$

$$\langle p, \beta\rangle = \sum_{j=1}^{n} p_j \beta_j$$

If the integer vector $(\beta_1, \ldots, \beta_n, \beta_o) \in \mathbb{Z}_+^{n+1}$ is perpendicular to one of the three edges of the Newton polyhedron $\Gamma \subset \mathbb{R}^{n+1}$, $\Gamma = \mathrm{conv}\{(p, q) \mid a_{pq} \neq 0\}$, which are seen from the origin of coordinates, then equation (19) is replaced by equation $\varphi(u) + \sum b_{pq} U^p \mu^q = o$ (where $\mu^m(\varphi(u) + \sum b_{pq} U^p \mu^q) = \tau(v, \lambda)$), whose every solution at $\mu = 0$ is the minor coefficient of the Puiseux series (i.e. of small solution).

REFERENCES

(Items I-3,6,7,IO,II,I3-I6,I9,20,22-32 are in Russian)

1. Krasnosel'sky, M.A. Topological Methods in the Theory of Nonlinear Integral Equations, Gostekhizdat, Moscow, 1956.
2. Wainberg, M.M. and Trenogin, V.A. Lyapunov-Schmidt Methods in the Theory of Nonlinear Equations and Their Further Development, Uspehi Math. Nauk, 1962, vol. 17, issue 2.
3. Arnold, V.I. Singularities of Smooth Mappings, Uspehi Math. Nauk, 1968, vol. 23, issue 1.
4. Thom, R. and Levine, H. Singularities of Differentiable Mappings, I.-Bonn, 1959.
5. Malgrange, B. Ideals of Differentiable Functions, Oxford University Press, 1966.
6. Wainberg, M.M. and Trenogin, V.A. Branching Theory of Solutions of Nonlinear Equations, Nauka, Moscow, 1969.
7. Krasnosel'sky, M.A., Vainikko, G.M., Zabreiko, P.P., Rutitsky, Ya.B., and Stetsenko, V.Ya. Approximate Solution of Operator Equations, Nauka, Moscow, 1969.
8. Pham, F. Introduction a l'etude topologique des singularites de Landau, Gauthier-Villars, Paris, 1967.
9. Milnor, J. Singular Points of Complex Hypersurfaces, Annals of Math. Studies 61, Princton University Press, New Jersey, 1968.
10. Palamodov, V.P. Remarks on Finite-to-One Differentiable Mappings, Functional Analysis, 1972, vol. 6, issue 2.
11. Arnold, V.I. Lectures on Bifurcations and Versal Families, Uspehi Math. Nauk, 1972, vol. 27, issue 5.
12. Bifurcation Theory and Nonlinear Eigenvalue Problems. Eds. J.B. Keller and S. Antman, W.A. Benjamin, Inc., New York - Amsterdam, 1969.
13. Shoshitaishvili, A.N. Structures of Ideals of Local Rings of Finite-to-One Mappings of Spaces with Equal Dimensions is Self-dual, Uspehi Math. Nauk, 1974, vol. 29, issue 3.
14. Kushnirenko, A.I. Newton Polyhedrons and Milnor Numbers, Functional Analysis, 1975, vol. 9, issue 1.

15. Botashev, A.I. Finite Methods in the Theory of Mutivariate Branching, Frunze, 1976.
16. Varchenko, A.N. Newton Polyhedrons and Estimates of Oscillating Integrals, Functional Analysis, 1976, vol. 10, issue 3.
17. Brecker, Th. and Lander, L. Differentiable Germs and Catastrophes, London Mathematical Society Lecture Notes 17, Cambridge, 1975.
18. Nirenberg, L. Topics in Nonlinear Functional Analysis, Courant Inst. of Math. Sci., New York, 1974.
19. Khimshiashvili, G.N. On Local Degree of a Smooth Mapping, Soobshch. AN GSSR, 1977, vol. 85, N 2.
20. Khimshiashvili, G.N. On Small Solutions of Nonlinear Fredholm Equations, Vestn. Moscow State University, ser. math. mech., 1977, N 2.
21. Golubitsky, M. and Guillemin, V. Stable Mappings and Their Singularities, Graduate Texts in Mathematics 14, Springer-Verlag, New York - Heidelberg - Berlin, 1973.
22. Borisovich, Yu.G., Zvyagin, V.G., and Sapronov, Yu.I. Nonlinear Fredholm Mappings and the Leray-Schauder Theory, Uspehi Math. Nauk, 1977, vol. 32, issue 4.
23. Zachepa, V.R. On Extension Along Parameter of Regularly Branching Solutions of Nonlinear Operator Equations, In : Methods of Solving Operator Equations, Voronezh University Press, Voronezh, 1978.
24. Bliznyakov, N.M. On Estimates of Topological Index of a Singular Point of Vector Field, Dep. in VINITI on Feb. 8, 1979, No. 589-79.
25. Arnold, V.I. Additional Chapters on the Theory of Differential Equations, Nauka, Moscow, 1978.
26. Khovansky, A.G. Newton Polyhedrons and the Euler-Jacobi Formula, Uspehi Math. Nauk, 1978, vol. 33, issue 6.
27. Bliznyakov, N.M. Computation and Estimation of Index of a Singular Point of Vector Field on a Plane, Deposited in VINITI on June 28, 1979, No. 3041-79.
28. Bruno, L.D. Local Methods of Nonlinear Analysis of Differential Equations, Nauka, Moscow, 1979.
29. Zachepa, V.R. and Sapronov, Yu.I. On Local Analysis of Nonlinear Fredholm Equations, Proceedings of the International Conference on Topology, Moscow, 1979.
30. Zachepa, V.R. On Finitely Defined Solutions to Karman's Equation, Deposited in VINITI on August 14, 1980, No. 3614-80.
31. Zachepa, V.R. Finitely Defined Equations, Deposited in VINITI on August 14, 1980, No. 3615-80.
32. Zachepa, V.R. and Sapronov, Yu.I. Regular Branching and Regular Deformations of Nonlinear Fredholm Equations, Deposited in VINITI

on August 14, 1980, No. 3617-80.

33. Chow, S.N., Hale, J.K., and Mallet-Paret, J. Applications of Generic Bifurcation, I- Arch. Rational Mech. Anal, 1975, vol. 59, N 1, II - item., 1976, vol. 62, N 2.
34. Ize, I.A. Bifurcation Theory for Fredholm Operators, Mem. Amer. Math. Soc., 1976, N 174.
35. Antman, S.S. Bifurcation Problems for Nonlinearly Elastic Structures, Applications of Bifurcation Theory, New York, 1977.
36. Knightly, G.H. Some Math. Problems from Plate and Shell Theory, Lecture Notes in Pure and Applied Mathematics, 1977, vol. 19.
37. Marsden, J.E. Qualitative Methods in Bifurcation Theory, Bull. Amer. Math. Soc., 1978, vol. 84.
38. Marsden, J.E. On the Geometry of Lyapunov-Schmidt Procedure, Lecture Notes Math., 1979, N 755.
39. Guimaraes, L.C. Contact Equivalence and Bifurcation Theory, Lecture Notes Math., 1980, N 799.
40. Hoyle, S.L. Local Solution Manifolds for Nonlinear Equations, Nonlinear Analysis: Theory, Methods, and Applications, 1980, vol. 4, N 2.
41. Thom, R. Local Topological Properties of Differentiable Mappings, Differential Analysis (Papers presented at the Bombay Colloquium, 1964), 191-202, Oxford University Press, Oxford and New York, 1964.
42. Mather, J.N. Stability of C^{∞}-Mappings, III : Finitely Determined Mappings, Publ. Math. IHES, 1968, vol. 35.
43. Chillingworth, D.A. A Global Genericity Theorem for Bifurcation in Variational Problems, J. of Functional Analysis, 1980, vol. 35.

Translated from the Russian by P.K. Dang

CHARACTERISTIC CAUCHY PROBLEM ON A COMPLEX-ANALYTIC MANIFOLD

B.Yu.Sternin and V.E.Shatalov
Moscow Institute of Civil Aviation Engineers
Department of Mechanics
125838 Moscow, USSR

This report is dedicated to the study of characteristic Cauchy problem on a complex analytic manifold in the class of analytic functions.

Fundamental studies on this topic have been made in the classical works of J. Leray (see [1] and the bibliography listed therein) who, using the uniformization concept, studied the qualitative nature of the solution to the Cauchy problem and obtained its asymptotic representation close to the initial surface. Unfortunately, the developed techniques proved to be unsuitable for global study of the problem; in particular, the solution in a neighbourhood of focal (caustic) points is not of the form proposed by J. Leray.

In the last few years the authors have made some progress in studying the characteristic Cauchy problem "in large" [3-5] . In these works, in particular, we introduced a new concept of uniformization based on the ideas and techniques of the theory of Feinman integrals and, the more so, constructed an asymptotics of the Cauchy problem solution in a neighbourhood of the singular points without assuming them to be close to the initial manifold.

This report consists of two parts. Part I, written by B.Yu. Sternin, contains practically the results of J. Leray concerning uniformization and asymptotic expansion of the Cauchy problem "in small". Our presentation and style differ largely from that of J. Leray.

Part II, written by V.E. Shatalov, contains new results concerning global theory of the Cauchy problem. It concerns mainly the study of the introduced concept of Legendre's uniformization (which includes J. Leray's uniformization as a particular case) and the construction of asymptotic solution to the Cauchy problem.

We shall briefly dwell on the main concepts and specific features of the Cauchy problem.

Consider the problem

$$\begin{cases} A(x, \frac{\partial}{\partial x}) \, u(x) = f(x) \\ u(x) - \varphi(x) \quad \text{has a zero of order } r \text{ on } S \end{cases} \tag{1}$$

for a differential operator of order r:

$$A(x, \frac{\partial}{\partial x}) = \sum_{|\alpha| \le r} a_\alpha(x) \left(\frac{\partial}{\partial x}\right)^\alpha$$

with holomorphic coefficients $a_\alpha(x)$ in the domain of n-dimensional complex space. We also assume that functions $f(x)$ and $\varphi(x)$ are holomorphic and the surface S on which the initial data are given is an analytic submanifold of co-dimensionality 1. The latter implies that locally the surface S may be written by equation $S(x) = 0$, and $dS_{S(x)=0} \ne 0$.

Note that problem (1), in general, cannot have an holomorphic solution, as it is shown by the following example:

$$\frac{\partial}{\partial x^1} u(x^1, x^2) = 1, \quad u(x^1, x^2)\big|_S = 0$$

Here $S = \{S(x) = 0\} = \{x^2 - (x^1)^2 = 0\}$

It is easily seen that function

$$u(x^1, x^2) = x^1 - \sqrt{x^2}$$

is a solution to this problem.

An analysis of this example shows that the solution to problem (1) is, in general, a multiple-valued analytic function, and, as a consequence of this, the solution is not smooth in a neighbourhood of the ramification point.

Naturally the statement of the problem concerning the study of Cauchy problem (1) includes in it the

1. determination of the nature of solution's ramification;
2. determination of the extent to which the solution is not smooth.

These problems can be solved by representing respectively the Riemannian surface of the solution and its asymptotics in a neighbourhood of the branching set.

I. Uniformization and Asymptotics of the Cauchy Problem "in small"

<u>The scalar case</u>. We include problem (1) into the one-parameter family of problems:

$$\begin{cases} A(x, \frac{\partial}{\partial x}) u(x,\xi) = f(x,\xi) & (2) \\ u(x,\xi) \equiv \varphi(x,\xi) \pmod{r} \quad \text{on } S_\xi & (3) \end{cases}$$

where $S = \{S(x) = \xi\}$, $f(x, \xi)$, $\varphi(x, \xi)$ are holomorphic functions of a set of variables, and the comparison of (3) indicates that the difference $u(x, \xi) - \varphi(x, \xi)$ has a zero of order r on S .

Obviously, without loss of generality, we may assume the function $\varphi = 0$. Therefore, condition (3) is written as :

$$u(x, \xi) \equiv 0 \quad \text{on } S \pmod{r}. \tag{3'}$$

We shall find the Riemannian surface of the solution in the form

$$\mathbb{C}^{n+1}_{(x,t)} \xrightarrow{\pi} \mathbb{C}^{n+1}_{(x,\xi)}, \quad (x,t) \overset{\pi}{\longmapsto} (x, S(x,t)) \tag{4}$$

where $S(x,t)$ is a holomorphic function. We shall choose function $S(x,t)$ in such a manner that on the Riemannian surface (with coordinates (x,t)) the solution $u(x, S(x,t)) = U(x,t)$ is holomorphic. For this, it is sufficient that function $U(x,t)$ should satisfy the Cauchy-Kovalevsky system with holomorphic initial data. Here we shall say that substitution

$$\xi = S(x,t) \tag{5}$$

uniformizes problems (2), (3). Obviously, if a uniformizing substitution is found, then problem (1) is solved. Indeed, the ramification points of the Riemannian surface (4) are determined in accord with the implicit-function theorem from equation

$$S_t(x,t) = 0$$

The solution $t = t(x, \xi)$ to equation (5) is then a multiple-valued function, and, hence, such is the solution $u(x, \xi) = U(x, t(x, \xi))$.

And what is more, the knowledge of uniformizing substitution enables us to determine the shape of solution's asymptotics in a neighbourhood of singular points.

Indeed,

$$\frac{\partial u(x,\xi)}{\partial x^j} = \frac{\partial U}{\partial x^j}(x,t(x,\xi)) + \frac{\partial U}{\partial t}(x,t(x,\xi))\,\frac{\partial t(x,\xi)}{\partial x^j} =$$

$$= \frac{\partial U}{\partial x^j}(x,t(x,\xi)) - \frac{S_{x^j}(x,t(x,\xi))}{S_t(x,t(x,\xi))}\,\frac{\partial U}{\partial t}(x,t(x,\xi))$$

so that $\frac{\partial u}{\partial x^j}$ (x, S(x,t)) has already a polar singularity at $S_t(x,t)=0$. Thus, we go over to finding a uniformizing substitution (i.e. a function $\xi = S(x,t)$ for which the function U(x,t) satisfies the Cauchy-Kovalevsky system).

By direct calculation we obtain

$$\left[A\left(x,\frac{\partial}{\partial x}\right)u(x,\xi)\right]_{\xi=S(x,t)} = \sum_{j=0}^{r}\hat{\mathcal{L}}_j[S]\,U_{r-j}(x,t) \qquad (6)$$

where

$$U_j(x,t) = \left(-\frac{\partial}{\partial \xi}\right)^j u(x,\xi)\Big|_{\xi=S(x,t)},$$

$$\hat{\mathcal{L}}_0[S] = A^{(r)}(x,S_x(x,t)),$$

$$\hat{\mathcal{L}}_1[S] = A^{(r)}_{p_i}(x,S_x(x,t))\frac{\partial}{\partial x^i} + \frac{1}{2}S_{x^ix^j}(x,t)\times$$

$$\times A^{(r)}_{p_ip_j}(x,S_x(x,t)) + A^{(r-1)}(x,S_x(x,t)),$$

$\hat{\mathcal{L}}_j[S]$ j = 1,2, . . . , r are functionals of S;

$A^{(r)}(x,p)$ is the principal symbol of operator $A(x, \frac{\partial}{\partial x})$:

$$A^{(r)}(x,p) = \sum_{|\alpha|=r} a_\alpha(x)p^\alpha;\quad A^{(r-1)}(x,p) = \sum_{|\alpha|=r-1} a_\alpha(x)p^\alpha.$$

Differentiating the identity

$$U_{k-1}(x,t) = \left(-\frac{\partial}{\partial \xi}\right)^{k-1} u(x,\xi)\Big|_{\xi=S(x,t)} \qquad (7)$$

we, in virtue of (5), obtain

$$\frac{\partial U_{k-1}}{\partial t} = -S_t U_k . \tag{8}$$

Thus, equalities (2), (6), and (8) yield a set of equations with respect to $U_0, \ldots, U_{r-1}$:

$$\begin{cases} -\dfrac{A^{(r)}(x, S_x(x,t))}{S_t(x,t)} \dfrac{\partial U^{(r-1)}}{\partial t} = -\sum_{j=1}^{r} \hat{\mathcal{L}}_j [S] U_{r-j} + f(x, S(x,t)), \\ \dfrac{\partial U_{k-1}}{\partial t} = - S_t(x,t) U_k , \quad k = 1, \ldots, r-1 \end{cases} \tag{9}$$

If in the first equation we choose S(x,t) in such a manner that

$$- \frac{A^{(r)}(x, S_x(x, t))}{S_t(x, t)} = 1$$

or

$$S_t + A(x, Sx) = 0 \tag{10}$$

then system (9) is converted into the Cauchy-Kovalevsky system if, of course, the initial data for it are given at $t = 0$. The latter can be obtained by assuming $S(x, 0) = S(x)$.

The initial data for functions U_j, $j = 0, \ldots, r-1$ are induced from the initial data of (3). Namely

$$U_j(x, 0) = 0, \quad j = 0, \ldots, r-1 \tag{11}$$

Thus, the following holds

<u>Theorem 1.0</u>. Functions $U_j(x,t)$ are holomorphic for $j = 0, \ldots, r-1$.

<u>Definition 1.1</u>. We say that function $u(x, \xi)$ is uniformizable up to order N, or briefly, N-uniformizable by substitution (5) (or projection (4)) if the functions

$$U_j(x,t) = \left(-\frac{\partial}{\partial \xi}\right)^j u(x, \xi) \Big|_{\xi = S(x,t)} \tag{12}$$

are holomorphic in a neighbourhood $t = 0$, $j = 0, \ldots, N$.

In these terms, Theorem 1.0 admits the following statement.

<u>Theorem 1.1</u>. (uniformization theorem). The solution $u(x, \xi)$ to problem (2), (3) is uniformizable by substitution $\xi = S(x,t)$ up to order r-1 if the function $S(x,t)$ satisfies the following Cauchy problem for the Hamilton-Jacobi equation:

$$\begin{cases} S_t + A^{(r)}(x, S_x) = 0 \\ S(x,0) = S(x) \end{cases} \tag{13}$$

Now the function $u(x, \xi)$ is not r-uniformizable by substitution (5). To know the nature of singularity of the r-th derivative function $u(x, \xi)$, we apply operator $(-\partial/\partial\xi)$ to equation (2) and resubstitute (5). Analogous to the above described, we obtain

$$\frac{\partial U_r}{\partial t} + \hat{\mathcal{L}}_1[S]U_r = -\sum_{j=2}^{r} \hat{\mathcal{L}}_j[S]U_{r-j+1} + \left(-\frac{\partial}{\partial\xi}\right) f(x,\xi)\Big|_{\xi = S(x,t)} \tag{14}$$

In virtue of the uniformization theorem, the right-hand side of (14) is a holomorphic function so that by module of holomorphic functions ($U_r = \tilde{U}_r$ + f. h.) equation (14) is rewritten as

$$\frac{\partial \tilde{U}_r}{\partial t} + \hat{\mathcal{L}}_1[S]U_r = 0. \tag{15}$$

From (2) and (6) we obtain the following initial data for equation (15):

$$A^{(r)}(x, S_x)U_r + \sum_{j=1}^{r} \hat{\mathcal{L}}_j[S]U_{r-j} = f(x, S(x,t)).$$

At t = 0 the second group of terms in the sum disappears in virtue of (11) and the fact that operators $\hat{\mathcal{L}}_j$ contain the derivatives only with respect to x. Thus

$$\tilde{U}_r\Big|_{t=0} = \frac{f(x, S(x))}{A^{(r)}(x, S_x(x))} = -\frac{f(x, S(x))}{S_t(x,0)}. \tag{16}$$

Now note that the function $A^{(r)}(x, S_x)$ is the first integral of the field

$$\frac{d}{dt} = \frac{\partial}{\partial t} + A^{(r)}_{p_i}(x, S_x(x,t))\frac{\partial}{\partial x^i}. \tag{17}$$

Here it may be assumed that

$$U_r = \frac{V_r(x, t)}{A^{(r)}(x, S_x)}$$

For V_r, from (15) and (16) we obtain

$$\begin{cases} \frac{\partial V_r}{\partial t} + \hat{\mathcal{L}}_1[S]V_r = 0, \\ V_r(x, 0) = f(x, S(x)) \end{cases} \tag{18}$$

This is an ordinary differential equation on the trajectories of field (17).

Theorem 1.2. (asymptotic expansion theorem). The equality holds

$$\left(-\frac{\partial}{\partial \xi}\right)^r u(x, \xi)\bigg|_{\xi = S(x,t)} = U_r(x,t) = \frac{V_r(x,t)}{A^{(r)}(x, S_x(x,t))} + f.h.$$

where the function $V_r(x,t)$ is the solution to problem (18).

Corollary 1.1. Function

$$u(x, \xi) - \frac{V_r(x,t)}{A^{(r)}(x, S_x(x,t))}\bigg|_{t = t(x, \xi)}$$

is uniformizable up to order r.

Note that the set of singularities of function $u(x, \xi)$ consists of points which are the images, when projected on (4), of those points (x,t) where $S_t = 0$, and hence, in virtue of (10), of those points where $A^{(r)}(x, S_x) = 0$. However, since $A^{(r)}$ is the first integral, the set of singularities is the projection on space (x, ξ) of the system of bicharacteristics that get out from those points of the initial surface S_ξ where

$$A^{(r)}(x, S_x) = 0$$

The points of the initial surface where the last equality is fulfilled are called the characteristic points.

From the above described it is seen that the presence of characteristic points in the initial surface gives rise to many-valued solutions and, as a result, to their non-smoothness in a neighbourhood of the rays that emerge from the characteristic points.

Sets of equations. Let $A^{(r)}(x,p)$ be a (quadratic) matrix of the principal symbol of operator $A(x, \frac{\partial}{\partial x})$. Assume the order of homogeneity in p of all matrix elements to be similar and equal to r *).

Suppose that

1. The characteristic points of the initial surface

$$S_\xi = \{\xi = S(x)\} \tag{19}$$

(points $x \in S_\xi$, where $\det A^{(r)}(x, S_x) = 0$) form an analytical set of co-dimensionality 1, i.e.

$$\det A^{(r)}(x, S_x) \not\equiv 0 \tag{20}$$

2. There exists a holomorphic non-zero matrix $\Sigma(x,p)$ such that

$$A^{(r)}(x,p)\,\Sigma(x,p) = \Sigma(x,p)A^{(r)}(x,p) = H(x,p)\cdot I \tag{21}$$

and

$$dH \neq 0 \text{ on } \operatorname{char} H = \{(x,p) \mid H(x,p) = 0\}. \tag{22}$$

Here I is a unit matrix.

The function $H(x,p)$ satisfying conditions (21) and (22) is called the Hamilton function of operator $A(x, \frac{\partial}{\partial x})$.

It can be verified that a set of zeros of the determinant of matrix $A^{(r)}(x,p)$ agrees with the zeros of the Hamilton function:

$$\{(x,p) \mid \det A^{(r)}(x,p) = 0\} = \operatorname{char} H$$

The role of the matrix appearing in (21), that is of the "scalarizator" of operator $A^{(r)}(x,p)$, is seen from the following algebraic lemma. We leave the proof of this lemma to the reader.

<u>Lemma 1.1</u>. On the set char H the following relations hold

$$\operatorname{Im} A^{(r)}(x,p) = \ker \Sigma(x,p),$$

$$\ker A^{(r)}(x,p) = \operatorname{Im} \Sigma(x,p)$$

Thus, we consider the Cauchy problem

$$\begin{cases} A(x, \frac{\partial}{\partial x})\, u(x,\xi) = f(x,\xi) \\ u(x,\xi) \equiv 0 \pmod{r} \quad \text{on } S_\xi \end{cases} \tag{23}$$

---

*) It is not difficult to go over to Douglas-Nirenberg systems.

for the matrix differential operator $A(x, \frac{\partial}{\partial x})$ of order r and dimension $m \times m$. Thus, $u(x, \xi)$ and $f(x, \xi)$ are holomorphic vector-functions. The surface S_ξ is described by equation (19) and $ds(x) \neq 0$ at the points of the surface. And (mod r) implies, as before, that on the surface S_ξ the function $u(x, \xi)$ has a zero of order r.

Following the scheme developed in the previous section we obtain on the Riemannian surface $x = x$, $\xi = S(x,t)$ the set of equations:

$$\begin{cases} -\dfrac{A^{(r)}(x, S_x)}{S_t} \dfrac{\partial U_{r-1}}{\partial t} + \sum\limits_{j=1}^{r} \hat{\mathcal{L}}_j [S] U_{r-j} = f(x, S(x,t)), \\ -\dfrac{\partial U_{k-1}}{\partial t} = S_t(x,t) U_k, \quad k = 1, \dots, r-1. \end{cases} \tag{24}$$

From condition (24) it follows that the matrix $A^{(r)}(x,p)/H(x,p)$ has a regular inverse $\Sigma(x,p)$. In this case it is sufficient to choose $S(x,t)$ as solution to equation

$$S_t + H(x, S_x) = 0 \tag{25}$$

and multiply the first equation of (24) by $\Sigma(x, S_x)$.

As a result we obtain

$$\begin{cases} \dfrac{\partial U_{r-1}}{\partial t} + \sum\limits_{j=1}^{r} \hat{P}_j [S] U_{r-j} = \Sigma(x, S_x) f(x, S(x,t)), \\ \dfrac{\partial U_{k-1}}{\partial t} = -S_t(x,t) U_k, \quad k = 1, \dots, r-1 \end{cases} \tag{26}$$

where $\hat{p}_j[S] = \Sigma(x, S_x) \hat{\mathcal{L}}_j[S]$.

The initial conditions for the Hamilton-Jacobi equation (25) and the Cauchy-Kovalevsky system (26) are chosen in the same manner as in the scalar case:

$$S(x,0) = S(x); \quad U_k(x, 0) = 0, \; k = 0, \; . \; . \; . \; , r-1$$

As before, from this follows the theorem on r-1 uniformization of solution $u(x, \xi)$ by substituting $\xi = S(x, t)$, and for obtaining (the first term) of the asymptotics of function $u(x, \xi)$ for function $U_r = \tilde{U}_r + f \cdot h$ we get the transport equation :

$$\begin{cases} \dfrac{\partial \tilde{U}_r}{\partial t} + \hat{P}_1 [S] \tilde{U}_r = 0, \\ \tilde{U}_r \Big|_{t=0} = \dfrac{1}{H(x, S_x(x))} \Sigma(x, S_x(x)) f(x, S(x)). \end{cases}$$

We shall now transform the operator $\hat{P}_1[S]$. Since $-1/S_t\ \partial/\partial t\ U_{r-1} = U_r$ (see Section 1) from (24) and the uniformization theorem it follows that $A^{(r)}(x, S_x)U_r(x,t)$ is the holomorphic function and, therefore, transformation of operator $\hat{P}_1[S]$ may be carried out modulo $\operatorname{Im} A^{(r)}$.

Computations yield

$$\hat{P}_1[S] \equiv H_{p_i}\frac{\partial}{\partial x^i} + \frac{1}{2} S_{x^i x^j} H_{p_i p_j} + A^{(r)}_{p_i} \Sigma_{x^i} + A^{(r-1)} \quad (\mathrm{mod}\ \operatorname{Im} A^{(r)}).$$

Further, since the function $H(x, S_x(x,t))$ is constant on the trajectories of the field d/dt, then assuming

$$\tilde{U}_r(x,t) = \frac{V_r(x,t)}{H(x, S_x(x,t))}$$

we find that the function $V_r(x,t)$ satisfies the transport equation

$$\begin{cases} \left\{\frac{d}{dt} + \frac{1}{2} S_{x^i x^j} H_{p_i p_j} + \Sigma_{p_i} A^{(r)}_{x^i} + \Sigma A^{(r-1)}\right\} V_r = 0, \\ V_r\big|_{t=0} = \Sigma(x, S_x(x)) f(x, S(x)). \end{cases}$$

Further, if $x = x(x_0,t)$, $p = p(x_0,t)$ are the solutions of the Hamilton system

$$\dot{x} = H_p, \quad \dot{p} = -H_x, \quad x(0) = x_0, \quad p(0) = S_x(x_0),$$

then assuming (for sufficiently small values of t)

$$V_r(x,t) = \varphi_r(x,t)\sqrt{\frac{Dx_0(x,t)}{Dx}}$$

for $\varphi_r(x,t)$ we obtain the following problem :

$$\begin{cases} \left[\frac{d}{dt} + \{\Sigma, A^{(r)}\} - \Sigma\left[\frac{1}{2} A^{(r)}_{x^i p_i} - A^{(r-1)}\right]\right] \varphi(x,t) = 0, \\ \varphi\big|_{t=0} = \Sigma(x, S_x(x)) f(x, S(x)). \end{cases}$$

where $\{\ ,\ \}$ are the Poisson brackets. This is how looks the transport operator for the system.

II. Uniformization and Asymptotics of Solution in "the large"

<u>Uniformization</u>. First we shall interpret the formula

$$u(x,\xi) = U(x, t(x,\xi)), \quad S(x, t(x,\xi)) = \xi, \tag{27}$$

which gives solution to the Cauchy problem (2), (3) (Section 1) with the use of parametric integrals. The Cauchy integral formula enables (27) to be rewritten as

$$u(x,\xi) = \frac{1}{2\pi i} \int_{h(x,\xi)} \frac{U(x,t) S_t(x,t)}{S(x,t) - \xi} dt, \tag{28}$$

where $h(x,\xi)$ is the homology class of cycle which includes the solution to equation $S(x,t) = \xi$. Note that if we are interested only in a particular part of the integral (28), there is no need to examine the whole contour $\gamma(x,\xi)$ which is the representative of the class $h(x,\xi)$. This is because the singularities of integral (28) lie in the set of points (x,ξ) for which the function $t(x,\xi)$ has ramification, i.e. the equation $S(x,t) = \xi$ has a multiple root:

$$S(x,t) - \xi = 0, \quad S_t(x,t) = 0. \tag{29}$$

We now isolate a particular part of integral (28). Let (x_0, ξ_0) be a point at which relations (29) hold, $t_i(x,\xi)$, $i = 1, \ldots, k$ -- corresponding solutions to equation $\xi = S(x,t)$, $t_i(x_0,\xi_0) = t_0$ agree for all $i = 1, \ldots, k$. Then there exists a neighbourhood V of the point t_0, which does not contain other roots of the equation $\xi_0 = S(x_0, t)$, and a neighbourhood U of the point (x_0, ξ_0) such that for $(x,\xi) \in U$ the boundary of the neighbourhood V does not contain solutions of $\xi = S(x,t)$. In a neighbourhood U by module of holomorphic functions we can replace integration in (28) with respect to cycle $\gamma(x,\xi)$ by integration over that part $\gamma(x,\xi)$ which lies in V, i.e. we assume

$$u(x,\xi) \equiv \frac{1}{2\pi i} \int_{h'(x,\xi)} \frac{U(x,t) S_t(x,t)}{S(x,t) - \xi} dt \pmod{\mathcal{O}(U)} \tag{30}$$

where $h'(x,\xi) \in H_*(V \setminus \Sigma_{(x,\xi)}, \partial V)$, $\Sigma_{(x,\xi)} = \{t \mid \xi = S(x,t)\}$, $O(U)$ is the ring of holomorphic functions in the domain U. Further, formula (30) has to be generalized because the solution $S(x,t)$ of the Hamilton-Jacobi equation (10) of Part I does not exist everywhere, and the representation of type (30) of the solution of problem (2), (3) is not valid in a neighbourhood of focal (caustic) points.

We shall now examine more general integrals of the type

$$f(x,\xi) = \frac{1}{2\pi i}\left(-i\frac{\partial}{\partial\xi}\right)^m \int_{h(x,\xi)} \frac{U(x,t,\tau)\,dt\wedge d\tau\wedge dy}{\Phi(x,t,\tau)+y^2-\xi} . \qquad (31)$$

Here $\tau = (\tau_1, \ldots, \tau_m)$, $y = (y_1, \ldots, y_m)$, $y^2 = \sum_{j=1}^{m} y_j^2$; U, Φ are holomorphic in a neighbourhood of the integration cycle.

Define the class $h(x, \xi)$. We require that

$$\operatorname{rank} \left\| \frac{\partial^2 \Phi}{\partial x^i \partial \tau_j} \; \frac{\partial^2 \Phi}{\partial \tau_k \partial \tau_j} \right\| = m \qquad (32)$$

In this case the set

$$K'_\Phi = \{(x,t,\tau) \mid \Phi_\tau(x,t,\tau) = 0\} \qquad (33)$$

is the analytic manifold. We shall also demand that the equations

$$\Phi_t(x,t,\tau) = 0, \; \Phi_\tau(x,t,\tau) = 0 \qquad (34)$$

for every fixed x determine a finite number of solutions $(t_1(x), \tau_1(x)), \ldots, (t_k(x), \tau_k(x))$. Let (x_0, t_0, τ_0) be a point that satisfies relations (34). Denote by

$$\Sigma_{(x,\xi)} = \{(t,\tau,y) \mid \xi = \Phi(x,t,\tau) + y^2\}$$

the set of singularities of the integrand in (31).

Lemma 2.1. In the space with coordinates (t, τ, y) there exists a sphere $K(r)$ with center at the point $(t_0, \tau_0, 0)$ and boundary $S(r)$ such that

(a) $\left[\Sigma_{(x_0, \xi_0)} \setminus \{(t_0, \tau_0, 0)\}\right] \cap K(r)$ is the analytic manifold;

(b) there exists a neighbourhood $V(x_0, \xi_0)$ of the point $(x_0, \xi_0) = (x_0, \Phi(x_0, t_0, \tau_0))$ such that for all $(x, \xi) \in V(x_0, \xi_0)$ the set $\Sigma_{(x,\xi)} \cap S(r)$ is the smooth manifold.

In this case, $h(x, \xi)$ is the class of homologies in $K(r)$ by module $S(r) \setminus \Sigma_{(x,\xi)}$. Integral (31) is completely defined in a neighbourhood $V_{(x_0, \xi_0)}$.

Using the Thom transversality theorem (see, for example, [2]), it can be shown that formula (31) defines the analytical function whose set of singularities is contained in the projection $\mathcal{L}$ on the space

$\mathbb{C}^{n+1}_{x,\xi}$ of set (34).

We shall now define the Legendre uniformizable functions.

Definition 2.1. The function $f(x, \xi)$ will be called the (Legendre) uniformizable function up to order N if all the derivatives $\partial^j f(x,\xi)/\partial \xi^j$, $0 \leq j \leq N$ can be represented as sum of integrals of type (31) in module of holomorphic functions in a neighbourhood of every point (x, ξ).

Denote the space of uniformizable up to order N functions by $U^{(N)}$.

They will play a leading role in determining the asymptotic expansion of solution. It is the asymptotic solution that is associated with filtration $U^{(0)} \supset U^{(1)} \supset U^{(2)} \supset \ldots$

It is therefore quite natural to consider the integrals (31) which will later define the asymptotics as elements of the factor-spaces $U^{(N)}/U^{(N+1)}$. For this we introduce two important concepts. By $\mathcal{J}(\Phi)$ we denote the ideal in a ring of holomorphic in (x, t, τ) functions, which is generated by generators $\Phi_{\tau_1}, \ldots, \Phi_{\tau_m}, \Phi_t$:

$$\mathcal{J}(\Phi) = \{\Phi_{\tau_1}, \ldots, \Phi_{\tau_m}, \Phi_t\}. \tag{35}$$

The ideal $\mathcal{J}(\Phi)$ will be called the gradient ideal.

We shall now define the mapping α by formulas:

$$\alpha(x, t, \tau) = (x, p, t, E, \xi) = (x, \Phi_x(x,t,\tau), t, \Phi_t(x,t,\tau), \Phi(x,t,\tau)) \tag{36}$$

Lemma 2.2. The image $\alpha(K'_\Phi)$ of the set (33) upon mapping is the submanifold $L(\Phi)$ in a space with coordinates (x, p, t, E, ξ). This is the Legendre submanifold, i.e.

$$d\xi - p\,dx - E\,dt \big|_{L(\Phi)} = 0. \tag{37}$$

Function Φ will be called the determining function for the manifold $L(\Phi)$.

Lemma 2.3. Suppose that the function $f(x, \xi)$ permits uniformization up to order N and, besides, $U_N(x, t, \tau) \in \mathcal{J}(\Phi)$. U_N is the function that appears in integral (31) in the representation for $\partial^N f/\partial \xi^N$. Then $f(x, \xi)$ permits uniformization up to order N+1.

Lemma 2.4. Suppose that two representations of type (35) are given, and are defined by functions $\{U_1(x, t, \tau_1, \ldots, \tau_m), \Phi_1(x, t, \tau_1, \ldots, \tau_m)\}$ and $\{U_2(x,t, \theta_1, \ldots, \theta_k), \Phi_2(x,t, \theta_1,\ldots,\theta_k)\}$, respectively. Let $K_1(r)$, $K_2(r)$ be spheres which are defined in Lemma 2.1 for these representations. If $L(\Phi_1) = L(\Phi_2)$, then there exists an isomorphism $\sigma : H_*(K_1(r) \setminus \Sigma_{1(x,\xi)}, S_1(r) \setminus \Sigma_{1(x,\xi)}) \longrightarrow$ $\longrightarrow H_*(K_2 \setminus \Sigma_2, S_2 \setminus \Sigma_2)$ such that for any function U_1 exists a function U_2 such that the two representations coincide in module $U^{(1)}$ for $\sigma(h_1(x,\xi)) = h_2(x,\xi)$.

More exact information on connection between U_1 and U_2 can be obtained in the following particular case. Let I be the subset of set $\{1,2,\dots,n\}$ and $\bar{I}$ be its complement. Let $(x^I, p_{\bar{I}}, t) = (x^{i_1}, \dots, x^{i_s}, p_{i_{s+1}}, \dots, p_{i_n}, t)$ be the coordinates on $L(\Phi)$ (here $I = \{i_1, \dots, i_s\}$, $\bar{I} = \{i_{s+1}, \dots, i_n\}$. Such systems of coordinates are always present on the Legendre manifold. And what is more, as a determining function we can take the function

$$\Phi(x, p_{\bar{I}}, t) = S_I(x^I, p_{\bar{I}}, t) + \sum_{j \in \bar{I}} x^j p_j .$$

Here, $S_I(x^I, p_I, t) = \xi - \sum_{j \in \bar{I}} x^j p_j \Big|_{L(\Phi)}$ is the action (generating function of the domain U_I with coordinates $(x^I, p_{\bar{I}}, t)$) on manifold L.

Let (x) and (y) be two different systems of coordinates on X. We introduce a measure μ on the manifold L. And let v be the measure on the base manifold X. If Φ_1 and Φ_2 are obtained in the manner described for the system of coordinates $(x^{I_1}, p_{\bar{I}_1}, t)$ and $(y^{I_2}, q_{\bar{I}_2}, t)$, respectively, then the mentioned connection between U_1 and U_2 is expressed as:

$$U_1\Big|_{K'_{\Phi_1}} \sqrt{\frac{v_x(x^{I_1}, p_{\bar{I}_1}, t)}{\mu_{I_1}(x^{I_1}, p_{\bar{I}_1}, t)}}\,(-1)^{\frac{|\bar{I}_1|(|\bar{I}_1|+1)}{2}} \left(\frac{\pi}{\sqrt{2}}\right)^{|\bar{I}_1|} =$$

$$= U_2\Big|_{K'_{\Phi_2}} \sqrt{\frac{v_y(y^{I_2}, q_{\bar{I}_2}, t)}{\mu_{I_2}(y^{I_2}, q_{\bar{I}_2}, t)}}\,(-1)^{\frac{|\bar{I}_2|(|\bar{I}_2|+1)}{2}} \left(\frac{\pi}{\sqrt{2}}\right)^{|\bar{I}_2|}$$

Here, v_x, v_y are the densities of the measure v on the base manifold X with respect to the systems of coordinates (x) and (y), respectively; μ_{I_1}, μ_{I_2} are the densities of the measure on the Legendre manifold L with respect to the coordinates (x^{I_1}, p_{I_1}, t) and (y^{I_2}, q_{I_2}, t).

Note that the function f belongs to the class $U^{(N)}$ if and only if it is representable as the sum of integrals of type

$$\frac{1}{2\pi i}\left(-i\frac{\partial}{\partial \xi}\right)^{m-N}\int\limits_{h(x,\xi)}\frac{U(x,t,\tau)\,dt\wedge d\tau\wedge dy}{\Phi(x,t,\tau)+y^2-\xi}. \tag{38}$$

For representations of this type, the analog of Lemma 1.4 is valid which determines necessary conditions for the coincidence of representations of type (38) of the functions from $U^{(N)}$ in module of the space $U^{(N+1)}$.

Let now L be the Legendre manifold in the space with coordinates (x, p, t, E, ξ), i.e.

$$d\xi - p\,dx - E\,dt|_L = 0.$$

Let $U(x_0, \xi_0)$ be a neighbourhood of the point (x_0, ξ_0) on which L decomposes into open sets $W_1, \ldots, W_k$. In each of these sets L is constructed by the determining function Φ_j, $j = 1, \ldots, k$. Denote by $K_j(r)$ the spheres constructed in Lemma 2.1 for each of the functions Φ_j; $\mathcal{L}$ is the join of sets $\mathcal{L}_j$ for $j = 1, \ldots, k$ (the definition of $\mathcal{L}$ is given after Lemma 2.1). In the set

$$\mathcal{O}\!\mathcal{L} = \bigcup_{(x,\xi)\in U(x_0,\xi_0)\setminus\mathcal{L}}\ \prod_{j=1}^{k} H_*\left(K_j(\tau)\setminus\Sigma_{j(x,\xi)},\, S_j(\tau)\setminus\Sigma_{j(x,\xi)}\right) \tag{39}$$

we shall introduce a topology in the following manner. If $V\subset U_{(x_0,\xi_0)}\setminus\mathcal{L}$ is an 1-connected open set, then the bundle

$$\bigcup_{(x,\xi)\in U(x_0,\xi_0)\setminus\mathcal{L}}\ \prod_{j=1}^{k}\left(K_j(\tau)\setminus\Sigma_{j(x,\xi)},\, S_j(\tau)\setminus\Sigma_{j(x,\xi)}\right)\to U_{(x_0,\xi_0)}\setminus\mathcal{L}$$

which according to Thom's theorem is locally trivial, becomes trivial over V. We take a trivialization. It gives the projection

$$\bigcup_{(x,\xi)\in U(x_0,\xi_0)\setminus\mathcal{L}}\ \prod_{j=1}^{k}\left(K_j(\tau)\setminus\Sigma_{j(x,\xi)},\, S_j(\tau)\setminus\Sigma_{j(x,\xi)}\right)\to$$

$$\to\prod_{j=1}^{\tau}\left(K_j(\tau)\setminus\Sigma_{j(\tilde{x},\tilde{\xi})},\, S_j(\tau)\setminus\Sigma_{j(\tilde{x},\tilde{\xi})}\right)$$

where $(\tilde{x}, \tilde{\xi})$ is a point from V. The pre-images of cycles from the right-hand side of the last relation give basis of the topology for different V.

Definition 2.2. An arbitrary component of the connection of space $\mathcal{O}\!\mathcal{L}$ will be called the ramified class $\{h_1(x,\xi),\ldots,h_k(x,\xi)\}$ over the set $U(x_o,\xi_o)$.

Note that further the ramified class will play two roles:

1. every ramified class will define the connected covering of the space $U_{(x_o,\xi_o)}\setminus\mathcal{L}$;

2. every point of this covering will be a set of homology classes $h_1(x,\xi),\ldots,h_k(x,\xi)$.

The isomorphism σ constructed in Lemma 2.4 enables us to remove the ambiguity in defining the ramified class, which appears due to lack of uniqueness in chosing the determining functions $\Phi_1,\ldots,\Phi_k$. Therefore, we shall not distinguish the ramified classes which change from one into another upon the isomorphism σ .

Definition 2.3. The sub-sheaf of a sheaf of germs of coverings, consisting of coverings whose every connection component is the ramified class, is called the index sheaf of the manifold L.

Lemma 2.4 enables us, using the set of three (L,μ,h), where L is the Legendre manifold, μ is a measure on it, and h is a section of the index sheaf, to construct the section of the sheaf $U^{(0)}/U^{(1)}$ whose local expression is

$$f(x,\xi)=\frac{1}{2\pi i}\sum_{j=1}^{k}(-1)^{\frac{|\bar{I}_j|(|\bar{I}_j|+1)}{2}}\left(\frac{\sqrt{2}}{\pi}\right)^{|\bar{I}_j|}\left(-i\frac{\partial}{\partial\xi}\right)^{|\bar{I}_j|}\int\limits_{h_j(x,\xi)}\Big[S_{I_j}(x^{I_j},p_{\bar{I}_j},t)+x^{\bar{I}_j}p_{\bar{I}_j}+ \tag{40}$$

$$+y^2_{\bar{I}_j}-\xi\Big]^{-1}\varphi(x^{I_j},p_{\bar{I}_j},t)\sqrt{\frac{\mu_{I_j}(x^{I_j},p_{\bar{I}_j},t)}{v_x(x^{I_j},p_{\bar{I}_j},t)}}\,dt\wedge dp_{\bar{I}_j}\wedge dy_{\bar{I}_j}$$

Here, φ is the function on L; I_j is the type of the chart on the Legendre manifold; μ_{I_j} denotes the density of the measure in the chart U_{I_j} in canonical coordinates $(x^{I_j},p_{\bar{I}_j},t)$; v_x stands for the density of the fixed chart on the manifold X raised on the Legendre manifold L. The section defined by formula (40) we shall denote by $K^{(0)}_{(L,\mu,h)}(\varphi)$. In a like manner the section $K^{(N)}_{(L,\mu,h)}(\varphi)$ of the sheaf $U^{(N)}/U^{(N+1)}$ is determined.

<u>Asymptotic solutions</u>. We shall search for the asymptotic solutions of problem (23) (Part I) in smoothness. To formulate the concept of asymptotic solution, we make use of the definition of classes $U^{(N)}$ given in Definition 2.1, namely:

The function $\bar{u}(x,\xi) \in U^{(r)}/U^{(r+N)}$ is called the asymptotic solution of problem (23) up to order N if

1. $\bar{u}(x,\xi)$ has a zero of order r on S ;

2. the difference $\hat{A}\bar{u}(x,\xi) - f(x,\xi)$ has on the manifold S a zero of order N; $\hat{A} = A(x, \partial/\partial x)$;

3. the inclusion $\hat{A}\bar{u}(x,\xi) \in U^{(N+1)}$ is valid.

Further the asymptotic solution $\bar{u}$ will be a function from the class $U^{(r)}$. In this case, the conditions (1), (2), and (3) for determining the asymptotic solution will be assumed to be fulfilled for a representation of the function $\bar{u}(x,\xi)$ in the class $U^{(r)}/U^{(r+N)}$.

We substitute the multiple-valued vector function, given by integral (38), into the equation of problem (23) of Part I. As we wish to be within the limits of the space $U^{(N)}$, $N \geq 0$, we must assume a priori that the vector function $u(x,\xi)$ belongs to the space $U^{(r)}$.

This function may be (locally) given by the integral

$$u(x,\xi) = \left(-i\frac{\partial}{\partial \xi}\right)^{m-r} \int_{h(x,\xi)} \frac{U(x,t,\tau)\,dt\wedge d\tau\wedge dy}{\Phi(x,t,\tau)+y^2-\xi} \tag{41}$$

and we assume, obviously, that $m = \dim[\tau] \geq r$*). Let us now operate on function (41) by operator $\hat{A}$:

$$\hat{A} = A\left(x, \frac{\partial}{\partial x}\right) = \sum_{k=0}^{r} A^{(k)}\left(x, \frac{\partial}{\partial x}\right)$$

(by $A^{(k)}$ is denoted the component of the operator $\hat{A}$ of order k). We obtain the equality

$$A\left(x,\frac{\partial}{\partial x}\right)u(x,\xi) = \left(-i\frac{\partial}{\partial\xi}\right)^{m-r}\int_{h(x,\xi)} A\left(x,\frac{\partial}{\partial x}\right)\left[\frac{U(x,t,\tau)}{\Phi(x,t,\tau)+y^2-\xi}\right]dt\wedge d\tau\wedge dy =$$

$$= \sum_{k=0}^{r}\left(-i\frac{\partial}{\partial\xi}\right)^{m-r+k}\int_{h(x,\xi)} \frac{\hat{\mathcal{L}}^k(x,t,\tau,\partial/\partial x)\,U(x,t,\tau)}{\Phi(x,t,\tau)+y^2-\xi}\,dt\wedge d\tau\wedge dy,$$

---

*) $[\tau]$ is the arithmetic vector space of points $(\tau_1, \ldots, \tau_m)$.

where the operators $\hat{\mathcal{L}}^k$ are the differential operators in variable x of order r-k, whose coefficients holomorphically depend on variables (x, t, τ). In this case the operators $\hat{\mathcal{L}}^r$ and $\hat{\mathcal{L}}^{r-1}$ which will be needed in further calculations take the form

$$\hat{\mathcal{L}}^r(x,t,\tau,\partial/\partial x) = A^{(r)}(x, \Phi_x(x,t,\tau)) \tag{42}$$

operator of zero order (operator of multiplication by function) and

$$\hat{\mathcal{L}}^{r-1}(x,t,\tau,\partial/\partial x) = A^{(r)}_{p_i}(x,\Phi_x(x,t,\tau))\frac{\partial}{\partial x^i} + \frac{1}{2}\Phi_{x^ix^j}(x,t,\tau)\times$$

$$\times A^{(r)}_{p_ip_j}(x,\Phi_x(x,t,\tau)) + A^{(r-1)}(x,\Phi_x(x,t,\tau)). \tag{43}$$

Note that, in general, the sum in the right-hand side of relation (28) is graduated according to the scale of spaces $U^{(N)}$. The higher term of this expansion, with consideration for formula (42), takes the form

$$\left(-i\frac{\partial}{\partial \xi}\right)^m \int_{h(x,\xi)} \frac{A^{(r)}(x,\Phi_x(x,t,\tau))\,U(x,t,\tau)}{\Phi(x,t,\tau)+y^2-\xi}\, dt\wedge d\tau\wedge dy \tag{44}$$

and, obviously, belongs to the space $U^{(0)}$. For the integral (44) to be a function from $U^{(1)}$ we must demand that

$$A^{(r)}(x,\Phi_x(x,t,\tau))U(x,t,\tau) \in J(\Phi)$$

in virtue of Lemma 2.3. Recall that $J(\Phi)$ is the gradient ideal defined by the function Φ and given by the generators $\{\Phi_t, \Phi_\tau\}$. In virtue of the maximality condition of the rank of matrix (32) it suffices that the function $A^{(r)}(x, \Phi_x(x,t,\tau))\, U(x,t,\tau)\big|_{K'_\Phi}$ be divisible by $\Phi_t\big|_{K'_\Phi}$ (the set K'_Φ is defined by formula (33)). Using the mapping α, from formula (36) we can rewrite the last relation in the form

$$A^{(r)}(x, p)\big|_{L(\Phi)}\, a \equiv E\big|_{L(\Phi)} \tag{45}$$

where by a we have denoted the image of the restriction of the function U on K'_Φ for the mapping α; a is the vector function on $L(\Phi)$. Comparing (45) means divisibility in holomorphic functions.

Taking into consideration the relation (21) of Part I, we see that (45) holds at $a = \Sigma(x, p)\big|_{L(\Phi)}\tilde{a}$, where $\tilde{a}$ is a new vector function on $L(\Phi)$ if we require that

$$E + H(x, p)\big|_{L(\Phi)} = 0 \tag{46}$$

Obviously, it can now be assumed that

$$U(x,t,\tau) = \Sigma(x, \Phi_x(x, t,\tau)) \widetilde{U}(x,t,\tau) \tag{47}$$

We have proved the following lemma.

Lemma 2.5. If the manifold $L(\Phi)$ lies on the Hamiltonnian $E + H(x,p)$ null surface, then the multiple-valued vector function

$$u(x,\xi) = \left(-i\frac{\partial}{\partial\xi}\right)^{m-\tau} \int\limits_{h(x,\xi)} \frac{\Sigma(x,\Phi_x(x,t,\tau))\widetilde{U}(x,t,\tau)}{\Phi(x,t,\tau)+y^2-\xi} dt\wedge d\tau\wedge dy \tag{48}$$

satisfies the comparison $\hat{A}u = 0(\mathrm{mod}\ U^{(1)})$.

For constructing the function $u(x,\xi)$ that satisfies requirement 3 of determining the asymptotic solution, it is necessary that the last relation should be made a comparison modulo $U^{(1)}$. In this case the following statement is valid.

Lemma 2.6. Under the conditions of Lemma 2.5 the multiple-valued vector function (48) satisfies modulo $U^{(2)}$ the comparison

$$\hat{A}u(x,\xi) \equiv \left(-i\frac{\partial}{\partial\xi}\right)^{m-1} \int\limits_{h(x,\xi)} \frac{[\hat{\mathcal{P}}_H + A^{(\tau)}(x,\Phi_x(x,t,\tau))\hat{\mathcal{P}}_\Sigma]\widetilde{U}(x,t,\tau)}{\Phi(x,t,\tau)+y^2-\xi} dt\wedge d\tau\wedge dy,$$

where

$$\hat{\mathcal{P}}_H = H_{p_i}(x,\Phi_x(x,t,\tau))\frac{\partial}{\partial x^i} + \frac{\partial}{\partial t} + \sum_{i=1}^m F_i(x,t,\tau)\frac{\partial}{\partial\tau_i} + \frac{1}{2}\Phi_{x^ix^j}(x,t,\tau)\times$$
$$\times H_{p_ip_j}(x,\Phi_x(x,t,\tau)) + \sum_{i=1}^m \frac{\partial F_i}{\partial\tau_i}(x,t,\tau) + A^{(\tau)}_{p_i}(x,\Phi_x(x,t,\tau))\times \tag{49}$$
$$\times \Sigma_{x^i}(x,\Phi_x(x,t,\tau)) + A^{(\tau-1)}(x,\Phi_x(x,t,\tau))\Sigma(x,\Phi_x(x,t,\tau))$$

and the functions $F_i(x,t,\tau)$ are defined by the relation

$$H(x,\Phi_x(x,t,\tau)) + \Phi_t(x,t,\tau) = -\sum_{i=1}^m F_i(x,t,\tau)\Phi_{\tau_i}(x,t,\tau);$$

$$\hat{\mathcal{P}}_\Sigma = \Sigma_{p_i}(x,\Phi_x(x,t,\tau))\frac{\partial}{\partial x^i} + \frac{1}{2}\Phi_{x^ix^j}(x,t,\tau)\Sigma_{p_ip_j}(x,\Phi_x(x,t,\tau)). \tag{50}$$

In virtue of the result of Lemma 2.6 the expression $\hat{A}u(x,\xi)$ modulo $U^{(2)}$ is split into two terms :

$$\hat{A}u(x,\xi) = f_1(x,\xi) + f_2(x,\xi) \tag{51}$$

where

$$f_1(x,\xi) = \left(-i\frac{\partial}{\partial\xi}\right)^{m-1}\int_{h(x,\xi)} \frac{\hat{\mathcal{P}}_H \tilde{U}(x,t,\tau)\, dt\wedge d\tau\wedge dy}{\Phi(x,t,\tau)+y^2-\xi}, \tag{52}$$

$$f_2(x,\xi) = \left(-i\frac{\partial}{\partial\xi}\right)^{m-1}\int_{h(x,\xi)} \frac{A^{(\tau)}(x,\Phi_x(x,t,\tau))\hat{\mathcal{P}}_\Sigma \tilde{U}(x,t,\tau)}{\Phi(x,t,\tau)+y^2-\xi}\, dt\wedge d\tau\wedge dy. \tag{53}$$

The expression (52) accurate to a function from $U^{(2)}$ is determined only by the values of the function $U(x, t, \tau)$ on the manifold

$$K'_\Phi = \{(x, t, \tau) \mid \Phi_\tau(x, t, \tau) = 0\}. \tag{54}$$

This statement follows from the following lemma.

Lemma 2.7. The vector field

$$\frac{\partial}{\partial t} + \frac{\partial H}{\partial p_i}(x, \Phi_x(x,t,\tau))\frac{\partial}{\partial x^i} + \sum_{j=1}^{m} F_j(x,t,\tau)\frac{\partial}{\partial \tau_j} \tag{55}$$

appearing in formula (54) that defines the operator $\mathcal{P}_H$ is tangent to the manifold K'_Φ.

For describing the manifold $L(\Phi)$, defined in Lemma 2.2, we consider the symplectic phase space $[x, t, p, E, \tau, q]$ with a structure form

$$\tilde{\Omega} = p\,dx + E\,dt + q\,d\tau$$

The function $\Phi(x, t, \tau)$ determines in this space the non-singular Lagrangian manifold $\tilde{L}(\Phi)$:

$$p = \frac{\partial\Phi(x,t,\tau)}{\partial x}, \quad E = \frac{\partial\Phi(x,t,\tau)}{\partial t}, \quad q = \frac{\partial\Phi(x,t,\tau)}{\partial\tau}. \tag{56}$$

Let us now consider the projection

$$\alpha : [x, p, t, E, \tau, q] \to [x, t, p, E]. \tag{57}$$

Its restriction on the manifold

$$K'_\Phi = \tilde{L}(\Phi) \cap \{q = 0\} \tag{58}$$

determines the embedding of manifold K'_Φ in the contact space $[x, t, p, E, \xi]$ with the structure form

$$d\xi - p\,dx - E\,dt \tag{59}$$

By Lemma 2.2, it is the Legendre imbedding.

Lemma 2.8. The field (55), contracted on the manifold K'_Φ, changes into the Legendre manifold $L(\Phi)$ in the field $\frac{\partial}{\partial t} + H_p\frac{\partial}{\partial x} - H_x\frac{\partial}{\partial p}$.

The lemma is proved by direct calculation.

Our further aim is to prove the global existence of the operator $\mathcal{P}_H$, given by formula (41), on the Legendre manifold. For this, we compute the comparison $\hat{A}u(x,\xi)$ if $u(x,\xi)$ is given by formulae (40). Obviously, the required expression can be computed separately for each term of the formula (40). Since our ultimate objective is to choose an amplitude φ such that the expression $\hat{A}u(x,\xi)$ should go into zero in module of the functions from $U^{(2)}$, there is no need to take care of the numerical coefficient. In virtue of Lemma 2.3, in integral (52) we can go over to canonical coordinates $(x^I, p_{\bar{I}}, t)$ of the Legendre manifold $L(\Phi)$. The function $f_1(x,\xi)$ given by formula (52) will now be written as

$$f_1(x,\xi) = \left(-i\frac{\partial}{\partial\xi}\right)^{|\bar{I}|-1} \int_{h_I(x,\xi)} \frac{\hat{\mathcal{P}}_H^I \tilde{U}(x^I, p_{\bar{I}}, t)\, dt \wedge dp_{\bar{I}} \wedge dy_{\bar{I}}}{S_I(x^I, p_{\bar{I}}, t) + x^{\bar{I}} p_{\bar{I}} + y_{\bar{I}}^2 - \xi} \tag{60}$$

where the operator $\mathcal{P}_H^I$ is constructed by formula (54) for the function

$$\Phi_I(x, p_{\bar{I}}, t) = S_I(x^I, p_{\bar{I}}, t) + x^{\bar{I}} p_{\bar{I}}.$$

For the global existence of operator $\mathcal{P}_H^I$ it is necessary to replace in formula (60) the function $U(x^I, p_{\bar{I}}, t)$ by function

$$\tilde{U}(x^I, p_{\bar{I}}, t) = \varphi(x^I, p_{\bar{I}}, t)\sqrt{\frac{\mu_I(x^I, p_{\bar{I}}, t)}{v_x(x^I, p_{\bar{I}}, t)}}. \tag{61}$$

Here, as before, $\mu_I(x^I, p_{\bar{I}}, t)$ is the density of the measure μ (of closed form of bidegree $(n+1, 0)$) on the Legendre manifold $L(\Phi)$, and $v_x(x^I, p_{\bar{I}}, t)$ is the rise on manifold $L(\Phi)$ of density of some measure (of closed form of bidegree $(n,0)$) v on the manifold X. For convenience we assume the measure μ to be invariant along the trajectories of the vector field $\frac{\partial}{\partial t} + V(H)$, i.e. $\mathcal{L}_{\frac{\partial}{\partial t}+V(H)}\mu = 0$, where $\mathcal{L}_{\frac{\partial}{\partial t}+V(H)}$ is the Lie derivative along the vector field $\frac{\partial}{\partial t} + V(H) = \frac{\partial}{\partial t} + H_p\frac{\partial}{\partial x} - H_x\frac{\partial}{\partial p}$.

<u>Lemma 2.9</u>. The equality holds

$$\hat{\mathcal{P}}_H^I \tilde{U}(x^I, p_{\bar{I}}, t) = \sqrt{\frac{\mu_I(x^I, p_{\bar{I}}, t)}{v_x(x^I, p_{\bar{I}}, t)}}\; \hat{P}_H^I \varphi(x^I, p_{\bar{I}}, t) \tag{62}$$

where

$$\hat{P}_H^I = \frac{\partial}{\partial t} + V(H) - \frac{1}{2}H_{xp}\Big|_{L(\Phi)} + \left[A_p^{(\tau)}\Sigma_x + A^{(\tau-1)}\Sigma\right]\Big|_{L(\Phi)} - \frac{1}{2}\left[\frac{\partial}{\partial t} + V(H)\right]\ln v_x. \tag{63}$$

Here we assume the functions $\tilde{U}$ and φ to be connected by relation (60). It is proved by direct computation, using the S.L. Sobolev lemma.

Note that, in virtue of the statement of Lemma 2.6, the operator $\hat{P}_H^I$ is indeed independent of the type of map I. We denote it by $\hat{P}_H$ and call it the transport operator.

Let now the multiple-valued vector function $u(x, \xi)$ be defined by formula

$$u(x, \xi) = K^{(r)}_{(L, \mu, h)} \left(\Sigma|_L \varphi\right) \tag{64}$$

and let the function φ on the Legendre manifold L satisfy the transport equation

$$\hat{P}_H \varphi = 0 \tag{65}$$

Proposition 2.1. Suppose the Legendre manifold L lies on the Hamiltonnian $E + H(x, p)$ null surface; the measure μ is invariant with respect to the trajectories of the vector field $\frac{\partial}{\partial t} + V(H)$; φ satisfies the transport equation (65) and h is the section of the index sheaf. Then in a neighbourhood of any point (x_0, ξ_0) there exists a function $\bar{u}(x, \xi)$ different from the function (64) on an element of the space $U^{(r+1)}$ and satisfying the comparison

$$\hat{A}\bar{u}(x, \xi) \equiv 0 \pmod{U^{(2)}}.$$

Proof. In a neighbourhood of any point (x_0, ξ_0) the function $u(x, \xi)$ is locally given by the sum of expressions of type (48)

$$u(x,\xi) = \left(-i\frac{\partial}{\partial \xi}\right)^{m-r} \int_{h(x,\xi)} \frac{\Sigma(x, \Phi_x(x,t,\tau))\tilde{U}(x,t,\tau)}{\Phi(x,t,\tau) + y^2 - \xi}\, dt \wedge d\tau \wedge dy.$$

We change this function by an element of the space $U^{(r+1)}$ after assuming

$$\bar{u}(x,\xi) = u(x,\xi) - \left(-i\frac{\partial}{\partial \xi}\right)^{m-r-1} \int_{h(x,\xi)} \frac{\hat{\mathcal{P}}_\Sigma \tilde{U}(x,t,\tau)}{\Phi(x,t,\tau) + y^2 - \xi}\, dt \wedge d\tau \wedge dy.$$

By Lemma 2.6 we see that the comparison holds

$$\hat{A}\bar{u}(x,\xi) \equiv \left(-i\frac{\partial}{\partial \xi}\right)^{m-1} \int_{h(x,\xi)} \frac{\hat{\mathcal{P}}_H \tilde{U}(x,t,\tau)\, dt \wedge d\tau \wedge dy}{\Phi(x,t,\tau) + y^2 - \xi} \pmod{\mathcal{U}^{(2)}}.$$

In virtue of condition (65) and the results of Lemma 2.5, the integral in the right-hand side of the last formula is a function belonging to the space $U^{(2)}$. The proposition is proved.

The initial conditions for the equation have been obtained in Part I.

We now go over to the next theorem.

Theorem 2.1. In the assumptions of Proposition 2.1 the function $K^{(r)}_{(L,\mu,h)}(\sum|_L \varphi)$ is the asymptotics of problem (23), of Part I, up to order 1 if the function φ satisfies additionally the initial condition

$$\varphi|_{t=0} = \sum(x, S_x(x))\, f(x, S(x))$$

References

1. Leray, J., Gårding, L., et al, Problème de Cauchy, Paris, 1964.
2. Pham, F., Introduction à l'étude topologique des singularités de Landau, Paris, 1967.
3. Sternin, B.Yu. and Shatalov, V.E. Analitic Lagrangian Manifolds and Feinman Integrals, Uspehi Math. Nauk, 1979, vol. 34, issue 6.
4. Shatalov, V.E. Global Asymptotic Expansions in the Characteristic Cauchy Problem for Complex Analytic Functions, Uspehi Math. Nauk, 1980, vol. 35, issue 4.
5. Sternin, B.Yu. and Shatalov, V.E. Legendre Uniformization of Multiple-Valued Analytic Functions, Math. sb., 1980, vol. 113, issue 2.

Translated from the Russian by P.K. Dang

CATEGORY OF NONLINEAR DIFFERENTIAL EQUATIONS

A.M.Vinogradov
Department of Mechanics and Mathematics
Moscow University
117234 Moscow, USSR

The aim of this report is to provide a categorical framework for the theory of nonlinear partial differential equations, and demonstrate the expedience of this step. To do this, we shall first list data from the invariant theory of differential equations, based on the geometry of jet spaces, which are necessary for invariant introduction of the category of differential equations (DE) in descriptive-geometric terms. Thereafter, it will be discussed from the algebraic view point that leads to the construction of effective differential calculus on the objects of the category DE. The meaning and significance of the formalism introduced will be illustrated with examples.

We shall restrict ourselves to the discussion of motives which lead to the category DE, to schematic description of basic elements of its structure, and to some examples. Proofs of necessary facts can be partially found in [1]. A detailed information on some of the problems considered elsewhere is available in [2-5].

1. Jet Spaces

Let N be a smooth manifold, $\dim N = m + n$. Call the class of n-dimensional submanifolds $L \subset N$ touching each other at point $x \in N$ with order k the k-jet of an n-dimensional submanifold in N. Denote by $[L]^k_x$ the k-jet of submanifold L at point x. Let $N^k_m(x) = \{[L]^k_x \mid \dim L = n, L \ni x\}$ and $N^k_m = \bigcup_x N^k_m(x)$. The set N^k_m is equipped in a natural way with the structure of a smooth manifold. Projections $\pi_{k,l}: N^k_m \to N^l_m$, $\pi_{k,l}([L]^k_x) = [L]^l_x$ are determined for $k \geqslant l$. If $L^n \subset N$, then the smooth map $j_k(L): L \to N^k_m$, $j_k(L)(x) = [L]^k_x$ is defined. Obviously, $\pi_{k,l} \circ j_k(L) = j_l(L)$. Denote by N_m the inverse limit of the chain of maps

$$\dots \to N^k_m \xrightarrow{\pi_{k,k-1}} \dots \xrightarrow{\pi_{1,0}} N^0_m = N$$

Denote by $j(L): L \to N^\infty_m$ the limit of the sequence of maps $j_k(L)$, $k \to \infty$.

The natural projection $N_m^\infty \to N_m^k$ is denoted by $\pi_{\infty,k}$. N_m^k, $0 \leq k \leq \infty$, is called the manifold of k-jets of n-dimensional submanifolds of the manifold N. Let $\mathcal{F}_m^k(N) = C^\infty(N_m^k)$. Denote by $\mathcal{F}_m(N)$ the direct limit of the chain of maps

$$\ldots \to \mathcal{F}_m^{k-1}(N) \xrightarrow{\pi^*_{k,k-1}} \mathcal{F}_m^k(N) \to \ldots$$

Let $\pi : E = E_\pi \to M^n$ be a submersion and dim E = n + m. Then the collection of k-jets of images of local sections of this submersion forms an open set $J^k(\pi)$ in E_m^k (called the manifold of k-jets of the fibering π, provided π is a fibering). Denote the collection of local sections of submersion π by $\Gamma_{loc}(\pi)$. For $\sigma \in \Gamma_{loc}(\pi)$, $\mathcal{U}_\sigma$ indicates the domain of σ. Let $j_k(\sigma) = j_k(L) \circ \sigma$, where $L = \sigma(\mathcal{U}_\sigma)$; $\pi_k = \pi \circ \pi_{k,0}$, $\mathcal{F}_k(\pi) = C^\infty(J^k(\pi))$, $\mathcal{F}_{-1}(\pi) =$

$= C^\infty(M)$, $\mathcal{F}(\pi) = \lim\limits_{k\to\infty} \operatorname{dir} \mathcal{F}_k(\pi)$, $k \to \infty$.

Let $U = V^n \times W^m$. Here V^n (correspondingly W^m) is a domain in R^n (correspondingly in R^m) and (x, u), where $x = (x_1, \ldots, x_n)$ and $u = (u_1, \ldots, u_m)$ is a corresponding system of coordinates. Then on the manifold $J^k(\alpha)$, $0 \leq k \leq \infty$, where $\alpha : U \to V$ is the natural projection, there appears a system of coordinates x_j, p_σ^i, $1 \leq j \leq n$, $1 \leq i \leq m$, $|\sigma| \leq k$. Here σ is an unordered system of integers $(i_1, \ldots, i_s)$, $1 \leq i_l \leq n$, and $|\sigma| = s$. Functions p_σ^i are uniquely determined by the following property:

$$j_k(f)^*(p_\sigma^i) = \frac{\partial^{|\sigma|} f_i}{\partial x_{i_1} \ldots \partial x_{i_s}}$$

where the section $f \in \Gamma(\alpha)$ is given by the equalities $u_i = f_i(x)$, $u^i = p_\emptyset^i$.

For any diffeomorphism $f: U \to U' \subset N$ the diffeomorphisms $f_{(k)}: J^k(\alpha) \to N_m^k$ are naturally defined; owing to this the aforementioned coordinates are transferred from $J^k(\alpha)$ to im $f_{(k)}$. Hereinafter they are called the canonical local coordinates, and the domain im $f_{(k)}$ equipped with coordinates is said to be a canonical chart.

By $\Lambda^i(M)$ we shall denote $C^\infty(M)$ - the module of smooth differential i-forms on manifold M. Let

$$C\Lambda^i(N_m^k) = \left\{ \omega \in \Lambda^i(N_m^k) \mid j_k(L)^*(\omega) = 0, \forall L^n \subset N \right\}.$$

As $\pi^*_{k,1}(\mathcal{C}\Lambda^i(N^1_m)) \subset \mathcal{C}\Lambda^i(N^k_m)$, the submodule $\mathcal{C}\Lambda^i(N^\infty_m) \subset \Lambda^i(N^\infty_m)$, where $\mathcal{C}\Lambda^i(N_m) = \lim\limits_{k\to\infty} \mathrm{dir}\, \mathcal{C}\Lambda^i(N^k_m)$ and $\Lambda^i(N_m) = \lim\limits_{k\to\infty} \mathrm{dir}\, \Lambda^i(N^k_m)$, is defined. The submodule $\mathcal{C}\Lambda^1(N^k_m) \subset \Lambda^1(N^k_m)$ is of a constant rank and, in a dual manner, determines the distribution $\theta \to C_\theta \subset T_\theta(N^k_m)$ known as the Cartan distribution. If $\theta \in N^\infty_m$ and $\theta = \{\theta_k\}$, where $\theta_k \in N^k_m$, $\pi_{k,1}(\theta_k) = \theta_1$, then we shall define the tangent space $T_\theta(N^\infty_m)$ as inverse limit of the chain of linear maps

$$\ldots \longrightarrow T_{\theta_k}(N^k_m) \xrightarrow{d\pi_{k,k-1}} T_{\theta_{k-1}}(N^{k-1}_m) \longrightarrow \ldots$$

Since $d\pi_{k,k-1}(C_{\theta_k}) \subset C_{\theta_{k-1}}$, the inverse limit of the chain

$$\ldots \longrightarrow C_{\theta_k} \xrightarrow{d\pi_{k,k-1}} C_{\theta_{k-1}} \longrightarrow \ldots$$

denoted by C_θ is defined. The distribution $\theta \mapsto C_\theta$ is called the Cartan distribution on N^∞_m. The module $\mathcal{C}\Lambda^1(N^k_m)$ that annihilates the Cartan distribution is generated, in the limits of a canonical chart, by forms

$$U(p^j_\sigma) = dp^j_\sigma - \sum_i p^j_{\sigma i}\, dx_i, \ |\sigma| < k, \ \sigma i = (i_1, \ldots, i_l, i)$$

if $\sigma = (i_1, \ldots, i_l)$.

An integral manifold of the Cartan distribution is said to be the maximal one, if locally in a neighbourhood of any of its point it does not lie in an integral manifold of higher dimension.

<u>Proposition 1</u>. (a) Maximal integral manifolds of the Cartan distribution on N^k_m, $0 < k < \infty$, are divided into types so that the manifolds of type $n-p$ are of dimension $m\, C^k_{p+k-1} + n - p$;

(b) Maximal integral manifolds of the Cartan distribution on N^∞_m are of dimension n and locally are of the form $\mathrm{im}\, j(L)$.

<u>Remark 1</u>. Excepting the cases $n = m = 1$ and $k = m = 1$, $\dim V > \dim V'$, where V and V' are maximal integral manifolds of the Cartan distribution on N^k_m and type $V <$ type V'. Here, the zero type integral manifolds coincide with the fibres of projection $\pi_{k,k-1}$.

This remark and the first statement of Proposition 1 reveal that the Cartan distribution on N^k_m, $0 < k < \infty$, is not a completely integrable distribution. On the contrary, for $k = \infty$, we have

Proposition 2. The Cartan distribution on N_m^∞ is completely integrable in the sense that

$$d\, C\Lambda^1(N_m^\infty) \subset C\Lambda^1(N_m^\infty) \wedge \Lambda^1(N_m^\infty).$$

Remark 2. The Frobenius theorem for Cartan's distribution on N_m^∞ is false in view of the infinite dimension of the latter. This is evident, for instance, from the fact that the submanifold $L^n \subset N$ is not defined locally by its infinite jet $[L]_x^\infty$ in a neighbourhood of the point $x \in L^n$. It may also be noted that, nonetheless, this theorem holds in the analytical variant of the theory described.

Maximal integral manifolds of the Cartan distribution of type n are of dimension n and are called the R-manifolds.

Proposition 3. Let $V \subset N_m^k$ be the R-manifold, $0 < k < \infty$ and/or $m > 1$, or $n + k > 2$. Then, with the exception of nowhere-dense close subset $V_o \subset V$, the manifold V is locally of the form $\mathrm{im}\, j_k(W)$. In this case, $V_o = \{ x \in V \mid \ker d\pi_{k,k-1} \mid_{T_x(V)} \neq 0 \}$.

2. Differential Equations

Elsewhere a system of nonlinear partial differential equations is interpretted as a closed submanifold (possibly, with some singularities) in N_m^k. Here, n and m stand respectively for the number of independent and dependent variables, and k denotes the order of the system. Further, instead of the "system of equations", we shall say "equation".

Let $\mathcal{Y} \subset N_m^k$ be an equation. The submanifold $L^n \subset N_m^\infty$ is called its solution in the usual sense, if $\mathrm{im}\, j_k(L) \subset \mathcal{Y}$. The classical interpretation of the concept of solution is however not very satisfactory, as it does not cover the useful "singular" and "generalized" cases. One of the possible ways to generalize the concept of solution resides in the following.

The R-manifold of $V \subset N_m^s$, $s \geqslant k$, is called the element of solution of equation $\mathcal{Y}$, if $\pi_{s,k}(V) \subset \mathcal{Y}$. Using the maps $\pi_{s,k}$, it is not difficult to determine the procedure of piecing together the elements of solution. We shall call the object obtained the generalized solution of $\mathcal{Y}$. For example, the Riemannian surfaces are the generalized solutions of the Cauchy-Riemann system.

We shall determine the s-th extension $\mathcal{Y}^{(s)} \subset N_m^{k+s}$ of the equation

$\mathcal{Y} \subset N^k_m$ assuming that $[L]^{k+s}_x \in \mathcal{Y}^{(s)} \Leftrightarrow$ im $j_k(L)$ touches $\mathcal{Y}$ at a point $j_k(L)(x)$ with order s. Obviously, $\pi_{s,t}(\mathcal{Y}^{(s)}) \subset \mathcal{Y}^{(t)}$, for $s \geq t$, and the equation $\mathcal{Y}^{(s)}$ has the same usual solutions as $\mathcal{Y}$. Let $\mathcal{Y}_{(s)} = \pi_{\infty,s}(\mathcal{Y}^{\infty})$, $\mathcal{Y}_{\infty} = \mathcal{Y}^{(\infty)}$, $\mathcal{F}_s(\mathcal{Y}) = C^{\infty}(N^s_m)|\mathcal{Y}_{(s)}$. We shall identify $\mathcal{F}_s(\mathcal{Y})$ with $\pi^*_{\infty,s}(\mathcal{F}_s(\mathcal{Y})) \subset \mathcal{F}_{\infty}(\mathcal{Y})$. Then the system of subalgebras $\mathcal{F}_s(\mathcal{Y})$ provides $\mathcal{F}_{\infty}(\mathcal{Y})$ with a structure of a filtered algebra.

Also we shall introduce the following system of notations. Let $\eta : E_{\eta} \to N$ be a smooth vector bundle and let $\eta_s = \pi^*_{s,0}(\eta)$ be a corresponding induced bundle and

$$\eta_s(\mathcal{Y}) = \eta_s|\mathcal{Y}_{(s)}, \quad \mathcal{F}_s(\mathcal{Y},\eta) = \Gamma(\eta_s(\mathcal{Y}))$$

Identifying as before

$$\mathcal{F}_s(\mathcal{Y},\eta) \text{ with } \pi^*_{\infty,s}(\mathcal{F}_s(\mathcal{Y},\eta)) \subset \mathcal{F}_{\infty}(\mathcal{Y},\eta)$$

we provide $\mathcal{F}_{\infty}(\mathcal{Y},\eta)$ with a structure of a filtered module over the filtered algebra $\mathcal{F}_{\infty}(\mathcal{Y})$.

An infinitely extended equation $\mathcal{Y}_{\infty}$ can be equipped with the Cartan distribution, if we restrict on it the Cartan distribution available on N^{∞}_m. It is convenient, however, to make use of dual construction. Let $C\Lambda^k(\mathcal{Y}_{\infty}) = C\Lambda^k(N^{\infty}_m)|_{\mathcal{Y}_{\infty}}$, considering $C\Lambda^1(\mathcal{Y}_{\infty})$ to be an annihalator of the Cartan distribution on $\mathcal{Y}_{\infty}$.

3. Objects of the Category DE (Preliminary Discussion)

In this section we shall introduce an important class of objects of the category DE. To do so, we start with the following motivations.

Let $C(\mathcal{Y})$ be the restriction of the Cartan distribution on N^k_m to $\mathcal{Y}$. The classical Cartan interpretation of a differential equation as a Pfaffian system is equivalent, obviously, to the consideration of a manifold $\mathcal{Y}$ equipped with the distribution $C(\mathcal{Y})$. This, useful in many respects, view point gives an idea of regarding pairs of type $(\mathcal{Y}, C(\mathcal{Y}))$ as objects of DE. In an implict way, this idea is accepted in many contemporary papers dedicated to various "principal" problems related to the theory of nonlinear equations. The following arguments are an indication of the inadequacy of this idea from several view points.

1. Proposition 1(a) reveals that maximal integral manifolds of the distribution $C(\mathcal{Y})$, which from the interior point of view should be regarded as solutions to equation $\mathcal{Y}$, differ in dimensions, in general. Of them, as solution of equation $\mathcal{Y}$ it is appropriate to consider only those which are R-manifolds (E. Cartan by-passed this difficulty by postulating a priori dimension of integral manifolds of the distribution $C(\mathcal{Y})$ which should be regarded as solutions).

2. Natural operations of differential calculus [1] on space N_m^k "jump" on N_m^l, $l \geqslant k$.

3. Questions like problem of compatibility (formal integrability) (see, for example, [5, 6]) or propagation of singularities of generalized solutions demand, in a nonlinear case, simultaneous consideration of all possible extensions of equation $\mathcal{Y}$.

4. Nonlinear differential operators should be morphisms in DE. This, however, cannot be achieved in the framework of Cartan's interpretation.

5. The distribution $C(\mathcal{Y})$ is not completely integrable.

All these drawbacks disappear on going over from the $(\mathcal{Y}, C(\mathcal{Y}))$ pair to the $(\mathcal{Y}_\infty, C(\mathcal{Y}_\infty))$ pair, where $C(\mathcal{Y}_\infty)$ is the Cartan distribution on $\mathcal{Y}_\infty$. For example, all integral manifolds of the Cartan distribution on N_m^∞ are of one type (and have the same dimension) and may be interpretted as solutions of equation $\mathcal{Y}_\infty$ and, hence, of $\mathcal{Y}$. This leads us to the following conclusion: as objects the category DE should contain $(\mathcal{Y}_\infty, C(\mathcal{Y}_\infty))$ pairs.

As regards the problems concerning the internal structure of objects of the category DE, this conclusion is quite satisfactory. However, having restricted the framework of the category DE with objects of type $\mathcal{Y}_\infty$ we would have deprived us of the possibility of realizing within its framework many important constructions which make the categorical consideration practically expedient. For example, the natural operation of factorization $\mathcal{Y}_\infty \to \mathcal{Y}_\infty/G$, where G is the Lie group of automorphism of equation $\mathcal{Y}_\infty$, as it is shown elsewhere, does not lead in general to objects of type $\mathcal{Y}_\infty$. It is therefore necessary to widen the just contemplated framework of the category DE. To do this, we shall take recourse to the algebraic view point described below.

4. Differential Calculus in Commutative Algebras

Let K be a commutative ring with identity and A be an unitary commutative K-algebra. With every $a \in A$ and $\Delta \in \mathrm{Hom}_K(P,Q)$, where P and Q are A-modules, one can connect K-homomorphism $\delta_a(\Delta): P \to Q$, assuming $\delta_a(\Delta)(p) = \Delta(ap) - a\Delta(p)$. Obviously, $\delta_a\delta_b = \delta_b\delta_a$. A K-homomorphism Δ is called a K-differential operator (d.o.) of order $\leq s$, if for any set of elements $a_0, a_1, \ldots, a_s \in A$ $(\delta_{a_0} \circ \ldots \circ \delta_{a_s})(\Delta) = 0$.

Let $K = \mathbb{R}$ and $A = C^\infty(M)$, where M is a smooth manifold. Let $P = \Gamma(\xi)$ and $Q = \Gamma(\eta)$, where ξ and η are smooth vector bundles on M. In this case the earlier introduced concept of differential operator agrees with the classical notion of linear differential operator acting from the section of bundle ξ into the section of bundle η.

The collection of all d.o. of order $\leq s$, from P to Q, forms an A-module relative to the operation of multiplication $(a, \Delta) \to a\Delta$, where $(a\Delta)(p) = a\Delta(p)$, $a \in A$, $p \in P$. We denote it by $\mathrm{Diff}_s(P,Q)$ and pose $\mathrm{Diff}_s P = \mathrm{Diff}_s(A,P)$. Obviously, $\mathrm{Diff}_0(P,Q) = \mathrm{Hom}_A(P,Q)$, $\mathrm{Diff}_0 P = P$. Let $D(P) = \{\Delta \in \mathrm{Diff}_1 P \mid \Delta(1) = 0\}$. Then A-module D(P) consists of all differentiations of algebra A with values in module P.

The differential operators over an algebra A have all common properties of "usual" differential operators, for example, $\Delta_1 \circ \Delta_2 \in \mathrm{Diff}_{s+t}(P,R)$, if $\Delta_1 \in \mathrm{Diff}_s(Q,R)$ and $\Delta_2 \in \mathrm{Diff}_t(P,Q)$. In further discussion we shall make use of these properties without specific reservations [1, 5].

The operations $P \mapsto \mathrm{Diff}_s P$ and $P \mapsto D(P)$ are examples of absolute functors of differential calculus [1, 5]; operation $Q \mapsto \mathrm{Diff}_s(P,Q)$ (P is fixed) is an example of a special functor of differential calculus. Hereinafter, these functors are denoted by Diff_s, D, and $\mathrm{Diff}_s(P,\cdot)$ respectively. It is now quite natural to raise the problem concerning representation of these functors. For example, call the A-module of $\mathcal{J}^s(P)$ together with the operator $j_s = j_s(P) \in \mathrm{Diff}_s(P, \mathcal{J}^s(P))$ the representing object for functor $\mathrm{Diff}_s(P, .)$, provided the operation $\varphi \mapsto \varphi \circ j_s(P)$, where $\varphi \in \mathrm{Hom}_A(\mathcal{J}(P), Q)$ realizes the isomorphism of A-modules $\mathrm{Diff}_s(P,Q)$ and $\mathrm{Hom}_A(\mathcal{J}(P), Q)$ for all Q. Similarly, the representing objects for functors Diff_s and D are determined. The corresponding pairs are denoted by $(\mathcal{J}^s = \mathcal{J}^s(A), j_s)$ and $(\Lambda^1 = \Lambda^1(A), d)$ where $j_s \in \mathrm{Diff}_s(\mathcal{J}^s)$ and $d \in D(\Lambda^1)$.

Proposition 4. The representing objects for functors $Diff_s$, $Diff_s(P,.)$, and D exist and are unique up to an isomorphism. They are called the modules of jets (for $Diff_s$ and $Diff_s(P, .)$ and differential forms (for D).

It is expedient to generalize the statement of the problem concerning representing objects. Let $\mathfrak{K}$ be a subcategory of the category of all A-modules, which is closed under the action of functors of the differential calculus. Then one can pose the representativity problem in $\mathfrak{K}$ for these functors. For the earlier discussed functors, we denote the representing objects in $\mathfrak{K}$ (if they exist) by $\mathcal{J}^s_{\mathfrak{K}}(P)$ and $\Lambda^1_{\mathfrak{K}}$, respectively.

Category G of geometric A-modules is an important example of the closed category in the above sense. Module P is called the geometric module, if $\bigcap_{\mathfrak{p}} \mathfrak{p} P = 0$ where intersection is taken over all simple ideals of algebra A.

Proposition 5. Representing objects $\mathcal{J}^s_G(P)$ and Λ^1_G exist. If, in this case, $K = \mathbb{R}$ and $A = C^\infty(M)$, then $\Lambda^1_G = \Lambda^1(M)$ and $\mathcal{J}^s_G(P) = \Gamma(\xi_s)$, where $P = \Gamma(\xi)$ and $\xi_s: J^s(\xi) \to M$ is a bundle of s-jets.

Assume now that algebra A is filtered by subalgebras $\ldots \subset A_i \subset A_{i+1} \subset \ldots$ and consider a category of filtered A-modules $P = \{P_i\}$, $P = \bigcup P_i$. Call the differential operator $\Delta \in Diff_s(P, Q)$, where P and Q are filtered A-modules, the filtered operator, if there exists $m \in \mathbb{Z}$ such that $\Delta(P_i) \subset Q_{i+m}$. Call the filtered module P the geometric module, if all A_i-modules P_i are geometric. Denote the category of geometrically filtered A-modules of A by FG, and the functors of the filtered differential calculus by $Diff_s^F(P,.)$, D^F, and so on.

Proposition 6. Functors $Diff_s^F(P, .)$ and D^F are representative in the FG category.

5. The Category DE

Let $\mathcal{Y} \subset N^k_m$ be an equation $\mathcal{F} = \mathcal{F}_\infty(\mathcal{Y})$, $\mathcal{F}_s = \mathcal{F}_s(\mathcal{Y})$. We shall operate in the FG category on the filtered algebra $\mathcal{F} = \{\mathcal{F}_s\}$. To simplify notations, we write $Diff_s(P, Q)$ and $D(P)$ instead of $Diff_s^F(P,Q)$ and $D^F(P)$, and denote by $\mathcal{J}^s(P)$, Λ^1, etc. the representing objects in the FG category over $\mathcal{F}$. If Φ is a representing object in the FG category over $\mathcal{F}$ for a functor of differential calculus $\mathfrak{F}$, then the submodule $C\Phi \subset \Phi$:

$$\mathcal{C}\Phi = \left\{ \varphi \in \Phi \;\middle|\; j(L)^*(\varphi)\big|_x = 0,\ [L]_x^\infty \in \mathcal{Y}_\infty \right\}$$

is defined. If now $\Delta : \Phi_1 \to \Phi_2$ is a natural differential operator connecting the representing objects (say, d or j_s), then, obviously, $\Delta(\mathcal{C}\Phi_1) \subset \mathcal{C}(\Phi_2)$.

Thus, on $\mathcal{Y}_\infty$ (more exactly over the algebra $\mathcal{F}$) is defined the operation of "Cartan distributions" $\mathcal{C} : \Phi \mapsto \mathcal{C}\Phi$ consistent with natural operators connecting the representing objects. Besides, $\mathcal{C}\Lambda^* = \mathcal{C}\Lambda^1 \wedge \Lambda^1$, where $\Lambda^i = \wedge^i \Lambda^1$, $\Lambda^* = \sum \Lambda^i$.

Similarly to the fact that an algebra $C^\infty(M)$ uniquely defines the manifold M (particularly, a set of points constituting M and the topology on it), the filtered algebra $\mathcal{F}_\infty(\mathcal{Y})$ defines $\mathcal{Y}_\infty$ so that transition from the consideration of $\mathcal{Y}_\infty$ (which is highly nonconstructive) to that of $\mathcal{F}_\infty(\mathcal{Y})$ does not cause a loss of information. Further, putting operation $\mathcal{C}$ in the differential calculus in the FG category over $\mathcal{F}_\infty(\mathcal{Y})$ is equivalent to equipping with Cartan distribution. Therefore, the filtered algebra $\mathcal{F}_\infty(\mathcal{Y})$ equipped with operation $\mathcal{C}$ in the aforementioned sense may be regarded as algebraic representation of the object $\mathcal{Y}_\infty$ of the category DE. In view of the fact that by this way one obtains in a constructive form everything necessary for studying $\mathcal{Y}_\infty$, we accept it as the base for describing the category DE.

Thus, we call the filtered algebra $A = \{A_i\}$, $A_i \subset A_{i+1}$ to be an object of the category DE, if the differential calculus in the FG category over A is equipped with a natural operation $\mathcal{C}$ and $\mathcal{C}\Lambda^* = \mathcal{C}\Lambda^1 \wedge \Lambda^*$.

<u>Remark 3</u>. As we are interested not in the filtration of A-algebra, but in the FG category generated by it, two filtrations may, of course, be thought to be equivalent if they lead to the same FG category. For this reason, in the earlier given definition of the category DE it would be more satisfactory to fix the class of equivalent filtrations. Here it must be stressed that automorphisms of the object $\mathcal{Y}_\infty$ cannot, indeed, conserve the above described filtration of the algebra $A = \mathcal{F}_\infty(\mathcal{Y})$.

<u>Remark 4</u>. It would be more correct to formulate the condition $\mathcal{C}\Lambda^* = \mathcal{C}\Lambda^1 \wedge \Lambda^*$, interpreted as a condition of "infinite prolongativity" of the object considered, as differential closeness of the ideal $\mathcal{C}\widetilde{\Lambda}^1 \wedge \widetilde{\Lambda}^* \subset \widetilde{\Lambda}^*$, where $\widetilde{\Lambda}^*$ is the algebra of higher differential forms and corresponds to the sequence $\sigma = (\infty, \infty, \ldots)$ (see [7]). Since, in the interesting cases, both these conditions are equivalent, we

have used the first to avoid describing the theory of higher order de Rham complexes.

A geometric image related to filtered algebra $A = \{A_i\}$ is an inverse limit $\mathrm{Spec}_\infty A$ of the chain of maps $\ldots \leftarrow \mathrm{Spec}\, A_i \leftarrow \mathrm{Spec}\, A_{i+1} \leftarrow \leftarrow \ldots$ corresponding to the chain of imbeddings $\ldots \rightarrow A_i \rightarrow \rightarrow A_{i+1} \rightarrow \ldots$ This image together with the "Cartan distribution" given on it may be regarded as geometric description of an object of DE. It is clear that spaces of type $\mathrm{Spec}_\infty A$ are highly nonconstructive to work directly with them. For the purposes of classical theory of differential equations and its natural generalizations it is sufficient to restrict our consideration to algebras $A = \{A_i\}$, where $A_i = C^\infty(M)$ and M_i is a smooth manifold (possibly with singularities). Then $M_i \subset \mathrm{Spec}\, A_i$ and the "projection" $M_{i+1} \twoheadrightarrow M_i$ corresponds to the imbedding $A_i \subset A_{i+1}$. If M_∞ is the inverse limit of the chain $\ldots \leftarrow M_i \leftarrow M_{i+1} \leftarrow \ldots$, then $M_\infty \subset \mathrm{Spec}\, A$ and $\overline{M}_\infty = \mathrm{Spec}_\infty A$. Since M_∞ contains all "good" points of $\mathrm{Spec}_\infty A$, we shall further use M_∞ as geometric image connected with object A of the category DE.

The earlier given definition of an object of the category DE obviously leads to a definition of a morphism in DE as an homomorphism of filtered algebras consisted of the operation $\mathcal{C}$. Using the geometric language, as it is accepted in algebraic geometry, we shall talk about an object M_∞ instead of A, and about a morphism $\varphi : M^1_\infty \twoheadrightarrow M^2_\infty$ instead of homomorphism $\varphi^* : A_2 \twoheadrightarrow A_1$ of corresponding algebras. We shall call algebra A the algebra of functions on M_∞.

It is useful to interpret $M_\infty \in \mathrm{Ob}\, DE$ as an infinitely prolongated generalized differential equation. The sense in which the adjective "generalized" is used is explained in the examples given later.

6. Dimension in DE

If $M_\infty \in \mathrm{Ob}\, DE$ is taken as a generalized differential equation, then its dimension can be determined as number of independent variables whose functions are the solutions to this equation. The constructive method of realizing this view point consists in the following.

Let $\Lambda^i = \Lambda^i(M_\infty)$ and $\bar{\Lambda}^i = \bar{\Lambda}^i(M_\infty) = \Lambda^i / \mathcal{C}\Lambda^i$. Call M_∞ the n-dimensional object, if for almost all points $\theta \in M_\infty$, $\bar{\Lambda}^n_\theta \neq 0$, but $\bar{\Lambda}^i_\theta = 0$, $i > n$. If $\mathcal{Y} \subset N^k_m$, then $\mathcal{Y}_\infty$ is an n-dimensional object. Henceforth, we shall use the usual notation of dimension :

$\dim M_\infty = n$.

As an object of DE we may represent a smooth manifold M^n by two different methods. One of them consists in the following. Let $A = C^\infty(M)$, $A_i = A$. Then the FG category on A coincides with the geometric category. Further we shall introduce operation $\mathcal{C}$ assuming $\mathcal{C} = 0$. This will transform M into an n-dimensional object of DE.

This permits the following generalization. Let N be a smooth manifold, $\dim N = n + m$, and a smooth foliation $\mathcal{H}$ be given on N. Assume $A = C^\infty(N)$, $A_i = A$ and

$\mathcal{C}\Phi = \{\varphi \in \Phi \mid$ restrictions of φ on the leaves of $\mathcal{H}$ are trivial$\}$.

Then the foliation $\mathcal{H}$ is an n-dimensional object in DE.

Every manifold M may be regarded as foliated by its own points. This gives the second method of interpretting a smooth manifold M as an object of DE. Here, obviously, $\mathcal{C}\Lambda^i = \Lambda^i$, $i > 0$, and $\mathcal{C}\Lambda^0 = 0$. In the given case we obtain a zero-dimensional object in DE.

Thus, we have described two ways of imbedding the category of smooth manifolds in the category DE. In the first method an n-dimensional manifold becomes an n-dimensional object. We shall call them the dimensional and non-dimensional methods, respectively. Both these methods are the sources of fruitful and unobvious analogies between the category of smooth manifolds and DE.

Let now $F: M_\infty \to M'_\infty$ be a morphism. We determine its rank at a point $\theta \in M_\infty$ as maximum number r such that $F^*(\Lambda^r_{F(\theta)}(M'_\infty)) \neq 0$. Say F is of rank r, if F has a rank r at all $\theta \in M_\infty$. It is not hard to check that fibres $F^{-1}(\theta')$, $\theta' \in M'_\infty$, of morphism F of rank r are (n-r)-dimensional objects of the category DE, if $\dim M_\infty = n$. An important class of morphisms in DE yields morphisms that conserve the dimension or morphisms with zero-dimensional fibres. Surjective morphisms of this kind are further called the coverings. Note that the Wahlquist-Estabrook "prolongation structures" [8] are a particular case of coverings in DE, and the theory of Backlund transformations is also closely connected with the theory of coverings in DE [13].

Call the smooth manifold M^n which is assumed to be an n-dimensional object in DE together with an imbedding $i: M \to M_\infty$, which is a morphism in DE, the integral manifold in M_∞. If $\dim M_\infty = n$, then it is relevant to interpret its n-dimensional integral manifolds as solutions of the generalized equation M_∞. Obviously, an n-dimensional object in DE does not have an integral manifold of dimension $> n$.

Finally, for objects of the category DE we consider a concept of symmetry. If $M_\infty \in$ Ob DE, then its isomorphism onto itself as an object of DE is called the finite symmetry or the $\mathcal{C}$ -transformation. It is obvious that every $\mathcal{C}$ -transformation generates a transformation of the set of all solutions of the generalized equation M_∞.

Now we shall consider this from the infinitesimal point of view. Call a vector field X on M_∞, or, exactly, the filtered differentiation of the algebra A of functions on M_∞ the $\mathcal{C}$-field, if $X(\mathcal{C}\Lambda^1) \subset \mathcal{C}\Lambda^1$ (X(.) denotes the Lie derivative operation). Call $D_\mathcal{C}(M_\infty)$ the set of all $\mathcal{C}$-fields on M . Let also $\mathcal{C}D(M_\infty) = \{ X \mid X \rfloor \mathcal{C}\Lambda^1 = 0 \}$. As the Cartan distribution on M is completely integrable, $\mathcal{C}D(M_\infty) \subset D_\mathcal{C}(M_\infty)$.

Commutation operation obviously yields $D_\mathcal{C}(M_\infty)$ with the Lie algebra structure and $\mathcal{C}D(M_\infty)$ is an ideal of this algebra. Hence, the Lie factor-algebra

$$\mathrm{Sym}\, M_\infty = D_\mathcal{C}(M_\infty) / \mathcal{C}D(M_\infty)$$

is defined.

Call the elements of this factor algebra the infinitesimal symmetries of object M_∞. Further, when the context permits, we shall drop the adjective "infinitesimal" and call the Sym M_∞ the algebra of symmetry of object M_∞.

The motivation of introducing the algebra Sym M_∞ consists in the following. If on M_∞ there exist currents corresponding to fields $X \in D_\mathcal{C}(M_\infty)$, then they in their turn will generate currents in the space of all solutions of the generalized equation M_∞. The currents corresponding to the fields $X \in \mathcal{C}D(M_\infty)$ will generate trivial currents in this space of solutions. In other words, the Sym M_∞ algebra can be understood as algebra of vector fields on "the manifolds of all solutions of the generalized equation M_∞".

In [1, 4] and [5] the descriptions of the algebra Sym M_∞ are given for that case when $M_\infty = J^\infty(\pi)$, N_m^∞, $\mathcal{Y}_\infty$, where $\mathcal{Y}$ is a differential equation. In many interesting cases the Sym M_∞ may be described by the methods discussed in these papers. Examples of this are given below.

7. On the Structure of Morphisms in DE

This section contains examples and results illustrating the concept of morphism in DE. Let first $U_i \subset J^{\infty}(\pi_i)$, $i = 1, 2$ are open domains, where $\pi_i: E_i \to M_i$ are smooth bundles. It is not hard to see that the mapping $F: U_1 \to U_2$ is a morphism in DE, if $F^*(\mathcal{C}\Lambda^1(U_2)) \subset \mathcal{C}\Lambda^1(U_1)$. We now consider the canonical coordinates (x_i, p_σ^j) in U_1 and $(\tilde{x}_k, \tilde{p}_\alpha^l)$ in U_2 (see Section 2). Assume that F is a non-degenerate morphism, i.e. its rank is almost everywhere not less than dim U_1. Then holds

<u>Proposition 7</u>. (a) Every non-degenerate morphism $F: U_1 \to U_2$ is uniquely defined by functions $F^*(\tilde{x}_k) = \tilde{x}_k(\ldots, x_i, \ldots, p_\sigma^j, \ldots)$ and $F^*(\tilde{p}_\varnothing^l) = \tilde{p}_\varnothing^l(\ldots, x_i, \ldots, p_\sigma^j, \ldots)$ which can be chosen arbitrarily.

(b) Every non-degenerate morphism $F: \mathcal{Y}_\infty^1 \to \mathcal{Y}_\infty^2$ is uniquely defined by the restriction of F^* on $\mathcal{F}_0(\mathcal{Y}_\infty^2)$.

Let now $\mathcal{Y}_\infty^\varepsilon \subset J^\infty(\pi_\varepsilon)$, $\varepsilon = 1, 2$. Then the practical method of defining a non-degenerate morphism $F: \mathcal{Y}_\infty^1 \to \mathcal{Y}_\infty^2$ consists in pointing to a morphism $\overline{F}: J^\infty(\pi_1) \longrightarrow J^\infty(\pi_2)$ such that $\overline{F}^*(\mathcal{Y}_2) \subset \mathcal{Y}_1$, where $\mathcal{I}_\varepsilon \subset \mathcal{F}_\infty(\pi_\varepsilon)$ is the ideal of the equation $\mathcal{Y}_\infty^\varepsilon$ in $J^\infty(\pi_\varepsilon)$. We shall also stress that a non-degenerate morphism $F: M_\infty^1 \longrightarrow M_\infty^2$ may be defined, as a rule, by indicating $F^*\big|_{A_k^2}$ for some k, where $A^\varepsilon = \{A_i^\varepsilon\}$ is the algebra of functions on M_∞^ε.

If under the conditions of Proposition 7 we assume $M_1 = M_2$ and $F^*(x_i) = x_i$, then the nonlinear differential operator $(\hat{u}^1(x), \ldots, \hat{u}^m(x)) \mapsto (u^1 = u^1(x, \ldots, p_\sigma^i, \ldots), \ldots, u^m = u^m(x, \ldots, p_\sigma^i, \ldots))$ corresponds to morphism F, and conversely. Thus, differential operators are morphisms in DE.

This enables us to construct examples of automorphisms in DE (i.e. of $\mathcal{C}$-transformations) which do not conserve the filtration of basic algebra. It is of interest in view of Remark 3 of Section 5. We shall now consider a differential operator $\Delta : \Gamma(\pi) \to \Gamma(\pi)$, where π is the linear bundle on M. Then it generates a morphism $\varphi_\Delta : J^\infty(\pi) \to J^\infty(\pi)$. If this operator is invertible, then φ_Δ is the $\mathcal{C}$-transformation of the object $J^\infty(\pi)$. If in this case the order of operator Δ is more than zero, then φ_Δ does not conserve the filtration $\mathcal{F}_i(\pi)$ of the algebra $\mathcal{F}_\infty(\pi)$. The following is the most simple example of this kind of operator.

Let π be a trivial two-dimensional bundle over $\mathbb{R}^1$. Then

$$\Delta = \begin{pmatrix} 1 & D \\ 0 & 1 \end{pmatrix}, \quad \Delta^{-1} = \begin{pmatrix} 1 & D \\ 0 & 1 \end{pmatrix}^{-1}, \quad D = d/dx$$

8. Factorization in DE

Let $M_\infty \in \mathrm{Ob}\, DE$, $A = \{A_i\}$ be the algebra of functions on M_∞, and G be a Lie group of $\mathcal{C}$-transformations of M_∞. The representation $\alpha: \mathcal{G} \to D_{\mathcal{C}}(M)$ of the Lie algebra $\mathcal{G}$ of G is defined in a standard manner. Assume that G acts without fixed points (this implies that $\ker \alpha = 0$) and $\alpha(\mathcal{G})_\theta \cap \mathcal{C}D(M_\infty)_\theta = 0, \quad \forall \theta \in M_\infty$. The latter means that the orbits $G(\theta)$ of the action considered are transversal to the Cartan distribution $\theta \longmapsto \mathcal{C}D(M_\infty)$ on M_∞. This implies that the restriction of Cartan distribution on $G(\theta)$ induces on $G(\theta)$ the structure of a zero-dimensional object of the category DE. It may be noted also that from the assumptions made it follows that the composition β

$$\mathcal{G} \xrightarrow{\alpha} D_{\mathcal{C}}(M_\infty) \longrightarrow D_{\mathcal{C}}(M_\infty)/\, \mathcal{C}D(M_\infty) = \mathrm{Sym}\, M_\infty$$

is also monomorphic.

We now assume that for the action considered the "space of orbits" is defined. By this we mean the following. Let

$$M_b' \xleftarrow{f_{b+1}'} M_{b+1}' \longleftarrow \cdots \longleftarrow M_i' \xleftarrow{f_i'} M_{i+1}' \longleftarrow \cdots$$

be a chain of surjections of smooth manifolds, $A_i' = C^\infty(M_i')$, $A' = \lim\limits_{i \to \infty} \mathrm{dir}\, A_i'$, $M_\infty' = \lim \mathrm{inv}\, M_i'$, and $\gamma : M_\infty/G \longrightarrow M_\infty'$ be a bijection of sets such that $(\gamma \circ \pi_G)^* (A_i') \subset A_{i+s}$, where $\pi_G : M_\infty \to M_\infty/G$ is the natural projection. If the mapping $\gamma \circ \pi_G : M_\infty \to M_\infty'$ has smooth local sections, then M_∞' is called the space of orbits of G. Further, M_∞' and M_∞/G are identified.

If $M_\infty = \lim \mathrm{inv}\, M_i$, $\cdots \leftarrow M_i \xleftarrow{f_{i+1}} M_{i+1} \leftarrow \cdots$, and some actions of G on M_i commuting with mappings f_j are given, then an action of G on M_∞ arises. Suppose that the orbit spaces M_k/G are defined for all $k \geqslant k_0$. Then the orbit space M_∞/G is defined and $M_k' = M_k/G$. This situation is realized when $M_\infty = \mathcal{Y}_\infty$, where $\mathcal{Y}$ is an equation and G is a group of diffeomorphisms of the manifold $\mathcal{Y}_{k_0}$, $k_0 \geqslant 0$, that conserves the Cartan distribution on it, since such diffeomorphisms are canonically raised to diffeomorphisms of the manifolds M_k, $k \geqslant k_0$, which conserve the Cartan distribution.

Now M_∞/G may be supplied with a structure of an object of DE. To this end, note that the algebra of functions on M_∞/G coincide with $A^G = \{a \in A \mid ga = a,\ g \in G\}$. If Φ' is the representing object for functor $\mathcal{F}$ of the differential calculus over $A = A^G$, then we define

$$C\Phi' = \{\varphi \in \Phi' \mid \pi_G^*(\varphi) \in C\Phi\}$$

where Φ is an analogous object on A. This operation C transforms M_∞/G into an object of the category DE and $\pi_G : M_\infty \to M_\infty/G$ is a covering in DE. From this, in particular, follows that $\pi_G^{-1}(L)$, where $L \subset M_\infty/G$ is an integral manifold, has a foliation, say $\mathcal{H}$, which is determined by the restriction of the Cartan distribution on M_∞, and codim $\mathcal{H}$ = dim G. In this sense we may say that to every solution L of the "factor-equation" M_∞/G corresponds dim G - parametric family of solutions of the "equation" M_∞.

9. Examples of Factorization

If $M_\infty = \mathcal{Y}_\infty$ and $M_\infty/G = \mathcal{Y}'_\infty$, where $\mathcal{Y}$ and $\mathcal{Y}$ are some equations, then $\mathcal{Y}'$ will be called the factor-equation of $\mathcal{Y}$. Now we shall give some examples. In this connection it may be recalled that infinitesimal symmetry of the object $J^\infty(\mathbb{1}_M)$, where $\mathbb{1}_M$ is the trivial one-dimensional fibering on M, is uniquely defined by its generating function $f \in \mathcal{F}_\infty(\mathbb{1}_M)$ (see [1, 4, and 5]). Further we assume that $M = \mathbb{R}^2$ and $\mathcal{Y} \subset J^2(\mathbb{1}_M)$.

<u>Example 1.</u> <u>The Laplace equation</u> $\mathcal{Y} = \{u_{xx} + u_{yy} = 0\}$. The group G of translations of $\mathbb{R}^2$ is a group of symmetries for $\mathcal{Y}$. The generators of the algebra $\beta(G) \subset$ Sym Y correspond in this case to the generating functions $p_1, p_2 \in \mathcal{F}_1(Y)$ (see Section 1). Then $\mathcal{Y}'$ is again the Laplace equation and we, thereby, obtain a method of assigning to every solution of the Laplace equation a two-parametric family of solutions to this equation. For example, the family

$$u = e^{x+c_1} \sin(y + c_2);\ c_1,\ c_2 = \text{const}$$

corresponds in this case to the solution u = x.

Further, we shall characterize group G by generators of the algebra $\beta(\mathcal{Y})$.

<u>Example 2.</u> $\mathcal{Y}$ is the Laplace equation, $\beta(\mathcal{Y}) = \{x, y\}$. Then, as

before, $\mathcal{Y}'$ is the Laplace equation and we obtain a rule which is different from that of Example 1, for assigning a two-parametric family of solutions of this equation to every solution of the equation. In this case the family of solutions

$$u = x(c_1 - \ln(x^2 + y^2)^{1/2} + y(c_2 + \operatorname{arc\,tg} \frac{x}{y}), \quad c_1, c_2 = \text{const}$$

corresponds to the solution $u = x$.

Example 3. Wave equation $\mathcal{Y} = \{u_{xx} - u_{yy} = 0\}$, $\beta(\mathcal{Y}) = (p_1, p_2)$. Here, $\mathcal{Y}' \approx \mathcal{Y}$ and to the solution $u = x$ of equation $\mathcal{Y}$ corresponds the two-parametric family $u = e^{x+c_1} \operatorname{ch}(y + c_2)$, $c_1 c_2 = \text{const}$ of solutions of this very equation.

Example 4. Heat equation $\mathcal{Y} = \{u_x - u_{yy} = 0\}$, $\beta(\mathcal{Y}) = (p_1, p_2)$. In this case, $\mathcal{Y}'$ is the equation

$$(xu_x)^2 u_{xx} - 2xu_x(u_y - xu_x)u_{xy} + (u_y - xu_x)u_{yy} = 0$$

Example 5. Burgers equation $\mathcal{Y} = \{u_y + uu_x + u_{xx} = 0\}$, $\beta(\mathcal{Y}) = \{p_1, p_2\}$. Then $\mathcal{Y}'$ is the equation

$$2\, uu_x - u_x^2 + x^2 u_x + u_y - 5yu = 0$$

In fact, the description of the object $\mathcal{Y}/G$ given in the above examples is valid only locally. Globally, the objects $\mathcal{Y}/G$ are, as a rule, not of the form $\mathcal{Y}'_\infty$. Now we give a very simple example explaining this phenomena.

Example 6. Trivial equation $\mathcal{Y} = J^1(\mathbb{1}_{\mathbb{R}^1})$, $\mathcal{Y}_\infty = J^\infty(\mathbb{1}_{\mathbb{R}^1})$. The group $G = \mathbb{R}^1$ acts on $\mathcal{Y}_\infty$ such that the element $c \in \mathbb{R}^1$ maps the point $\theta = (x, u, \ldots, p_\sigma, \ldots) \in J^\infty(\mathbb{1}_{\mathbb{R}^1})$ into the point $(x+c, u, \ldots, p_\sigma, \ldots)$. Then the algebra $A' = A^G$ is composed of all smooth functions of variables $u, \ldots, p_\sigma, \ldots$. Besides, $\dim \mathcal{Y}/G = 1$. It is not difficult to check that

$$\mathcal{C}\Lambda^1(J^\infty(\mathbb{1}_{\mathbb{R}^1})/G) = \Big\{\sum_\sigma \lambda_\sigma \omega_\sigma \,\Big|\, \lambda_\sigma \in A', \ \omega_\sigma = p_1 dp_\sigma - p_{\sigma 1} du$$

and hence

$$\mathcal{C}D(J^\infty(\mathbb{1}_{\mathbb{R}^1})/G) = \Big\{ fD \,\Big|\, f \in A', \ D = \sum_\sigma p_{\sigma 1} \frac{\partial}{\partial p_\sigma} \Big\}$$

From the latter it immediately follows that the one-dimensional submanifold $L_{\omega,A} \subset J^\infty(\mathbb{1}_{\mathbb{R}^1})/G$, given by the system of equations

$$\begin{cases} u^2/A^2 + p_1^2/\omega^2 A^2 = 1 \\ p_{\sigma_{11}} + \omega^2 p_\sigma = 0\,, \quad \forall \sigma \end{cases}$$

is integral and diffeomorphic with the circle. Picking up sufficiently small ω and A we can place $L_{\omega,A}$ in any prescribed neighbourhood of the point $\theta = (0, 0, \ldots) \in J^\infty(\mathbb{1}_{\mathbb{R}^1})/G$. However one-dimensional objects of the category DE of type $\mathcal{Y}'_\infty$, where $\mathcal{Y}$ is an equation, do not contain arbitrarily small integral submanifolds which are diffeomorphic with the circle. This follows immediately from Proposition 1 (b). Therefore, $J^\infty(\mathbb{1}_{\mathbb{R}^1})/G$ is not of the form $\mathcal{Y}'_\infty$.

10. Partially Invariant Solutions

Suppose that a Lie group G of $\mathcal{C}$-transformations acts as before on $M_\infty \in$ Ob DE without fixed points, dim M_∞ = n, dim G = m. Let us consider a subset $M_{(k)} \subset M_\infty$:

$$M_{(k)} = \left\{ \theta \in M_\infty \,\middle|\, \dim\left(T_\theta(G(\theta)) \cap \mathcal{C}D(M_\infty)_\theta\right) = k \right\}$$

Obviously, $M_{(k)}$ is G-invariant. When the orbit space M_∞/G is defined, $M_{(k)}$ may be interpreted as "submanifold" of singular points of the projection $\pi_G: M_\infty \to M_\infty/G$ where its rank is equal to n-k. If $\theta \in M_{(k)}$, then $G(\theta) \subset M_{(k)}$ and the Pfaffian system on $G(\theta)$ corresponding to submodule $P_\theta = \mathcal{C}\Lambda^1|_{G(\theta)}$ is of dimension m-k and completely integrable.

Further we assume that $M_{(k)}$ is an n-dimensional subobject of the object M_∞ (as a rule, it is so) and the factor-object $M_{(k)}/G$ Ob DE is defined. Then, dim $M_{(k)}/G$ = n - k.

Remark 5. Having restricted ourselves to local considerations, we can take away the problem concerning the existence of the factor-object $M_{(k)}/G$.

If $M_\infty = \mathcal{Y}_\infty$, where $\mathcal{Y}$ is a differential equation, then it may happen that $M_{(k)}/G = \mathcal{Y}'_\infty$. Here, $\mathcal{Y}'$ is also a differential equation. In this case, $\mathcal{Y}'$ is an equation in n-k independent variables. If $L \subset \mathcal{Y}'_\infty = M_{(k)}/G$ is a solution of $\mathcal{Y}'$, then the Pfaffian system $P_L = \mathcal{C}\Lambda^1(\mathcal{Y}_\infty)|_{\pi_G^{-1}(L)}$ on $\pi_G^{-1}(L)$ is of dimension m-k and is completely integrable. Thus, on $\pi_G^{-1}(L)$ appears an n-dimensional foliation whose leaves are solutions to the original equation $\mathcal{Y}$. We have therefore specified a class of solutions to $\mathcal{Y}$. These solutions can be found by solving a differential equation in n-k independent variables and integrating a completely integrable Pfaffian system. We call them k-invariant (with respect to G) solutions.

Let us now discuss the case when k = m. Then every k-invariant solution $\mathcal{Y}$ is composed of some orbits of G, which are completely contained in it. Therefore, if we consider the corresponding action of G on the space of solutions of $\mathcal{Y}$, then m-invariant solutions will be the fixed points of this action. In view of this, further they are called the G-invariants or simply invariants.

The practical significance of k-invariants and, in particular, invariant solutions of differential equations we come across in geometry, mathematical physics, mechanics of continua, etc. derives first of all from the fact that they can be often found in an explicit manner, whereas this cannot be done by any other known methods. Besides, these solutions enable one to obtain an idea of most important and characteristic features of the phenomena described by the equation considered.

Now we shall describe in detail the procedure of finding $M_{(k)}$ if $M_{\infty} = \mathcal{Y}_\infty$, where $\mathcal{Y} \subset J^s(\pi)$ is an equation, $\dim \pi = 1$. Let $X_1, \ldots, X_m \in D_{\mathcal{C}}(\mathcal{Y}_\infty)($)) be the generators of the algebra $\alpha(G)$; $\widetilde{X}_1 \in D_{\mathcal{C}}(J^\infty(\pi))$ are such that $\widetilde{X}_i\big|_{\mathcal{Y}_\infty} = X_i$ and f_i is the generating function (see [1], [4], and [5]) of the field X_i. The set $J^\infty(\pi)_{(k)}$ of the points, where the linear span of the system of vectors $\widetilde{X}_{1,\theta}, \ldots, \widetilde{X}_{m,\theta}$ intersects the Cartan plane $\mathcal{C}D(J^\infty(\pi))$ in a subspace of dimension k, consists of such points where the system of evolution fields $\partial_{f_1}, \ldots, \partial_{f_m}$ [1], [4], and [5] is of the rank m-k. In canonical coordinates $\partial_f = \sum_{\sigma, i} D_\sigma(f^i) \frac{\partial}{\partial p^i_\sigma}$, where $f = (f^1, \ldots, f^l)$, $D_\sigma = D_{i_1} \circ \ldots \circ D_{i_s}$ if $\sigma = (i_1, \ldots, i_s)$ and $D_j = \frac{\partial}{\partial x_j} + \sum_{\sigma, i} p^i_{\sigma j} \frac{\partial}{\partial p^i_\sigma}$.

We now introduce vectors v(f) of infinite length and with components $D_\sigma(f_i)$. Then, from what has been said it follows that an infinite system of equations defining $J^\infty(\pi)_{(k)}$ is written as :

$\{$the rank of the system of vectors $v(f_1), \ldots, v(f_m)$ is equal to n-k$\}$.

Let us now consider a particular case illustrating the structure of a system of this kind.

<u>Example 7</u>. m = 2, $\dim \pi = l = 1$, k = 1. In this case the system of interest takes the form

$$\Delta(\sigma, \tau) \overset{\text{def}}{=} \begin{vmatrix} D_\sigma(f_1) & D_\sigma(f_1) \\ D_\sigma(f_2) & D_\sigma(f_2) \end{vmatrix} = 0,$$

Indeed, this system is the infinite prolongation of the system $\Delta(\emptyset, (i)) = 0$, $i = 1, \ldots, n$. Really, from the fact that $\Delta(\emptyset, (i)) = 0$ and $\Delta(\emptyset, (j)) = 0$ it follows that $\Delta((i), (j))=0$. Further, $D_i \Delta(\emptyset, (j)) = \Delta((i), (j)) + \Delta(\emptyset, (i, j))$. Hence, $\Delta(\emptyset, (i, j)) = 0$. Continuing this process we see that equations $\Delta(\sigma, \tau) = 0$ are differential corollaries of equations $\Delta(\emptyset, (i)) = 0$.

Since $M_{(k)} = J^\infty(\pi)_{(k)} \cap \mathcal{Y}_\infty$, then in the example considered $M_{(k)} = \mathcal{E}_\infty$, where $\mathcal{E}$ is the system $\{F_j = 0, \Delta(\emptyset, (i)) = 0\}$, where $\mathcal{Y} = \{F_j = 0\}$.

Arguing in the same manner, one can prove

<u>Proposition 8</u>. If $M_\infty = \mathcal{Y}_\infty$, where $\mathcal{Y} = \{F_j = 0\}$, then $M_{(m)} = \mathcal{E}_\infty$, where $\mathcal{E} = \{F_j = 0, f_i = 0\}$.

<u>Remark 6</u>. It may be noted that the given theory of partially invariant solutions differs essentially from that of partially invariant solutions in the sense of the book [9].

11. Coverings of $J^\infty(1_{\mathbb{R}^1})$

With the aim of giving simple examples of generalized differential equations, about which we talked earlier, we shall describe a class of morphisms that cover the object $M_\infty = J^\infty(1_{\mathbb{R}^1})$.

Let $\widetilde{M}_\infty = M_\infty \times \mathbb{R}^s$ and $F: \widetilde{M}_\infty \to M$ be the natural projection. Introduce in $\widetilde{M}_\infty$ the coordinates $(w_1, \ldots, w_s, x, u, \ldots, p_\sigma, \ldots)$, where $w = (w_1, \ldots, w_s)$ are standard coordinates in $\mathbb{R}^s$ and $(x, u, \ldots, p_\sigma, \ldots)$ are canonical coordinates in $J^\infty(1_{\mathbb{R}^1})$. Determine the algebra $\widetilde{A}$ of functions on $\widetilde{M}_\infty$ as the set of all smooth functions which depend on any finite number of the coordinates indicated. After assigning some filtrations to functions w_i, we obtain a filtration in $\widetilde{A}$. As the Cartan distribution on $\widetilde{M}_\infty$ should be one-dimensional, it can be defined after indicating a field $\widetilde{D} \in CD(\widetilde{M}_\infty)$ which is nonzero everywhere. As this field, we choose the field that projects into the field

$$D = \frac{\partial}{\partial x} + \sum_{\sigma} p_{\sigma 1} \frac{\partial}{\partial p_\sigma} \quad \text{on } M_\infty .$$

Then

$$\widetilde{D} = D + \sum_{i=1}^{s} H_i \frac{\partial}{\partial w_i} = D + H \frac{\partial}{\partial w}, \quad H_i \in \widetilde{A}, \; H = (H_1, \ldots H_s)$$

In a dual manner, $\widetilde{A}$-module $C\Lambda^1(\widetilde{M}_\infty)$ is freely generated by forms $U_1(p_\sigma) = dp_\sigma - p_{\sigma 1}\, dx, \forall \sigma$, $\omega_i = dw_i - H_i dx$, $i = 1, \ldots, s$. It is not hard to see that we thus transform $\widetilde{M}_\infty$ into an object of the category DE.

Let $\widetilde{L} \subset \widetilde{M}_\infty$ be an one-dimensional integral manifold. Then $L = F(\widetilde{L})$ is an integral manifold in M_∞ and therefore locally it takes the form $L = L_f$, where $L_f = \operatorname{im} j(f)$, $f \in C^\infty(\mathcal{U})$, $\mathcal{U} \in \mathbb{R}^1$. In other words, locally $\widetilde{L} = F^{-1}(L_f)$. Let us now identify $\mathcal{U} \times \mathbb{R}^s$ and $F^{-1}(L_f)$ via diffeomorphism $\pi_f = j(f) \times 1_{\mathbb{R}^s} : \mathcal{U} \times \mathbb{R}^s \longrightarrow F^{-1}(L_f)$. In so doing, the integral manifolds of the object $\widetilde{M}_\infty$ lying in $F^{-1}(L_f)$ are identified with integral manifolds of the Pfaffian system $\pi^*_f(C\Lambda^1(\widetilde{M}_\infty)) = 0$ on $\mathcal{U} \times \mathbb{R}^s$. Obviously, $\pi^*_f(U_1(p_\sigma)) = 0$ so that the submodule $\pi^*_f(C\Lambda^1(\widetilde{M}_\infty)) \subset \Lambda^1(\mathcal{U} \times \mathbb{R}^s)$ is generated by forms $\omega_i^f = \pi^*_f(\omega_i) = dw_i - H_i^f\, dx$, where $H_i^f = \pi^*_f(H_i)$. Thus, finding of integral manifolds lying in $F^{-1}(L_f)$ amounts to solving the Pfaffian system $\omega_i^f = 0$, $i = 1, \ldots, s$, which is equivalent to the following system of ordinary differential equations:

$$\frac{dw_i}{dx} = H_i^f(x, w_1, \ldots, w_s), \quad i = 1, \ldots, s$$

Denote this system by S_f. From what has been said it follows that locally the one-dimensional integral manifolds on M_∞ are in bijective correspondence with the collection of pairs of type (f,g), where $f \in C^\infty(\mathcal{U})$, $\mathcal{U} \subset \mathbb{R}^1$, and g is a solution of S_f.

12. Generalized Differential Equations

We shall describe a class of "generalized ordinary differential equations", which contains, in particular, integro-differential equations also. To this end, it may first of all be noted that $\widetilde{M}_\infty = \lim\operatorname{inv} \widetilde{M}_i$. Actually, let a_i be a filtration of the function w_i and $r(j)$ be the number of indexes $1 \leq i \leq s$ such that $a_i \leq j$. Assume $\widetilde{M}_j = J^j(1_{\mathbb{R}^1}) \times \mathbb{R}^{r(j)}$.

We determine the projection $\widetilde{f}_{j+1}\colon \widetilde{M}_{j+1} \longrightarrow \widetilde{M}_j$ as direct product of projections $\pi_{j+1,j}\colon J^{j+1}(\mathbb{1}_{\mathbb{R}'}) \longrightarrow J^j(\mathbb{1}_{\mathbb{R}'})$ and $\mathbb{R}^{r(j+1)} \longrightarrow \mathbb{R}^{r(j)}$. Then $\widetilde{M}_\infty$ is the inverse limit of the chain $\ldots \leftarrow \widetilde{M}_j \xleftarrow{f_{j+1}} \widetilde{M}_{j+1} \leftarrow \ldots$ For $j \geqslant \max a_i$, $\widetilde{M}_j = J^j(\mathbb{1}_{\mathbb{R}'}) \times \mathbb{R}^s$ and $\widetilde{f}_{j+1} = \pi_{j+1,j} \times \mathrm{id}_{\mathbb{R}^s}$.

Assuming the manifolds $\widetilde{M}_j$ in the theory of generalized differential equations to be the analogs of $J^j(\pi)$ (or N^j_m) spaces, we determine such a generalized equation as the submanifold $\mathcal{Y} \subset \widetilde{M}_1$. Let $\varphi = 0$, $\varphi \in C^\infty(\widetilde{M}_1)$ be an equation of submanifold $\mathcal{Y}$. Then the least radical ideal close with respect to the action of operator $\widetilde{D}$ and containing φ in an ideal in $\widetilde{A}$ of infinite prolongation $\mathcal{Y}_\infty$ of the generalized equation $\mathcal{Y}$. If, for example, $\partial\varphi/\partial p_\sigma \neq 0$ on $\mathcal{Y}$, where $|\sigma| = 1$, then $\mathcal{Y}_\infty$ is given by a set of equations $\widetilde{D}^k(\varphi) = 0$, $k = 0, 1, 2, \ldots$

Further we assume for brevity that $a_i = -1$, $\forall i$, and let k be a maximum of filtrations of functions H_i and $\forall i$. The Cartan distribution on $\widetilde{M}_j$ is determined by the Pfaffian system

$$\begin{cases} U_1(p_\sigma) = 0 & |\sigma| < j \\ \omega_i = 0, & \text{for such } i \text{ that filtration } H_i \leq j. \end{cases}$$

In particular, for $j \geqslant k$ the Cartan planes on $\widetilde{M}_j$ are two-dimensional. In this case, the restriction of Cartan distribution with $\widetilde{M}_j$ on $\mathcal{Y}_j = \widetilde{\pi}_{\infty,j}(\mathcal{Y}_\infty)$, where $\widetilde{\pi}_{\infty,j}\colon \widetilde{M}_\infty \longrightarrow \widetilde{M}_j$ is the natural projection, determines a field of directions on $\mathcal{Y}_j$ if the singular points are excluded. If $1 < k$, then, after prolongation of equation $\mathcal{Y}$, we arrive at the situation indicated. Geometrically this shows that the solution of generalized ordinary differential equations is reduced to the solution of ordinary differential equations.

If all functions H_i are independent of w, then the generalized equation is an integro-differential equation. Let us explain this. Suppose an integro-differential equation of the type

$$A(x,u,u', \ldots) + \int_0^x B(x,u,u', \ldots)\, dx$$

is given. Consider a covering $\widetilde{M}_\infty$ of the type described above, assuming $s = 1$ and $H = H_1 = B(x, u, p, \ldots)$. Then the solutions of the generalized equation $w + A(x, u, p, \ldots) = 0$ are the pairs $(f(x), w(x))$, where $w(x)$ is the solution of the equation

$$\frac{d\,w(x)}{dx} = B(x, f, f', \ldots)$$

and, besides, $w(x) + A(x, f, f', \ldots) = 0$. Hence, the solutions of the generalized equation considered, which satisfy the condition $w(0) = 0$, are the solutions to the starting integro-differential equation.

13. C-Spectral Sequence

Let $M_\infty \in \mathrm{Ob}\, DE$ and let $\Lambda^* = \Lambda^*(M_\infty)$ be the algebra of differential forms on M_∞. From the definition of an object of DE it follows that $C\Lambda^* \subset \Lambda^*$ is an ideal in Λ^* closed with respect to the operator d. In view of this, the degrees $C^k\Lambda^*$, $k = 1, 2, \ldots$, of ideal $C\Lambda$ are closed with respect to d. Because of this the complex $\{\Lambda^*, d\}$ is found to be filtered by subcomplexes $\{C^k\Lambda^*, d\}$: $\Lambda^* = C^0\Lambda^* \supset C\Lambda^* \supset C^2\Lambda^* \supset \ldots$, and we can consider the spectral sequence $\{E_r^{p,q}(M_\infty), d_r^{p,q}\}$ determined by this filtration. Here, p denotes, as usually, the filtration index, and p+q stands for the degree. Further, this special sequence is called the C-spectral sequence of the object M_∞.

Proposition 9. (1) C-spectral sequence is natural in the category DE, that is to every morphism $F : M_\infty \longrightarrow M'_\infty$ corresponds the homomorphism $F^*: E_r^{p,q}(M'_\infty) \longrightarrow E_r^{p,q}(M_\infty)$ of C-spectral sequences;

(2) $E_0^{p,q} = 0$ if $q < 0$;

(3) $E_0^{p,q} = 0$ for $q > n$, if $\dim M_\infty = n$;

(4) if $\dim M_\infty = n$, then the C-spectral sequence converges and its term E_∞ is attached to the de Rham cohomology algebra $H^*(M_\infty)$ of the object M_∞.

In order to understand the meaning of the C-spectral sequence we describe it on a category of smooth manifolds imbedded in the category DE by any of the two methods described in Section 6. First we shall consider the dimensional imbedding. In this case, $C \equiv 0$ and hence $E_0^{p,q}(M) = 0$, $p > 0$, $E_0^{p,q}(M) = \Lambda^q(M)$, where M is the considered manifold. Thus, in dimensional imbedding of the category of smooth manifolds in DE the complex $\{E_0^{p,q}, d_0^{0,q}\}$ proves to be an analog of the de Rham complex. From the definition it follows that $E_1^{0,q} = \bar{\Lambda}^q \overset{\text{def}}{=} \Lambda^q / C\Lambda^q$, and the operator $d_0^{0,q} = \bar{d} : \bar{\Lambda}^q \longrightarrow \bar{\Lambda}^{q+1}$ is the factor-operator of the operator $d : \Lambda^q \longrightarrow \Lambda^{q+1}$ by mod $C\Lambda^*$. Call the complex $0 \to A = \bar{\Lambda}^0 \xrightarrow{\bar{d}} \bar{\Lambda}^1 \xrightarrow{\bar{d}} \ldots \xrightarrow{\bar{d}} \bar{\Lambda}^n \; 0$ the de Rham $\bar{d}$-complex, and denote its cohomology by $\bar{H}^i$, in particular, $\bar{H}^i(M_\infty) = E_1^{0,i}(M_\infty)$.

Let us consider a manifold M which is regarded as zero-dimension object of DE. Then $\mathcal{C}^k\Lambda^*(M) = \sum_{i\geq k}\Lambda^i(M)$. Therefore, $E_0^{p,q}(M) = 0$ if $q > 0$ and $E_0^{p,o}(M) = \Lambda^p(M)$. In this case, $E_1^{p,o}(M) = \Lambda^p(M)$ and the complex $\{E_1^{p,o}(M), d_1^{p,o}\}$ coincides with the de Rham complex of the manifold M. Thus, the term E_1 of $\mathcal{C}$-spectral sequence is the "zero-dimensional" analog of the de Rham complex, and the term E_2 is the "zero-dimensional" analog of the algebra $H^*(M)$. It may be noted also that the Lie algebra Sym M_∞ is the "zero-dimensional" analog of the Lie algebra D(M) of smooth vector fields on manifold M. Naturally we assume that on the term $E_1(M_\infty)$ are carried out the basic facts of the calculus of vector fields and the differential forms on smooth manifolds; this is true, of course. We shall illustrate this by considering the Stokes infinitesimal formula : $X(\omega) = X \rfloor d\omega + d(X \rfloor \omega)$, where $X(\omega)$ is the Lie derivative of the form $\omega \in \Lambda^*(M)$ with respect to $X \in D(M)$.

Proposition 10. Natural operations of the Lie derivative and the substitution of a vector field in forms on $\Lambda^*(M_\infty)$ are reduced to a term $E_1(M_\infty)$. The following "Stokes infinitesimal formula" holds :

$$\mathcal{X}\{e\} = \mathcal{X} \rfloor d_1^{p,q}(e) + d^{p-1,q}(\mathcal{X} \rfloor e), \mathcal{X} \in \text{Sym } M_\infty ,$$

$$e \in E_1^{p,q}(M_\infty)$$

where $\mathcal{X}\{\cdot\}$ is the reduced Lie derivative, and $\mathcal{X} \rfloor$ is the reduced operation of substitution.

The direct product $M_\infty \times [0, 1]$ may be supplied with the structure of an object of the category DE so that all of its integral manifolds take the form L x t, where $L \subset M_\infty$ is an integral manifold. Here, the mappings of the form $id_{M_\infty} \times F : M_\infty \times [0,1] \to M_\infty \times [0,1]$, where $F: [0,1] \to [0,1]$ is a smooth function, are morphisms in DE. This enables the concept of homotopy, homotopic equivalence, etc. to be introduced in DE in a standard manner. From the existence of the "Stokes infinitesimal formula" follows

Proposition 11. If morphisms $F_i: M_\infty \to M'_\infty$, i = 1, 2 are homotopic, then the homomorphisms $F_i^* : E_2^{p,q}(M'_\infty) \longrightarrow E_2^{p,q}(M_\infty)$ coincide. In particular, the term $E_2^{p,q}(M_\infty)$ is a homotopic invariant in DE.

Let $\pi : E_\pi \to M$ be a vector bundle and $\alpha_t: E_\pi \longrightarrow E_\pi$ be its fibre homothety with coefficient 1-t. In a natural manner the mappings α_t are raised to morphisms $\bar{\alpha}_t : J^\infty(\pi) \longrightarrow J^\infty(\pi)$. From this it is seen that the morphism $j(0) : M \to J^\infty(\pi)$, $\quad 0 \in \Gamma(\pi)$, where M is understood to be an n-dimensional object of DE, is an homotopic equivalence in DE. Call the equation $\mathcal{Y} \in J^k(\pi)$ to be conic if $\operatorname{im} j_k(0) \subset \mathcal{Y}$ and $\mathcal{Y}_\infty$ is invariant under $\mathcal{Z}_t$. The following statement, which immediately follows from Proposition 11, is an analog of the Poincare lemma.

Proposition 12. If $\mathcal{Y}$ is a conic equation, then $E_2^{p,q}(\mathcal{Y}_\infty) = 0$ if $p > 0$, and $E_2^{o,q}(\mathcal{Y}_\infty) = H^q(M)$. In particular, the complexes

$$(E_1^{*,q}) : 0 \to E_1^{o,q}(\mathcal{Y}_\infty) \xrightarrow{d_1^{o,q}} E_1^{1,q}(\mathcal{Y}_\infty) \xrightarrow{d^{1,q}} \cdots$$

are acyclic in positive dimensions.

In view of the facts given in the following section, computation of $\mathcal{C}$-spectral sequence and, particularly, its terms (E_1, E_2, and E_∞) is an important problem. Propositions 11 and 12 are simple examples of those most general methods which can be developed in that direction (some methods may be found in [2], [5] and others we shall discuss in a separate article).

14. Meaning of $\mathcal{C}$-Spectral Sequence

Let $M_\infty \in \operatorname{Ob} DE$. We shall pose on M_∞ a variational problem of finding an extremum of the functional $L \mapsto \int i_L^*(\omega)$, where $\omega \in \bar{\Lambda}^n(M_\infty)$ is the "Lagrangian density" and L is a solution of the "equation" M (we omit the discussion of boundary and other type of conditions). Here, $i_L: L \to M_\infty$ is the imbedding map, L being "dimensionally" imbedded in DE, so that i_L is a morphism in DE and i_L^* is regarded as the corresponding homomorphism $E_o^{o,n}(M_\infty) = \bar{\Lambda}^n(M_\infty) \to E_o^{o,n}(L) = \bar{\Lambda}^n(L) = \Lambda^n(L)$ of $\mathcal{C}$-spectral sequences. In this case, the integral $\int_L i_L^*(\omega)$ may be identified a cohomology class $\langle i_L^*(\omega)\rangle \in H^n(L) = E_1^{o,n}(L)$ of the form $i_L^*(\omega) \in \Lambda^n(L)$. Hence the cohomology class $\langle\omega\rangle \in \bar{H}^n(M_\infty) = E_1^{o,n}(M_\infty)$ may be interpreted as Lagrangian of the considered variational problem.

Proposition 13. For an integral manifold $L \subset M_\infty$ to have an extremum it is necessary that $d_1^{o,n}\langle\omega\rangle|_L = 0$

In other words, $d_1^{o,n}\langle\omega\rangle = 0$ is the "Euler-Lagrange equation" of the problem considered. If, for example, $M_\infty = J^\infty(\pi)$, then this equation coincides with the classical Euler-Lagrange equation. If $M_\infty = \mathcal{Y}_\infty$, where $\mathcal{Y} \subset J^k(\pi)$ is a constraint equation in the general Lagrange problem, then the extremality condition $d_1^{o,n}\langle\omega\rangle = 0$ is the general form of the Lagrange multiplier theorem. Hence, the differential $d_1^{o,n}$ is the most general form of the Euler operator which attaches with Lagrangians the corresponding Euler-Lagrange equations.

Let us consider the complex $\{E_1^{*,n}, d_1\}$ containing the Euler operator $d_1^{o,n}$. As $d_1^{1,n} \circ d_1^{o,n} = 0$, we can study the inverse problem of the calculus of variations. In particular, in a situation without connections, i.e. when $M_\infty = J^\infty(\pi)$ or N_m^∞, or when $M_\infty = \mathcal{Y}_\infty$ and the equation of connections is conic, the general global solution of this old problem follows from Proposition [12] [2], [5]. Consideration of this very complex leads also to a global solution of other classical problem, i.e. of "degenerated" Lagrangians and, in addition, to generalization of Schwartz's formula from the theory of minimal surfaces on a very wide class of Lagrangians [2], [5], and [10].

The group $\bar{H}^{n-1}(\mathcal{Y}_\infty) = E_1^{o,n-1}(\mathcal{Y}_\infty)$ is the group of the conservation laws of equation $\mathcal{Y}$ and $\bar{\Lambda}^{n-1}(\mathcal{Y}_\infty) = E_o^{o,n-1}(\mathcal{Y}_\infty)$ is the group of conserved densities. If the equation $\mathcal{Y}$ is not overdetermined and regular, and $\mathcal{Y} = \{\varphi = 0\}$, then $E_1^{1,n-1}(\mathcal{Y}_\infty) = \ker \bar{l}_\varphi^*$, where l_φ is operator of universal linearization [1, 5] and $\bar{l}_\varphi$ is its restriction on $\mathcal{Y}_\infty$. The operator $d_1^{o,n-1}$ in this case has a kernel of purely topological nature, and therefore the evaluation of conservation laws for an equation reduces to the solution of equation $\bar{l}_\varphi^* = 0$. This, in most cases, is a very effective procedure. The discussion yields, in particular, a general solution of the problem of non-triviality of conserved densities. No general methods were earlier known for solving this problem [2, 5].

Further we shall point to very interesting observations made by T. Tsujishita [11]. In the first place he has shown that the Gelfand-Fuchs cohomology [12] is naturally put in the theory of $\mathcal{C}$-spectral sequences. Secondly, he has interpreted the theory of characteristic classes (primary and secondary), the theory of characteristic classes of deformations, Bott's topological obstructions, and others in terms of $\mathcal{C}$-spectral sequence for G-structure integrability type equations.

Finally, it is worth to point out that Hamiltonian formalism in the field theory is intimately related with $\mathcal{C}$-spectral sequences [3].

Thus, in a framework of the category DE we can extend the earlier described results to the "generalized equations" M_∞. This is another justification of the earlier accepted view point.

References

1. Vinogradov, A.M., Krasil'shchik, I.S., and Lychagin, V.V. Application of Nonlinear Differential Equations in Civil Aviation, MIIGA, Moscow, 1977.
2. Vinogradov, A.M. A Spectral Sequence Associated with a Nonlinear Differential Equation and Algebro-Geometric Foundations of Lagrangian Theory with Constraints, Soviet Math. Dokl., 1978, vol. 19, N 1, 144-148.
3. Vinogradov, A.M. Hamiltonian Structures in Field Theory, Soviet Math. Dokl., 1978, vol. 19, N 4, 790-794.
4. Vinogradov, A.M. Theory of Higher Infinitesimal Symmetries of Nonlinear Partial Differential Equations, Soviet Math. Dokl., 1979, vol. 20, N 5, 985-990.
5. Vinogradov, A.M. Geometry of Nonlinear Differential Equations, J. Soviet Math., vol. 17, N 1, 1624-1649.
6. Goldschmidt, H. Integrability Criteria for Systems of Nonlinear Differential Equations, J. Diff. Geom., 1967, vol. 1, N 3.
7. Vinogradov, A.M. Some Homological Systems Connected with Differential Calculus in Commutative Algebras, Uspehi Mat. Nauk, 1979, vol. 34, N 6.
8. Wahlquist, H.D. and Estabrook, F.B. Prolongation Structures of Nonlinear Evolution Equations, J. Math. Phys., 1975, vol. 16, N 1.
9. Ovsyannikov, A.B. Group Analysis of Differential Equations, Nauka, Moscow, 1978.
10. Dedecker, P. Généralisation d'une formule de H.A. Schwartz relative aux surfaces minima, J. C.R. Acad. Sci., 1977, vol. 285, IL-item, 1977, vol. 285.
11. Tsujishita, T. On Variation Bicomplexes Associated to Differential Equations, Osaka Journ. of Math., 1980.
12. Gelfand, I.M. and Fuchs, D.B. Cohomologies of Lie Algebras of Formal Vector Fields, Izv. AN SSSR, ser. mat., 1970, vol. 34, N 1.
13. Krasil'shchik, I.S. and Vinogradov, A.M. Nonlocal Symmetries and the Theory of Coverings, Acta Appl. Math., 1984, vol. 2, N 1.

Translated from the Russian by P.K. Dang

ALGEBRAIC STRUCTURE OF CERTAIN INTEGRABLE HAMILTONIAN SYSTEMS

A.T. Fomenko

Department of Mechanics and Mathematics,
Moscow State University, 119899, Moscow, USSR

Among the Hamiltonian systems that are naturally related to finite-dimensional Lie algebras of special interest are the so-called Euler's equations or their generalizations, some of which may be of importance in physics (for example, the equations of inertial motion of a rigid body in an ideal liquid). Complete integrability of certain such system is a consequence of profound algebraic properties of those Lie algebras on which these systems are naturally defined.

First, we recall several basic concepts.

Let there be defined a symplectic manifold M^{2n} and the corresponding exterior non-degenerate closed 2-form ω on M^{2n}. A vector field $\dot{X}$ on M^{2n} is called Hamiltonian if it preserves the form ω (i.e. if the derivative of the form along this field vanishes). In this case the field $\dot{X}$ can be represented locally as sgrad f, where f is a smooth function on the manifold, and sgrad is a skew-symmetric gradient (uniquely defined by ω). If f and g are two smooth functions on the manifold, their Poisson bracket $\{f, g\}$ is defined as ω(sgrad f, sgrad g), i.e. the skew-symmetric scalar product (with respect to the form ω) of the fields sgrad f and sgrad g. The functions f and g are said to be in involution (i.e. they commute) if $\{f, g\} = 0$. Since a Hamiltonian system cannot always be integrated explicitly (and the very concept of "explicit integration" often requires a non-trivial definition), the Liouville theorem is a very important tool for describing the solution of a Hamiltonian system. This theorem is formulated as follows. Let, on a symplectic manifold M^{2n}, there be defined n functions $f_1, \ldots, f_n$ such that $\{f_i, f_j\} = 0$, $i,j = 1, \ldots, n$, and let all these functions be functionally independent on a level surface $T_c:(f_1=c_1, \ldots, f_n=c_n)$, where c_i are constants. Then T_c is a smooth submanifold in M^{2n} which is invariant relative to any vector field sgrad f_i, $1 \leq i \leq n$. If T_c is compact and connected, then it is diffeomorphic to a torus T^n (if the level surface is not compact, it is diffeomorphic to a "cylinder" -- a factor of $\mathbb{R}^n$ with respect to standard action of a cycle free group of order not higher than n), and each flow sgrad f_i defines on the torus a quasiperiodic motion $\dot{\varphi}_i = \omega_i(c_1, \ldots, c_n)$; $\varphi_1, \ldots,$ φ_n are angular coordinates of the torus. Furthermore, the vector fields sgrad f_i define on the torus T^n a system which is integrable in quadratures (see [1]).

An important example of symplectic manifolds are the orbits O(X) of the (co)adjoint representation of a Lie group $\mathcal{G}$ on its (co)algebra G^*. If the group is compact, G^* can, apparently, be identified (by the Cartan-Killing form) with a Lie algebra G. A canonical symplectic structure is naturally defined on the orbits of the (co) adjoint representation. Consider a point X on the orbit $O(X) = \mathcal{G}/C(X)$, where C(X) is the centralizer of the point X, and let ξ_1, ξ_2 be vectors in the tangent space $T_X O(X)$ to the orbit O(X). Then, $T_X O(X) = \{[X, y]\}$, where y runs the Lie algebra G (in particular, there exsist $y_1, y_2 \in G$ such that $\xi_i = [X, y_i]$, i = 1,2; the vectors y_1 and y_2 are not defined uniquely). In the compact case we set $\omega(\xi_1, \xi_2) = \langle X, [y_1, y_2] \rangle = \langle \xi_1, y_2 \rangle$, where $\langle \ , \ \rangle$ denotes the Cartan-Killing form; in the non-compact case, $\omega(\xi_1, \xi_2) = X([y_1, y_2])$, $X \in G^*$, $[y_1, y_2] \in G$.

Many of classical Hamiltonian systems can be represented as systems on orbits in the corresponding Lie (co)algebras. Recent investigations have shown that a natural approach to the problem of integrating such systems is as follows. We fix on a Lie (co)algebra G a symplectic structure (for example, the one described above) and seek a family F(G) of smooth functions on the orbits (algebra) of the (co)adjoint representation that satisfy the requirements: (1) the space of functions F(G) forms a commutative Lie algebra (with respect to the Poisson bracket); (2) an additive basis of functionally independent functions $f_1, \ldots, f_k$, $k = \frac{1}{2}\dim O$, can be chosen in the space F(G). In other words, we seek a family of functions the number of which is equal to half the dimension of the orbit; first of all, for the orbits of maximal dimension (orbits of general position). For a compact semisimple, the family in question must have the dimension $\frac{1}{2}(n-r)$, where n is the dimension of G and r is the rank of the algebra G (i.e. the dimension of the maximal commutative subalgebra, the so-called Cartan subalgebra). It is reasonable to seek the functions $f_1, \ldots, f_k$ that are defined on the whole algebra G and such that their restrictions on the orbits of general position give the commutative family of maximal dimension in question. For a semisimple group, it is possible (see [4,5]) to find a set of $\frac{1}{2}(n-r) + r$ independent functions which are commutative with respect to the Poisson bracket defined in the whole Lie algebra (and not only on the orbits). In this case, r functions of the set are invariants of the adjoint representation (i.e. their common level surface determines the orbit of general position), and the remaining $\frac{1}{2}(n-r)$ functions are no longer constant on the orbits. We recall that the symplectic form just described is, generally, degenerate on the whole algebra (unlike the orbits).

Clearly, if for a certain Lie algebra G we can construct a set F(G) (which is commutative and maximal), then taking any function of this set as a Hamiltonian f, we may consider the Hamiltonian vector field sgrad f, which is completely integrable (in the sense of Liouville) on the orbits, because all the other functions of the set F(G) are first integrals of this Hamiltonian system.

In such an approach, the major problem lies in constructing a set, as large as possible, of commutative and maximal families of functions on the orbits of the (co)adjoint representation. The larger this set, the wider is the class of Hamiltonian systems that are completely integrable in the sense of Liouville. The inverse problem naturally arises: let there be given a particular system, and it is desirable to integrate this system completely, say, in terms of the Liouville theorem. Then, one should try to represent this system as a Hamiltonian vector field on an orbit of the (co)adjoint representation of some Lie (co)algebra. Particular examples show (see, e.g., [2,3,4,6]) that the same system can be represented by Hamiltonian fields on the orbits of different Lie (co)algebras, the representations possessing, in general, quite distinct algebraic properties. By varying the Lie (co)algebra, one may attempt at finding a representation of the system in which it is of the form sgrad f, where $f \in F(G)$, for a certain commutative and maximal family of functions on G and, therefore, at obtaining a complete set of integrals of the system. It is sometimes possible to algebrize a given system, i.e. to represent it as the field sgrad f not on all the orbits of general position, but only on one such orbit (in particular, on a singular orbit). Once a system has been algebrized (if it is possible), search is made for a complete set of integrals (i.e. for a maximal commutative family of independent functions). Several methods have been suggested for constructing integrals on the orbits of the co-adjoint representation. A rather effective method is that for finite-dimensional Lie algebras [4] . This methods rests upon the idea of shifting invariants of the co-adjoint representation by a covector of general position. The method also offers a possibility of constructing integrals via the shift (by a covector of general position) of semi-invariants of a Lie (co)algebra, i.e. functions multiplied by the character of the (co)adjoint representation [7-9] . Some of the presently existing methods of constructing the integrals (maximal commutative families of functions) are based on the algebraic properties of chains of subalgebras in the Lie algebra under study [8,9] .

Investigation of particular problems has shown that among Liouville-integrable systems of great importance are systems related to left-invariant metrics of the form $\varphi_{a,b,D}$ [4]. Although an invariant metric characteristic of these metrics, which distinguishes them in the class of all left-invariant metrics (e.g., on semisimple groups), has not so far been found, complete integrability of the corresponding equations of motion (with these metrics taken as Hamiltonians) should imply that common level surfaces of the integrals possess homogeneity properties. For example, it is not excepted that some of these surfaces (which are half-dimension tori in the orbits of the co-adjoint representation) are minimal submanifolds or even completely geodesic submanifolds with respect to the metric $\varphi_{a,b,D}$. If a group (algebra) is not semisiple, no operators of the form $\varphi_{a,b,D}$ are involved in the formation of the metric [4] . Thus, in the case of an arbitrary finite-dimensional Lie algebra, the question of the existence of a maximal commutative set of functions has not so far been solved (nevertheless, such a set is most likely to exist). It should be noted that this question is also related to the reduction of the non-commutative Liouville theorem (it is proved in [5]) to the commutative Liouville theorem. In other words, if a maximal commutative set of functions does exist on an arbitrary finite-dimensional Lie algebra, the existence of a maximal non-commutative algebra of the integrals of a Hamiltonian system on a semisimple manifold automatically implies the existence of a maximal commutative algebra of the integrals.

This paper deals mainly with serveying the results **concerning** complete integration of new Hamiltonian systems related to (not necessarily compact) Lie algebras. These results were obtained by the author and his pupils in 1980-1981. We succeeded in integrating (in the sense of Liouville) the equations of inertial motion of a multidimensional rigid body in an ideal liquid, the analogues of these equations on Lie algebras $(U(p, q) \times \mathbb{C}^{p+q})^*$, the equations of magnetic hydrodynamics, etc. Many of these results were obtained by analysing the algebraic structure of integrable Hamiltonian systems on Lie algebras and using a general scheme of section operators that was suggested by the author for an arbitrary linear representation in order to construct a special class of systems that includes the main cases of complete integrability.

The idea of constructing a general form of section operators is as follows. Let H be a Lie algebra, φ_y the corresponding Lie group, $\rho : H \to \mathrm{End}\, V$ the representation of H in a linear space V, $\alpha = \exp(\rho)$ the corresponding representation of the group, and O(X) orbits of the action of the group φ_y in the space V, $X \in V$. Once a linear operator $C : V \to H$ (called a section operator) is defined, a natural vector field $\dot{X}_c = \rho(CX)(X)$ arises on the orbits. If a section operator is given, one can define a symplectic structure on an orbit. In view of applications, of great importance is a special class of section operators that form a many-parameter family, the principal parameters being $a \in V$ and $b \in \mathrm{Ker}\, \Phi_a$, $\Phi_a(h) = (\rho h)a$. For example, in the case of $H = so(n)$, $\rho = ad$ and the field $\dot{X}_c$ coincides with the equations of motion of a rigid body with a fixed point (in the absence of gravity). An important property of this many-parameter family of section operators is that it permits a unified construction of multidimensional analogues of systems related to the motion of a rigid body (and admitting complete integration). Precisely this identification reveals a general algebraic cause of appearing integrals for such a system.

In a particular case where $H + V = G$ coincides with a standard decomposition of a compact symmetric space (H is a standard subalgebra), the field $\dot{X}_c$ turns to a flow on the orbit of action of a stationary group. Complete classification of all Hamiltonian flows of the form $\dot{X}_c$ has been developed, and necessary and sufficient conditions have been established that ensure the existence on the orbit of symplectic structures invariant relative to the fields $\dot{X}_c$. This permits the study of multidimensional analogues of the equations of motion of a "singular rigid body" when the inertia tensor satisfies a system of linear equations. For the motion group $\mathbb{R}^n$, it was found that the system $\dot{X}_c$ turns (for a special series of section operators) to the equations of inertial motion of a rigid body in an ideal liquid that admit a complete set of commuting integrals on the orbit of the co-adjoint representation of this group (see [3]). The integrals can be calculated explicitly (in the form of polynomials).

We should mention here the algorithm by V.V. Trofimov, which enables one to construct, proceeding from a Lie algebra with a complete (maximal) commutative set of functions, a set of new Lie algebras (not isomorphic to the initial one) that also possess a maximal commutative set of functions. A.V. Brailov has developed a procedure of finding a maximal set of commuting functions on certain Lie algebras, which are "contractions" of large Lie algebras. This procedure has enabled complete integration (in the sense of Liouville) of the equations of the form X_c on all Lie algebras $(U(p, q)\times \mathbb{C}^{p+q})^*$, where $U(p, q)\times \mathbb{C}^{p+q}$ is the semidirect product of the algebra $U(p, q)$ and its standard representation of minimal dimension. Advancing the results of V.V. Trofimov, A.V. Brailov succeeded in constructing a maximal commutative set of functions on Lie algebras that are tensor products of Lie algebras (say, semisimple) on which a complete set already exists and algebras provided with Poincaré duality. This gives new arguments in favour of the hypothesis that a complete commutative set of functions exists on any finite-dimensional Lie algebra.

§ 1. The construction of a special many-parameter family of section operators

Here we outline the algorithm of constructing section operators for an arbitrary finite-dimensional linear representation (the algorithm was suggested by the author of this paper). Let H be an arbitrary finite-dimensional Lie algebra, $\rho : H \to \mathrm{End}\, V$ a finite-dimensional representation of H in V, φ_y a simply connected group corresponding to H, and $\alpha : \varphi_y \to \mathrm{Aut}\, V$ the corresponding representation of this group, $\rho = d\alpha$. When acting on V, the group φ_y foliates V into orbits (we denote them by O(X)). Let $a \in V$ be an arbitrary point of general position (i.e. the orbit through this point has a maximal dimension) and let $K \subset H$ be an annulator of the element $a \in V$, $K = \mathrm{Ker}\, \varphi_a$, where $\varphi_a(h) = (\rho h)a$. If a is an element of general position, K has a minimal dimension. For an arbitrary element $b \in K$ we consider the action of ρb on V; M stands for $\mathrm{Ker}(\rho b) \subset V$. Let K' be an arbitrary algebraic complement to K in H, i.e. $H = K + K'$, $K \cap K' = 0$. The choice is not unique; that is to say, the possibility of varying this complement gives rise to a family of parameters.

Clearly, $a \in M$. By the definition of K', the mapping $\varphi_a: H \to V$ transforms K' homeomorphically into a plane $\varphi_a K' \subset V$. Since $\varphi_a K' = \mathrm{Im}(\varphi_a: H \to V)$, the plane does not depend on the choice of $K' \subset H$ and is uniquely determined by the element a and representation ρ. Let there exist $b \in K$ such that V can be decomposed into the sum of two subspaces M and $\mathrm{Im}(\rho b)$, i.e. $V = M + \mathrm{Im}(\rho b)$, $M \cap \mathrm{Im}(\rho b) = 0$. For example, as b we can take semisimple elements of K. The plane $\varphi_a K'$ intersects M and $\mathrm{Im}(\rho b)$ along planes, which are denoted by $B = M \cap \varphi_a K'$ and $R' = \varphi_a K' \cap \mathrm{Im}(\rho b)$, respectively. Thus, we obtain a decomposition of $\varphi_a K'$ into the direct sum $B + R' + P$, where B and R' are defined uniquely, and the additional plane P is not, thereby giving rise to its own set of the parameters. Consider the action of ρb on $\mathrm{Im}(\rho b)$. Apparently, ρb maps $\mathrm{Im}(\rho b)$ isomorphically onto itself; in particular, ρb is invertible on $\mathrm{Im}(\rho b)$. Let $(\rho b)^{-1}$ be an operator inverse to ρb on $\mathrm{Im}(\rho b)$. Setting $R = (\rho b)^{-1} R'$, we have $(\rho b) R = R'$. The plane $R \subset \mathrm{Im}(\rho b)$ is defined uniquely. Consider in $\mathrm{Im}(\rho b)$ a plane Z, which is the algebraic complement to R in $\mathrm{Im}(\rho b)$. Clearly, $\mathrm{Im}(\rho b) = Z + R'$. Take the algebraic complement of T to B in M.

Thus, we have constructed a decomposition of the representation space into the direct sum of four spaces: $V = T + B + R + Z$, a $M = T + B$, $\mathrm{Im}(\rho b) = Z + R$. The planes R, B, M, and $\mathrm{Im}(\rho b)$ are defined uniquely, and Z and T can be varied, thereby providing extra parameters, which should be chosen, depending on the problem. If a scalar product is given in V, Z and T can be defined uniquely as orthogonal complements.

Since K' is isomorphic to $\varphi_a K'$, we obtain the decomposition $K' = \tilde{B} + \tilde{R} + \tilde{P}$, where $\tilde{B} = \varphi_a^{-1} B$, $\tilde{R} = \varphi_a^{-1} R'$, and $\tilde{P} = \varphi_a^{-1} P$. Hence, one arrives at a many-parameter decomposition of the algebra H into the direct sum of four spaces: $H = K + \tilde{B} + \tilde{R} + \tilde{P}$. If a non-degenerate Killing's form is defined in H (the semisimple case), among these decompositions we can single out the one constructed with orthogonal complements.

Let us now define a section operator $Q: V \to H$, $Q: T + B + R + Z \to K + \tilde{B} + \tilde{R} + \tilde{P}$ by setting

$$Q = \begin{pmatrix} D & 0 & 0 & 0 \\ 0 & \varphi_a^{-1} & 0 & 0 \\ 0 & 0 & \varphi_a^{-1}(\rho b) & 0 \\ 0 & 0 & 0 & D' \end{pmatrix}$$

where $D: T \to K$ is an arbitrary linear operator, $\varphi_a^{-1} B \to \tilde{B}$ is the operator inverse to φ_a on the plane B, $\varphi_a^{-1}(\rho b): R \to \tilde{R}$, $\rho b: R \to R'$,

$\varphi_a^{-1}: R' \to \widetilde{R}$, $D': Z \to \widetilde{P}$. Thus, the operator $Q: V \to H$ is of the form $Q(a, b, D, D')$. The point of general position in V is chosen as a because only in this case K' has a maximal dimension, i.e. the operators $\varphi_a^{-1}(\rho b)$ and φ_a^{-1} have a largest domain in V. Let us consider important particular cases of this construction.

If $V = H^*$ and $\rho = ad^*: H \to \mathrm{End}\, H^*$, then $\varphi_a^{-1}(\rho b) = \varphi_a^{-1}(ad_b)^*$. Taking as H a non-compact Lie algebra $so(n) \oplus \mathbb{R}^n$, i.e. the Lie algebra of the group of motions of a Euclidean space, we observe (see [3]) that the system $\dot{X}_Q = ad^*_{Q(X)}(X)$ turns to the equations of inertial motion of a rigid body in an ideal liquid. Here $K \cong K^*$ under natural identification of H and H^*), $Z = \widetilde{Z} = 0$, and $R \cong \widetilde{R}$.

Let $\mathcal{V} = \mathcal{G} / \varphi_y$ be a compact homogeneous space. In this case, the Lie algebra G of the group $\mathcal{G}$ is decomposed into the sum of subspaces, $G = H + V$, where $H = T_e \varphi_y$ is a stationary subalgebra, V is a tangent space to $\mathcal{V}$. The subalgebra H has adjoint action on V, $\rho = ad$.

<u>Proposition 1.1</u>. Let $\mathcal{V} = \mathcal{G} / \varphi_y$ be a compact symmetric space and $a \in V$ an element of general position. Then, the decomposition $V = T + B + R + Z$ determining the section operator is of the form: T is a maximal commutative subspace in V, $a \in T$, $R = R'$, $Z = 0$, $b \in K$, $\varphi_a K' + T = V = T + B + R$. If $C: V \to H$ is a section operator, the exterior 2-form $F_c(X; \xi, \eta) = \langle CX, [\xi, \eta] \rangle$ arises on the orbits $O(X) = V$, where $X \in V$ and $\xi, \eta \in T_X O(X)$.

There exist rich series of symmetric spaces and operators C for which F_c determines (almost everywhere on the orbit) a symplectic structure on orbits (which is non-invariant under the action of the group).

Let us again consider a general case. Suppose $\xi, \eta \in T_X O$ are arbitrary tangent vectors; then there exist uniquely defined vectors $\xi', \eta' \in K'(X)$ such that $\rho \xi' = \xi$ and $\rho \eta' = \eta$. Let there be given a section operator $C: V \to H$. We define a bilinear form $F_c = \langle CX, [\xi', \eta'] \rangle$, where $[\xi', \eta'] \in H$, $CX \in H$. The form F_c (of degree 2) is defined on the orbits and is skew-symmetric. The flow $\dot{X}_Q$ is also defined on the orbits, $\dot{X}_Q = (\rho Q X) X$.

<u>Question A</u>. What are the operators C for which the form F_c is closed on the orbits and non-degenerate? <u>Question B</u>. What are the operators C and Q for which the flow $\dot{X}_Q$ is Hamiltonian relative to the closed form F_c? Such operators are solutions of the equation $\dot{X}_Q F_c = 0$.

It turns out that in the case of symmetric spaces these equations have interesting non-trivial solutions. Let us now analyse the form of a section operator in a symmetric space. Since $ad : V \to H$, the operator C is induced by operators of the form ad_v, $v \in V$. The spaces V and H are represented as $V = T + R + B$ and $H = K + \tilde{R} + \tilde{B}$, the section operator being defined by the matrix

$$C = ad_{a'} + ad_a^{-1} ad_b + D\,,\quad C = \begin{pmatrix} ad_{a'} & 0 & 0 \\ 0 & ad_a^{-1} ad_b & 0 \\ 0 & 0 & D \end{pmatrix}$$

where a, a' are elements of general position in T; $b \in K$ is not necessarily an element of general position; $D: T \to K$ is an arbitrary linear operator; $ad_a^{-1} ad_b : R \to \tilde{R}$; $ad_{a'} : B \to \tilde{B}$. The exterior 2-form F_c on V is defined as $F_c(X, \xi, \eta) = \langle C\lambda, [\xi, \eta] \rangle$, where $X, \xi, \eta \in V$ and C was defined earlier. Following the general scheme, one should consider the operation $[\xi', \eta']$ instead of $[\xi, \eta]$, where $\xi = [X, \xi']$ and $\eta = [X, \eta']$. Since we are now dealing with a semisimple case, the forms $\langle CX, [\xi, \eta] \rangle$ and $\langle C\dot{X}, [\xi', \eta'] \rangle$ are equivalent. The vector field $\dot{X}_Q$ on V is given by $\dot{X}_Q = [X, QX]$, where $Q = Q(\bar{a}, \bar{b}, \bar{a}', \bar{D})$.

It can easily be seen that if f is a smooth function on V and the field sgrad $f \in V$ is such that $Y \in V$ ($Y(f)$ is the derivative of f along the field Y) for $F_c(\text{sgrad } f, Y) = Y(f)$ and for an arbitrary vector field, then the equality sgrad $f = [CX, \text{sgrad } f]$ is valid. In a general case, the 2-form defined above may be degenerate.

We note that if the rank of a symmetric space is maximal, the 2-form F_c on V is induced by the Riemannian curvature tensor of this symmetric space, i.e. the equality $F_c(X; \xi, \eta) = 4 \langle a', R(X, \xi)\eta \rangle$ is valid, where R is the curvature tensor, $a' \in T$ is a fixed vector.

<u>Proposition 1.2</u>. Consider a symmetric space of maximal rank SU(3)/SO(3). In this case, the equations $\dot{X}_c = (\rho QX)X$ on B (Q is a section operator) coincide with Euler's equations for the motion of a rigid body with fixed point in $\mathbb{R}^3$ with an arbitrary inertia tensor.

Straightforward calculation shows that if the section operator C for the symmetric space SU(n)/SO(n) of maximal rank is defined by the elements $a' \in T$, then the 2-form F_c is closed on V and on the orbits of the action of the group.

§ 2. Classification of Hamiltonian flows of the form $\dot{X}_C$ on a standard sphere

Consider a sphere $S^{n-1} = SO(n)/SO(n-1)$ and the decomposition of the Lie algebra $so(n) = G = V + H$, where $H = so(n-1)$ (the embedding is standard) and $V = T_e S^{n-1}$. In the algebra G we choose a standard basis formed by skew-symmetric matrices E_{ij} in which the element at the place (i,j) is $+1$, that at the place (j,i) is -1, and the other elements are equal to zero (i is the row number, j is the column number). Let us write (i,j) for E_{ij}; a basis in V is the matrices $(1k)$, $2 \leq k \leq n$, i.e. $X = \sum_{k=2}^{n} x_k(1k)$. Clearly, $T = \lambda(12)$, $T' = \sum_{k=3}^{n} x_k(1k)$,

$K = so(n-2)$, $K' = \sum_{k=3}^{n} q_k(2k)$. Thus, $K = so(n-2)$, i.e. $B = 0$, $V = T + R$, and $H = K + \tilde{R}$. The operator $C: V \to H$ is of the form $CX = C(\pi X + Y) = D\pi X + ad_a^{-1} ad_b Y$, where $D: T \to K$. Any element $a \in T$ is of the form $\lambda(12)$, so that (12) can be taken as a. The operator D is defined by only one element $d = D(12)$.

<u>Theorem 2.1</u>. The form F_C, where $C = C(a, b, 0, D)$, $a = (12)$, $b \in K = so(n-2)$, and $D(a) = d \in K$, is closed on V (in particular, on the orbits as well) if and only if $d + 2b = 0$. In this case the form F_C is represented by $d\ell$, where the 1-form ℓ is defined by the vector field $\vec{\ell}(X) = (0, x_2, b(Y))$, where $Y = (x_3, \ldots, x_n)$ and $b \in so(n-2)$. The form F_C is not invariant under the action of $SO(n-2)$ and $SO(n-1)$. If the orbit S^{n-2} is even-dimensional and the skew-symmetric matrix $b \in K$ is non-degenerate (e.g., b is an element of general position in K), the form f_C (the restriction of F_C on the orbit) is non-degenerate almost everywhere on the orbit. Thus, if $d + 2b = 0$, b is non-degenerate, and n is even, the form f_C defines a symplectic structure (which is not invariant under the action of the group) on an open subset which is everywhere dense in the orbit (i.e. in the sphere).

The fact that the symplectic structure of f_C on the orbits is not invariant under the action of the group can be explained by the absence of invariant symplectic structure on spheres of dimension higher than 2. Let us define the flow $\dot{X} = [C'X, X]$ on orbits. The operator C' is given by the parameters a', b', and D'. Since $\dim T = 1$, $c' = \lambda'(12)$ and we may assume that $a' = a = (12)$. The element b is arbitrary and the operator D' is uniquely defined by the element $d = D'(12)$.

<u>Lemma 2.1</u>. Let $a = (12)$ and $b', d' \in K = so(n-2)$. Then the flow $\dot{X} = [C'X, X]$ is defined on V as follows: $\dot{x}_2 = 0$, $\dot{x}_k = x_2 \sum_{\alpha=3}^{n} x_\alpha (b'_{\alpha k} - d'_{\alpha k})$, $3 \leqslant k \leqslant n$. This means that the flow $\dot{X}$ preserves the foliation of the sphere S^{n-2} into spheres $S^{n-3} = \{x_2 = \text{const}\}$, and if $X = (x_2, Y)$, $Y \in \mathbb{R}^{n-2}$, the flow $\dot{X}$ on S^{n-3} coincides with the flow $Y = x_2 pY$, where $x_2 = \text{const}$, $p = b' - d' \in K = so(n-2)$, and pY is the image of the vector Y under the standard action of $so(n-2)$ on $R(x_3, \ldots, x_n)$.

We now discuss the question whether the flow X (where $X = (x_2, Y)$) is Hamiltonian relative to the symplectic structure of f_c on the orbit. Let the 2-form F_c be defined by the operator $CX = x_2 D(12) + ad^{-1}_{(12)} ad_b X$ and let the flow $\dot{X} = [C'X, X]$ be defined by the element $p \in K$, $\dot{X} = (0, pY)$, $X = x_2(12) + Y$. The form F_c on V is invariant with respect to $\dot{X}$ if and only if $\dot{X}F_c = 0$. This gives conditions on the elements b, p, and d.

<u>Theorem 2.2</u>. Let the 2-form F_c be defined by the operator $CX = x_2 D(12) + ad^{-1}_{(12)} ad_b X$ and let the flow $\dot{X}_{c'} = [C'X, X]$ be represented by the element $p \in K = so(n-2)$, $\dot{X} = (0, p\ Y)$, $X = x_2(12) + Y$, $Y \in R$. The form F_c on V is invariant with respect to $\dot{X}$ if and only if the elements b, p, and $d = D(12)$ satisfy the relations $[p, d] = 0$ and $[p, b] = 0$. If $p, b, d \in K$ are elements of general position, the condition that they are commutative is equivalent to the condition that p, b, and d belong to the same Cartan's subalgebra in $K = so(n-2)$. In particular, if $n = 2q$, $d + 2b = 0$, and the matrix b is non-degenerate (i.e. the form f_c on the orbit defines (almost everywhere) a symplectic structure), the flow $\dot{X}$ is Hamiltonian if and only if $[p, b] = 0$. In particular, if the element p is fixed, then by varying b in such a fashion that $[p, b] = 0$, we obtain a series of distinct symplectic forms with respect to which the flow is Hamiltonian.

Here we should mention a relationship between Hamiltonian flows $\dot{X}_Q$ and the equations of motion of a multidimensional rigid body. Consider a compact symmetric space $\mathcal{V} = \mathcal{G} / \varphi_y$ and an orbit $\mathcal{G}(X)$ in $G = H + V$, where $X \in V$. Then the intersection of the orbit with the plane V coincides with the orbit $O = \varphi_y(X) \subset V$.

Let $a, b \in G$ be two commuting elements of general position. Consider a Cartan's subalgebra $S(a, b) \subset G$ which contains a and b, and construct operators $\varphi_{a,b,D}: G \to G$ defined in [4], where $\varphi = ad_a^{-1} ad_b + D$. In this case the flow $\dot{X} = [X, \varphi X]$ describes the motion of a multidimensional rigid body with a fixed point. Let $K \neq 0$, then the elements a, b tend to V and H, and eventually become singular and find themselves in T and K. Then, the flow $\dot{X}$ rotates on the orbit $\mathcal{O}(X)$ and preserves, in the "limiting case", the intersection $\mathcal{O}(X) \cap V = O$. On O, this limiting flow of a "rigid body" may be considered as coinciding with the flow $\dot{X}_Q$, where $Q = ad_{a'} + ad_a^{-1} ad_b + D$. The flow $\dot{X}_Q$ is not a limit of the flow of a "rigid body" in the literal sense, since for some roots α we have $\alpha(a) = 0$. Let us turn back to the sphere.

Consider a sphere S^{n-1} and the corresponding embedding $so(n-1) = V + H$. Embed $so(n-1)$ standardly in $U(n)$ (on a subspace of real matrices). Unlike the prece ing case, the elements a and b a e now taken not from the algebra $so(n-1)$, but from the surrounding algebra $U(n)$. Let T_{ij} be a matrix the only non-zero element of which is equal to 1 and stands at the place (ij). Then, $(ij) = T_{ij} - T_{ji}$. We set $|ij| = T_{ij} + T_{ji}$. As $\bar{a}$ we take the matrix $i|12|$, $i^2 = -1$, and as $\bar{b}$ we take $i \sum_{3 \le p < q \le n} b_{pq} |pq|$, where $\bar{b} = ib$, $b_{pq} = b_{qp}$. Clearly, $\bar{a}, \bar{b} \in U(n)$ and $[\bar{a}, \bar{b}] = 0$. It can be assumed that $\bar{a}$ and $\bar{b}$ belong to the same Cartan's subalgebra in $U(n)$. In order that b belong to $sU(n)$, we require that $\sum_{k=3}^{n} b_{kk} = 0$.

<u>Lemmas 2.2</u>. The form $F_{\bar{c}}$, where $\bar{c} = ad_{\bar{a}}^{-1} ad_{\bar{b}} + D$ (the elements $\bar{a}$ and $\bar{b}$ were introduced above) is defined on the space V by the formula

$$F_{\bar{c}} = F_1 + F_2 = -x_2 \sum_{3 \le i < j \le n} d_{ij}\, dx_i \wedge dx_j + \sum_{k=3}^{n} \Big(\sum_{\alpha=3}^{n} x_\alpha b_{\alpha k} \Big) dx_i \wedge dx_k$$

where $b_{\alpha k} = b_{k\alpha}$. The form F_2 is exact on the space V, i.e. $F_2 = d\ell$, where the 1-form ℓ is given by $x_2 \sum_{(\alpha,k)=3}^{n} b_{\alpha k} x \;\; dx_k$. The orbits of adjoint action are spheres S^{n-2}, the restriction $f_{\bar{c}}$ of the form $F_{\bar{c}}$ on the orbit being expressed as

$$f_{\bar{c}} = f_1 + f_2 = \sum \omega_{ij}\, dx_i \wedge dx_j =$$

$$= -x_2 \sum_{3\le i<j\le n} d_{ij}\, dx_i \wedge dx_j - \frac{1}{x_2} \sum_{3\le i<j\le n} \Big[\sum_{\alpha=3}^{n} x_\alpha (x_i b_{\alpha j} - x_j b_{\alpha i}) \Big] dx_i \wedge dx_j$$

where $x_2{}^2 = R^2 - \sum_{k=3}^{n} x_k{}^2$, or in other form

$$\omega_{ij} = -x_2 d_{ij} - \frac{1}{x_2} \Big[\sum_{\substack{\alpha=3 \\ \alpha\ne(i,j)}}^{n} (x_\alpha (x_i b_{\alpha j} - x_j b_{\alpha i}) +$$

$$+ x_i x_j (b_{jj} - b_{ii}) + b_{ij} (x_i^2 - x_j^2)) \Big]$$

where $x_3, \ldots, x_n$ are coordinates on the sphere S^{n-2}.

<u>Theorem 2.3</u>. The form $F_{\bar{c}}$, $\bar{C} = ad_{\bar{a}}^{-1} ad_{\bar{b}} + D$, where the elements $\bar{a}$ and $\bar{b}$ were described above and $d = D(12) \in K = so(n-2)$, is closed on $V = T_e S^{n-1}$ and, in particular, on the orbits $O = S^{n-2}$ if and only if $d = 0$. In this case, the form $F_{\bar{c}}$ is exact on V and can be represented as $d\ell$, where $\ell = x_2 \sum_{k=3}^{n} (\sum_{\alpha=3}^{n} b_{\alpha k} x_\alpha) dx_k$. The form $F_{\bar{c}} = \sum_{k=3}^{n} (\sum_{\alpha=3}^{n} x_\alpha b_{\alpha k}) dx_2 \wedge dx_k$ is not invariant on V and on the orbits under the action of stationary groups. If the orbit S^{n-2} is even-dimensional ($n = 2r + 2$, $r = \text{rank } SO(2r)$) and if the symmetric operator $b = \| b_{pq} \|$ is an element of general position (as the element $\| ib_{pq} \|$ of the algebra $U(n-2)$), then the form $f_{\bar{c}}$ (restriction of F_c on the orbit) is non-degenerate almost everywhere on the orbit. Thus, if $d = 0$, n is even, and the operator b is an element of general position, the form $f_{\bar{c}}$ defines a non-degenerate symplectic structure on an open subset which is everywhere dense in the orbit.

Proof. Since $F_{\bar{c}} = F_1 + F_2$, we have $dF_{\bar{c}} = dF_1$. If $F_1 = \sum_{i<j} \omega_{ij} dx_i \wedge dx_j$, then

$$dF_1 = \frac{1}{x_2} \sum_{i<j<k} (x_i d_{jk} - x_j d_{ik} + x_k d_{ij}) dx_i \wedge dx_j \wedge dx_k$$

because $x_2{}^2 = R^2 - \sum_{\alpha=3}^{n} x_\alpha^2$. Thus, $dF_{\bar{c}} = 0$ if and only if $d = 0$.

We now prove that $f_{\bar{c}}$ is non-degenerate on the orbit, provided n is even and b is an element of general position. Clearly, the form f_c may be considered on the space R^{n-2}

$$f_{ij} = -\frac{1}{x_2} \Big[\sum_{\substack{\alpha=3 \\ \alpha\ne(i,j)}}^{n} x_\alpha (x_i b_{\alpha j} - x_j b_{\alpha i}) + x_i x_j (b_{jj} - b_{ii}) + b_{ij} (x_i^2 - x_j^2) \Big] .$$

Suffice it to prove that $\|f_{ij}\|$ is non-degenerate for some operator b. Let this operator be close to a diagonal one; then the matrix $\|f_{ij}\|$ turns to $\|(b_{ii} - b_{jj})\frac{X_i X_j}{X_2}\|$. Put $b_{ii} = b_i$; then it is sufficient to prove that the matrix is non-degenerate under the condition that the vector $b = (b_3, \ldots, b_n)$ is an element of general position. Non-degeneracy of the matrix $\|b_i - b_j\|$ (for an even n) can be verified by considering the point $x_3 = \ldots = x_n = 1$ in R^{n-2}. Putting $b_\alpha = b^\alpha$ (the degree), where $b \neq 0$, we have

$$\|b_i - b_j\| = b \begin{pmatrix} 0 & 1-b & 1-b^2 & \ldots & 1-b^{2r-1} \\ b-1 & 0 & 1-b & \ldots & 1-b^{2r-2} \\ \ldots & \ldots & \ldots & \ldots & \ldots \\ b^{2r-1}-1 & b^{2r-2}-1 & b^{2r-3}-1 & \ldots & 0 \end{pmatrix}$$

Letting b tend to zero, we see that to prove non-degeneracy of $\|b_i - b_j\|$, it is sufficient to demonstrate non-degeneracy of the matrix

$$\begin{pmatrix} 0 & 1 & 1 & \ldots & 1 \\ -1 & 0 & 1 & \ldots & 1 \\ \ldots & \ldots & \ldots & \ldots & \ldots \\ -1 & -1 & -1 & \ldots & 0 \end{pmatrix}$$

which is obvious. The theorem is proved.

Let us calculate the flow $\dot{X}_{\bar{Q}}$ on the orbit. The operator $\bar{Q}: V \to H$ is of the form $ad_{a'}^{-1} ad_{b'} + D'$. Since $a' = \lambda i\,|12| = \lambda \bar{a}$, we set $a' = \bar{a}$, $b' = c = i \sum_{3 \leq p < q \leq n} c_{pj}\,|pj|$; $c_{pj} = c_{jp}$. Let c be a symmetric operator of general position. The operator $D': T \to K$ is uniquely determined by the element $d' = D'(12)$. According to the above considerations, the element d taking part in the construction of a symplectic structure of $f_{\bar{c}}$ on the orbit is zero, so that while calculating the flow $\dot{X}_{\bar{Q}}$, we denote d' by d. Then, straightforward calculation shows that if the elements $\bar{a}$, $\bar{c}$, $c^T = c$, and $d \in K = so(n-2)$ are chosen as it was described above, the flow $\dot{X}_{\bar{Q}} = [\bar{Q}X, X]$, where $\bar{Q} = ad_{\bar{a}}^{-1} ad_{\bar{c}} + D$, is defined on V as follows:

$$\dot{x}_2 = -\sum_{3 \leq \alpha, \kappa \leq n} x_\alpha x_\kappa c_{\alpha\kappa} ; \qquad \dot{x}_\kappa = x_2 \sum_{\alpha=3}^{n} x_\alpha (c_{\alpha\kappa} - d_{\alpha\kappa}) .$$

Here $c^T = c$ and $d^T = -d$. Since the flow $\dot{X}_{\bar{Q}}$ has the integral $x_2^2 + \ldots + x_n^2$, it touches the orbits $S^{n-2} \subset V$. Let us discuss the question whether the flow $\dot{X}_{\bar{Q}}$ is Hamiltonian with respect to the structure of $f_{\bar{c}}$ on the orbit.

Theorem 2.4. Let the 2-form $f_{\bar{c}}$ (restriction on the orbit) be defined on the orbit by the operator $\bar{c} = ad_{\bar{a}}^{-1} ad_{\bar{b}}$, where the elements $\bar{a}$ and $\bar{b}$ were described above; $\bar{b} = ib$, $b^T = b$. Let the flow $\dot{X}_{\bar{Q}}$ on the orbit be defined by the operator $\bar{Q} = ad_{\bar{a}}^{-1} ad_{\bar{c}} + D$, $\bar{c} = ic$, $c^T = c$, $d = D(12)$. Along with symmetric operators c and b that map R^{n-2} into itself, we consider symmetric operator $h = tb + bt^T$, where $t = c-d$, i.e. $h = (bc + cb) + [b, d]$, $h^T = h$, $h: R^{n-2} \to R^{n-2}$. On R^{n-2}, we take four vector fields $eX = X$, hX, bX, and cX, and the corresponding 1-forms:

$$EX = \sum_{\alpha=3}^{n} x_\alpha dx_\alpha \quad , \quad HX = \sum_{\alpha=3}^{n} (hX)_\alpha dx_\alpha \ ,$$

$$BX = \sum_{\alpha=3}^{n} (bX)_\alpha dx_\alpha \quad , \quad CX = \sum_{\alpha=3}^{n} (cX)_\alpha dx_\alpha \ .$$

In this case, the form $f_{\bar{c}}$ is invariant on the orbits with respect to the flow $\dot{X}_{\bar{Q}}$ if and only if the 1-forms E, H, B, and C satisfy the relation $E \wedge H = 2B \wedge C$. where $\wedge$ denotes exterior product. If $n = 2r + 2$ is even and the operator b is an element of general position, $f_{\bar{c}}$ is a non-degenerate symplectic structure almost everywhere on the sphere S^{2r}, and the relation $E \wedge H = 2B \wedge C$ describes all Hamiltonian flows $\dot{X}_{\bar{Q}}$ with respect to this form. In particular, if $\dot{X}_{\bar{Q}}$ is fixed, then by varying b and d in such a manner that $E \wedge H = 2B \wedge C$, we obtain a series of distinct symplectic forms relative to which the flow $\dot{X}_{\bar{Q}}$ is Hamiltonian. If the operators b and c are both diagonal and $b_{ii} = b_i$, $c_{ii} = c_i$, the condition $E \wedge H = 2B \wedge C$ turns to $[b, d] = 0$ and $[b, \tilde{c}] = 0$, where $\tilde{c} = \|\tilde{c}_{ij}\|$, $\tilde{c}_{ij} = c_i + c_j$. This means that if $b_k \neq b_p$, then $c_k + c_p = 0$ and $d_{pk} = 0$; if $b_k = b_p$, then $c_k + c_p$ and d_{pk} are arbitrary.

Remark. The vector fields eX, hX, bX, and cX on R^{n-2} can be identified with the gradients of the corresponding quadratic functions f_e, f_h, f_b, and f_c. Then, the necessary and sufficient condition that the flow $\dot{X}_{\bar{Q}}$ is Hamiltonian can be expressed as

$$df_e \wedge df_h = 2\, df_b \wedge df_h \quad , \quad E \wedge (BC + CB + [B, d]) = 2B \wedge C \ .$$

Thus, the freedom in varying the flow $\dot{X}_{\bar{Q}}$, which is Hamiltonian with respect to the form $f_{\bar{C}}$, depends on to what extent the decomposition of the form $E \wedge H = 2B \wedge C$ in the algebra of exterior forms is ambiguous.

§ 3. Complete integration of the equations of inertial motion of a multidimensional rigid body in an ideal liquid

These results are reported in [3] . The Lie algebra E(n) of the group of motions of a Euclidean space is a semidirect sum $so(n) \oplus_\varphi \mathbb{R}$, where $\varphi : so(n) \to End(\mathbb{R}^n)$ is the differential of the standard representation (with minimal dimension) of the group SO(n) in $\mathbb{R}^n$. The Lie algebra E(n) admits the following matrix realization

$$\left(\begin{array}{c|c} so(n) & \begin{matrix} y_1 \\ \vdots \\ y_n \end{matrix} \\ \hline 0 \dots 0 & 0 \end{array}\right)$$

and the space $E(n)^*$ is of the form

$$\left(\begin{array}{c|c} so(n) & \begin{matrix} 0 \\ \vdots \\ 0 \end{matrix} \\ \hline y_1 \dots y_n & 0 \end{array}\right)$$

The minor of the matrix X formed by the elements at the intersection of rows $i_1, \dots, i_s$ and columns $j_1, \dots, j_s$ will be denoted by $M^{i_1 \dots i_s}_{j_1 \dots j_s}(X)$. Let us define, on E(n) and $E(n)^*$, functions $F^{i_1 \dots i_s}_{j_1 \dots j_s}(X)$ by the relation $F^{i_1 \dots i_s}_{j_1 \dots j_s}(X) = M^{i_1 \dots i_s}_{j_1 \dots j_s}(\frac{1}{2}(X - X^T))$, where T denotes transposition relative to the principal diagonal.

<u>Proposition 3.1</u>. The functions $J_k(X)$ on $E(n)^*$, where

$$J_k(X) = \sum_{1 \le i_1 < \dots < i_k \le n} F^{i_1 \dots i_k, n+1}_{j_1 \dots j_k, n+1}(X)$$

are invariants of the co-adjoint representation of the group of motions of a Euclidean space.

<u>Proposition 3.2</u>. Let O be an orbit of maximal dimension of the co-adjoint representation Ad^* of the Lie group E(n) of motions of an n-dimensional Euclidean space. Then, $\operatorname{codim} O = \left[\frac{n+1}{2}\right]$. The set of invariants of the co-adjoint representation Ad^* of the Lie group E(n)

(see Proposition 3.1) is a complete set of invariants, i.e. any other invariant can be expressed functionally in terms of J_k.

<u>Proposition 3.3</u>. Let $\xi \in so(n)$, $a \in \mathbb{R}^n$, $S \in so(n)^* \cong so(n)$, $M \in (\mathbb{R}^n)^*$, $a + \xi \in so(n) \oplus_\varphi \mathbb{R}^n = E(n)$, $S + M \in E(n)^* =$
$= (so(n) \oplus \mathbb{R}^n)^* \cong so(n)^* \oplus (\mathbb{R}^n)^* \cong so(n) \oplus \mathbb{R}^n$.

Then we assert that

$$\{\xi + a, S + M\}|_{so(n)} = [S, \xi] + \frac{1}{2}(Ma^T - aM^T),$$

$$\{\xi + a, S + M\}|_{\mathbb{R}^n} = -\xi M.$$

Here $M, a \in \mathbb{R}^n$ are written as coordinate columns, T denotes transposition.

Consider a section operator C(a, b, D). It turns out that the corresponding Euler's equations not only coincide with the equations of inertial motion of a rigid body in an ideal liquid, but also admit a complete set of integrals. These integrals are polynomials (of different degrees).

<u>Theorem 3.1.</u> The system of differential equations $\dot{x} = \{C(a,b,D)(x), x\}$ on $E(n)^*$ is completely integrable on the orbits (which are in general position) of Ad^* of the Lie group $\mathcal{E}(n)$. Let F be a function on $E(n)^*$ which is invariant relative to the co-adjoint representation $Ad^*(\mathcal{E}(n))$. Then the functions $F_\lambda(x) = F(x + \lambda a)$ are first integrals of motion for arbitrary numbers λ. Any two integrals $F_\lambda(x)$ and $G_\mu(x)$ are in involution on the orbits of the representation Ad^* of the Lie group $\mathcal{E}(n)$, the number of independent integrals of this type being equal to half the dimension of the orbit of general position.

Let us remind the classical equations of inertial motion of a rigid body in an ideal liquid (for details see, e.g., [10,11]). Let a coordinate system be fixed to the moving body. We write u_i for the components of the translational motion of the origin and ω_i, for the components of the angular speed of rotation of the body. Then the kinetic energy of the whole system (liquid-body) is $T = \frac{1}{2}(A_{ij}\omega_i\omega_j + B_{ij}u_iu_j + C_{ij}\omega_iu_j)$, where A_{ij}, B_{ij}, and C_{ij} are constants dependent on the body shape and density, and on liquid density; summation is assumed over doubly repeated indices. Let $N = (y_1, y_2, y_3)$, where $y_i = \partial T/\partial\omega_i$, and $K = (x_1, x_2, x_3)$, where $x_i = \partial T/\partial u_i$. Then the inertial motion of the body in the liquid is described by the equations

$$\begin{cases} \frac{dN}{dt} = N \times \omega + K \times U \\ \frac{dK}{dt} = K \times \omega \end{cases},$$

where $U = (u_1, u_2, u_3)$; $\omega = (\omega_1, \omega_2, \omega_3)$.

The kinetic energy of a rigid body is a homogeneous quadratic form of six variables u_i, ω_i (in the classical cases it is positive definite); this form is determined by 21 coefficients A_{ij}, B_{ij}, and C_{ij}. In a three-dimensional case, the equations of motion have three classical Kirchhoff's integrals: T = const, $x_1^2 + x_2^2 + x_3^2$ = const, and $x_1y_1 + x_2y_2 + x_3y_3$ = const. Complete integrability of the equations in three dimensions requires the existence of four functionally independent integrals. The corresponding Hamiltonian flow is represented by a vector field on a four-dimensional orbit. Two integrals "cut away" this orbit from the surrounding six-dimensional algebra, and two other integrals must be in involution on the orbit to foliate it into "Liouville's tori" along which the system is moving. Three classical cases are known in which there exists an additional, fourth integral, i.e. these equations can be integrated explicitly (see [12]).

The first general solution of the equation of motion of a rigid body in a liquid was obtained by Kirchhoff for a body of revolution. In 1871, Clebsch invented two new types of kinetic energy for which there exists an additional fourth integral, so that the problem can be reduced to quadratures. In the first Clebsch's case, where the fourth integral is, generally, a linear homogeneous function of x_i and y_i, the problem was solved by Halphen in 1888 (see [13,14]). In 1878, Weber analysed the second Clebsch's case, where the fourth integral is a homogeneous quadratic function of x_i and y_i, under a particular assumption about the choice of arbitrary constants (see [15]). The third type of kinetic energy, for which the equations can be integrated explicitly, was discovered by V.A. Steklov (see [11]). A specific feature of all these cases is that certain restrictions are imposed on the "body configuration", and the body is assumed invariant under special symmetry operations.

By virtue of straightforward calculation, we arrive at the following result.

Theorem 3.2. For $n = 3$, the system of differential equations $x = \{C(a, b, D)(X), X\}$ on $E(n)^*$ constructed on the basis of a section operator coincides with the equations of inertial motion of a rigid body in an ideal liquid (in a three-dimensional space).

Usinf this fact and proceeding from the explicit form of Euler's equations constructed above, one can assume that in the case of an arbitrary dimension these equations describe the motion of a multidimensional body in a liquid.

Interestingly, even for $n = 3$ we obtain a new example of integrability in which kinetic energy is of definite sign and has an additional linear fourth integral. It should be noted that for all dimensions (starting with the fourth and higher) additional integrals are in the form of polynomials of different degrees. Explicit calculation of kinetic energy shows that although this energy is not positive definite, this case of integrability is rather similar (in its algebraic structure) to the first Clebsch's case for a positive form. In the case where integration is possible, the explicit form of the sign-indefinite matrix of kinetic energy is

$$A = \begin{pmatrix} -2\alpha & 0 & 0 & 0 & 0 & \frac{\gamma}{2}-\beta \\ 0 & 0 & 0 & 0 & -\frac{2a_1}{b_2} & 0 \\ 0 & 0 & 0 & \frac{a_1}{b_2} & \frac{a_1}{b_2} & 0 \\ 0 & 0 & \frac{a_1}{b_2} & b_2^{-2}(-b_2a_2+2b_1a_1) & 0 & 0 \\ 0 & -\frac{2a_1}{b_2} & \frac{a_1}{b_2} & 0 & b_2^{-2}(-b_2a_2+2b_1a_1) & 0 \\ \frac{\gamma}{2}-\beta & 0 & 0 & 0 & 0 & \delta \end{pmatrix}$$

Here α, β, γ and δ are constants determining the operator $D: K^* = \mathbb{R}^2 \to K \cong \mathbb{R}^2$ used to define the section operator. Elementary calculation shows that the kinetic energy form T is reduced to

$$T = -2\alpha\left(f_1 - \frac{\gamma-2\beta}{4\alpha}u_3\right)^2 + \left[\frac{(\gamma-2\beta)^2}{8\alpha} + \delta\right]u_3^2 - \lambda\left(u_1 - \frac{a_1}{b_2\lambda}f_3\right)^2 - \lambda\left(u_2 + \frac{2a_1}{b_2\lambda}f_2 - \frac{a_1}{b_2\lambda}f_3\right)^2 + \frac{(2a_1)^2}{b_2\lambda}\left(f_2 - \frac{f_3}{2b_2}\right)^2 + \frac{2a_1^2(b_2-1)}{b_2^3\lambda}f_3^2 .$$

The corresponding system of equations on an orbit is integrated explicitly because the additional fourth integral is linear. In the cases of high dimensions (starting with the fourth), explicit integration becomes a much more complicated procedure.

§ 4. Construction of Lie algebras with a complete involutary set of functions on orbits of general position

As was already said, the existence of a maximal commutative set of functions on orbits of general position has been proved for many important series of Lie algebras (semisimple, Borel, and others). Besides an analysis of important series of algebras encountered in various applications, there is another approach to constructing Lie algebras with a maximal commutative set of functions. Consider a Lie algebra on which this set does exist. A question arises: how can new Lie algebras with the same properties be constructed proceeding from the initial algebra? It turns out that there exists a natural algorithm for constructing series of new "extended" Lie algebras with a maximal commutative set of functions. In what follows we briefly describe new results obtained by A. Brailov. Some of these results rest upon the scheme suggested by V. Trofimov.

Definition. An algebra with a Poincaré duality is said to be a finite-dimensional, graded, and commutative algebra over the field of real numbers $A = A_o \oplus \dots \oplus A_n$, $A_i \cdot A_j \subset A_{i+j}$ $(i+j \leqslant n)$ such that $\dim A_n = 1$ and the bilinear symmetric pairing $\langle\ ,\ \rangle_s : A \times A \to \mathbb{R}$ given by the formulas $\langle a, b\rangle_s = 0$, $a \in A_i$, $b \in A_j$, $i+j \neq n$; $\langle a, b\rangle_s\ a_n = ab$, $ab \in A_n$, is non-degenerate for a nonzero element $a_n \in A_n$.

Let $\dim A = N$. Then we choose in the algebra A a basis $B = \{\varepsilon_i\}_{i = 1, \dots, N}$ as follows: set $\varepsilon_1 = 1$ and choose bases in subspaces A_i passing from a smaller graduation to a larger one. If $i < \frac{n}{2}$, the bases are chosen arbitrarily; if $i = \frac{n}{2}$, we choose a basis in such a way that the bilinear form described above should be diagonal in this basis; if $i > \frac{n}{2}$, we choose bases that are conjugate to the bases of smaller graduations in the subspaces A_{n-i}. Instead of the field $\mathbb{R}$ we can, of course, take any field k.

Such algebras are convenient to extend Lie algebras with a maximal commutative set of functions. These extensions are constructed by the following simple procedure.

Let G be a Lie algebra over the field k and let A be an algebra with a Poincaré duality (over the field k). Consider the tensor product $G_A = G \otimes A$. Clearly, this algebra turns to a Lie algebra over the field k , and dim A = N, dim G = m.

Let $P \in k[x_1, \ldots, x_m]$ be an arbitrary polynomial considered as a function on the co-algebra G^* (the bases fixed in the algebras G and G^* are conjugate). Let $e_1, \ldots, e_m$ be a basis in G^*. Then as a basis in the new Lie co-algebra G_A^* we take the following set of elements: $e_1^{\varepsilon_1}, \ldots, e_m^{\varepsilon_1}, e_1^{\varepsilon_2}, \ldots, e_m^{\varepsilon_2}, e_1^{\varepsilon_3}, \ldots, e_m^{\varepsilon_3}, \ldots, e_1^{\varepsilon_N}, \ldots, e_m^{\varepsilon_N}$, where $\varepsilon_i \in B$ is a basis in A.

Remark. Here the notation for the tensor product is replaced, for convenience, by the symbol $e_i^{\varepsilon_i}$.

The corresponding coordinates in the co-algebra G_A^* are denoted by $x_1^{\varepsilon_1}, \ldots, x_m^{\varepsilon_N}$ and it is assumed that in the algebra A there is fixed a uniquely defined operator $* : A \to A$ induced by the above bilinear pairing, i.e. $\langle * \varepsilon_i, \varepsilon_j \rangle = \delta_{ij}$; $* \varepsilon_i \in B$, $*^2 = 1$.

Now we describe the algorithm that enables one to construct a function (a polynomial) on G_A^* with the aid of a function (a polyn mial) defined on G^*. Precisely this algorithm permits the construction of new integrals of an extended algebra proceeding from the integrals of the initial Lie algebra.

Let $P \in k[x_1, \ldots, x_m]$ be a polynomial on G^*. Then the polynomial $\nu P \in A[x_1^{\varepsilon_1}, \ldots, x_m^{\varepsilon_N}]$ is defined as follows ($\varepsilon_N = *1$):

$$\nu P(x_1^{\varepsilon_1}, \ldots, x_m^{\varepsilon_N}) = \sum_{k=0}^{\infty} \sum_{i_1 \ldots i_k} \sum_{j_1 \ldots j_k} P_{i_1 \ldots i_k} \varepsilon_{j_1 \ldots j_k} x_{i_1}^{*\varepsilon_1} \ldots x_{i_k}^{*\varepsilon_{j_k}}$$

or (in the short conventional form) $\sum_{I,J} P_I \varepsilon_J x_I^{*\varepsilon_J}$.

Straightforward calculation shows that the mapping $\nu : k[x_1, \ldots x_m] \to A[x_1^{\varepsilon_1}, \ldots, x_m^{\varepsilon_N}]$ is a homomorphism of k-algebras. We note that the symmetric pairing of elements of the algebra A induces a natural scalar product, which we denote by $\langle \ , \ \rangle$. Indeed, since $\langle a, b \rangle_s a_n = ab$ for $ab \in A_n$, then we obtain the scalar product $A \times A \to \mathbb{R}$ (because dim $A_n = 1$).

We now describe the operation which is a generalization of the "shift by a covector of general position", i.e. a generalization of the main operation introduced in [4] for an arbitrary Lie algebra and leading in many situations to a maximal commutative set of functions.

For an element $a \in A_n$, we define $\nu_a P \in k[x_1^{\varepsilon_1}, \ldots, x_m^{\varepsilon_N}]$;

$$P^a(x_1^{\varepsilon_1},\ldots,x_m^{\varepsilon_N}) = \nu_a P(x_1^{\varepsilon_1},\ldots,x_m^{\varepsilon_N}) = \sum_{I,J} P_I \langle a, \varepsilon_J \rangle x_I^{*\varepsilon_J} \quad .$$

<u>Theorem 4.1</u>. If the polynomials $P_1, \ldots, P_r \in k[x_1, \ldots, x_m]$ are independent (functionally) at a point $y \in G^*$, $y = (y_1, \ldots, y_m)$, then the polynomials $P_1^{\varepsilon_1}, \ldots, P_r^{\varepsilon_N} \in k[x_1^{\varepsilon_1}, \ldots, x_m^{\varepsilon_N}]$ are independent at any point $x \in G_A^*$ the higher coordinates of which are of the form $x_i^{*\varepsilon_1} = y_i$, $i = 1, \ldots, m$.

The theorem is proved by straightforward calculation.

As was established in [4] , the shifts of invariants are in involution. A similar statement holds in the present situation. To be more exact, the following is valid: if $a, b \in A$ and $P, Q \in k[x_1, \ldots, x_m]$, then the Poisson bracket is defined by

$$\{\nu_a P, \nu_b Q\} = \langle a \otimes b, \frac{\partial \nu P}{\partial x_i^{\varepsilon_j}} \otimes \frac{\partial \nu Q}{\partial x_q^{\varepsilon_i}} \{x_i^{\varepsilon_j}, x_q^{\varepsilon_i}\} \rangle \ .$$

We recall that the co-dimension of the orbit of general position (i.e. of the orbit of maximal dimension) is called by the index of a Lie algebra. The following statement permits the calculation of the index of an extended algebra G_A^*, using information on the initial algebra G^* and on the algebra A.

<u>Theorem 4.2</u>. Let the number r of polynomial invariants of the algebra G^* be equal to the index of G^*. Then the equality $\operatorname{ind}(G_A^*) = \dim A \operatorname{ind}(G^*)$ is valid.

<u>Remark</u>. Not for each Lie algebra one can find a set of polynomial invariants of this algebra the number of which is equal to the index. At any rate, this is true for algebraic Lie algebras.

Now we formulate the principal result of this section.

<u>Theorem 4.3</u>. Let $P_1, \ldots, P_s \in k[x_1, \ldots, x_m]$ form a complete (i.e. maximal) commutative set of functions (polynomials) on G^*. Then the polynomials $P_i^{\varepsilon_j}$ form a complete commutative set of functions on the extended algebra G_A^*.

In other words, to construct a complete commutative set of functions, the invariants of the algebra G_A^* must be shifted by the elements $a = \varepsilon_j$.

Let there be given a Lie algebra G and its decomposition $G = H + V$ such that $[H, H] \subset H$, $[H, V] \subset V$, $[V, V] \subset H$ (this decomposition corresponds to a symmetric space with the group G). Such Lie algebras are sometimes called $\mathbb{Z}_2$-graded algebras. The results presented below were obtained by A. Brailov.

Let $G^* = H^* \oplus V^*$ be a dual decomposition. If $x \in G^*$, $g \in G$, then by x_H, x_V, g_H, and g_V we d note the H- and V-components of x and H^*- and V^*-components of g, respectively. Define on G a bilinear operation $[g, g']_\lambda = [g_H, g'_H] + [g_V, g'_H] + [g_H, g'_V] + \lambda^2[g_V, g'_V]$, $g, g' \in G$. This operation satisfies the Jacobi identity, so that we obtain a new Lie algebra (it is denoted by G_λ). By definition, a Lie algebra G_0 is called the contraction of the $\mathbb{Z}_2$-graded Lie algebra G.

Along with the contraction of an algebra, we define the contraction of a function.

Let F be a smooth function on G^*. Suppose the expansion of F in the powers of the variables from V terminates at a certain term (for instance, if F is a polynomial), we have $F(x) = F^0(x_H, x_V) + F^1(x_H, x_V) + \dots + F^n(x_H, x_V)$. Let $F_\lambda(x) = \lambda^n F(x_H, \lambda^{-1}x_V)$. Since $\lambda^n F^n(x_H, \lambda^{-1}x_V) = F^n(x_H, x_V)$, then $F_\lambda(x)$ is also defined for $\lambda = 0$. The function $F_0(x)$ is called the contraction of the function F.

<u>Lemma</u>. Let F and F' be two functions on G^*, which are in involution. If the contractions F_0 and F'_0 of these functions are defined (this is always the case if the initial functions are polynomials), these contractions are also in involution on the "contracted algebra" G_0^*, i.e. $[F_0, F'_0]_0 = 0$.

<u>Proposition 4.1</u>. If F is a polynomial on G^*, which is invariant relative to the co-adjoint representation (i.e. it is an invariant of the algebra G), the contraction of the polynomial F_0 is an invariant of the co-adjoint representation of the Lie group corresponding to G_0 (i.e. an invariant of G_0).

Let G be a real Lie algebra, $G^{\mathbb{C}}$ its complexification, Σ a group acting via automorphisms on the algebra $G^{\mathbb{C}}$, and G_n a $\sum$-fixed subalgebra in G, $a \in G^*$. Suppose a is an eigencovector of weight relative to the representation of the compact group $\sum$ in the algebra $(G^{\mathbb{C}})^*$. Any symmetric operator $\varphi : G_n^* \to G_n$, for which there exists an element $b \in G$ such that the identities $\{a, b\} = 0$ and $\{a, \varphi(x)\} \equiv \{x, b\}$ hold, $(x \in G_n^*)$, is called an (a,b)-operator. As the algebra G, we now take $u(p, q) \times \mathbb{C}^{p+q}$, i.e. a semidirect product of the algebra u(p, q) and standard representation of minimal dimension. Then $G^{\mathbb{C}}$ is obtained from the algebra $sl(n, \mathbb{C})$ via contraction.

Theorem 4.4. For any covector $a \in G^*$ of general position, Euler's equations $\dot{x} = \{x, \varphi(x)\}$ are completely integrable (in the sense of Liouville) for any (a,b)-operator, such operators existing for any covector a of general position.

This result is an extension of the result obtained in [3] for a semidirect product of an orthogonal Lie algebra and its standard representation of minimal dimension. Operators of the (a,b)-series are analogues of operators used to construct systems of the form $\dot{X}_{\bar{C}}$ (i.e. they are analogues of section operators). A similar result is also valid for Lie algebras $so(p, q) \times \mathbb{R}^{p+q}$.

REFERENCES

1. Arnold V.I. Mathematical methods of classical mechanics.- Moscow, 1974 (in Russian).

2. Fomenko A.T. Group symplectic structures on homogeneous spaces.- Dokl. Akad. Nauk SSSR, 1980, 253, No. 5 (in Russian).

3. Trofimov V.V. and Fomenko A.T. The methods of constructing Hamiltonian flows on symmetric spaces and integrability of dynamical systems.- Dokl. Akad. Nauk SSSR, 1980, 254, No. 6 (in Russian).

4. Mishchenko A.S. and Fomenko A.T. Euler's equations on finite-dimensional Lie groups.- Izv. Akad. Nauk SSSR, 1978, 42, No. 2 (in Russian)

5. Mishchenko A.S. and Fomenko A.T. The generalized Liouville method for integrating Hamiltonian systems. Funktsionalny analiz i yego prilozheniya, 1978, 12, No. 2 (in Russian)

6. Vishik S.V. and Dolzhansky F.V. Analogues of the Euler-Poisson equations of magnetic hydrodynamics related to Lie groups.- Dokl. Akad. Nauk SSSR, 1978, 238, No. 5 (in Russian).

7. Arkhangelsky A.A. Completely integrable Hamiltonian systems on a group of triangular matrices.- Matem. sbornik, 1979, 108 (150) (in Russian).

8. Trofimov V.V. Euler's equations on Borel subalgebras of semisimple Lie algebras.- Izv. Akad. Nauk SSSR. Ser. Matem., 1979, 43 (in Russian).

9. Trofimov V.V. Finite-dimensional representations of Lie algebras and completely integrable systems.- Matem. sbornik, 1980, 111 (153) (in Russian).

10. Kochin N.E., Kibel I.A., and Roze N.V. Theoretical hydromechanics, Moscow, 1963, Part I (in Russian).

11. Steklov V.A. On the motion of a rigid body in a liquid, Kharkov, 1893 (in Russian).

12. Gorr G.V., Kudryashova L.V., and Stepanova L.A. Classical problems in solid-state dynamics, Kiev, 1978 (in Russian).

13. Halphen G.H. Sur le mouvement d'un solide dans un liquide.- J. Math., 1888, vol. 4, ser. 4.

14. Halphen G.H. Traité des fonctions elliptiques et de leurs applications. Paris, 1888, vol. 2.

15. Weber H. Anwendung der Thetafunctionen zweiter veränderlicher und die Theorie der Bewegung eines festen Körpers in einer Flüssigkeit.- Math. Ann., 1879, Bd. 14.

RIEMANNIAN PARALLEL TRANSLATION IN NON-LINEAR MECHANICS

Yu.E. Gliklikh

Department of Mathematics, Voronezh State University
394693, Voronezh, USSR

The paper deals with the description and investigation of mechanical systems within the framework of the differential-geometric formalism, which is natural for mechanics. While considering a mechanical system, it is customary to use such geometric objects as symplectic form, covariant derivative, geodesic, and others (see, for example, [28,21,1]). In modern differential geometry, however, there is a large variety of geometric objects related to mechanics which, though studied sufficiently well, have not been used to describe the motion of a mechanical system. This refers, in particular, to a Riemannian parallel translation. The mechanical meaning of the translation was discovered in 1926 (see [23]), and it was not until 1975 that a question was raised whether the concept of parallel translation is needed to describe the motion of a mechanical system [8] .

For several years, the author have been engaged in studying some problems related to Riemannian parallel translation and to constructions based on this translation. The present paper is a survey of the results relevant to mechanics. The Riemannian parallel translation is used here to construct integral-type operators that make it possible to write the equations of motion in geometrically invariant integral form, and to obtain some statements concerning the global behaviour of their solutions. Furthermore, using the concept of parallel translation, we have succeeded in describing certain classes of mechanical systems, which have extensively been studied in recent years on a flat configuration space, in terms of the geometrically invariant language.

Among the systems that have been "geometrized" via parallel translation, one may note systems with random force fields [19,20] , and also systems with a retarded control force or a control force varying in a "fixed" coordinate system [16-18] . The latter geometrization is based on the Radonian mechanical interpretation of parallel translation, which is described in [23] . In plain terms, Radon's results can be described as follows: while in a flat configuration space the direction of pendular oscillations remains constant, in a "curved" configuration space this direction undergoes a Riemannian parallel translation along the motion trajectory (e.g., the Foucault pendulum).

This paper is focused on the integral-type operators constructed on the basis of parallel translation. These operators are a geometrically invarianr analogue of the well-known Volterra's integral operators in the theory of ordinary differential equations [13-15] . Irrespective of the choice of the complete Riemannian metric on a manifold, the fixed points of these operators are integral curves of a given vector field. It is worth noting that these operators are defined globally (on the entire manifold), and not only in a local map.

Mechanical systems considered in this paper are characterized by quadratic kinetic energy defined by the complete Riemannian metric on a configuration space. The equation of motion of such systems is Newton's second law written in a geometric-invariant form in terms of the covariant derivative of the Levi-Civita connection of this metric. The integral operators thus constructed are used then to derive integral equations of motion naturally written in terms of the Riemannian parallel translation of the same connection [14,15] (random force fields [19,20] , many-valued and discontinuous fields [10,17]). A two-point boundary-value problem is considered as an example of applications.

This paper is an attempt to give a consistent presentation of the basic constructions. A somewhat brief exposition is compensated, to some extent, by numerous references.

§ 1. Integral operators with Riemannian parallel translation

1. Let M be a complete Riemannian manifold, $m_0 \in M$, $I = [0, T] \subset R$ a segment, and $v: I \rightarrow T_{m_0}M$ a continuous curve.

Theorem 1.1. There exists only one C^1-curve $\gamma: I \rightarrow M$ such that $\gamma(0) = m_0$ and the tangent vector $\dot{\gamma}(t)$ at each point $t \in I$ is parallel along γ to the vector $v(t) \in T_{m_0}M$.

Proof. Let $\theta = (e_1, \ldots, e_n)$ be an orthonormal basis in the tangent space $T_{m_0}M$. Using this basis, one can define an isomorphism of a standard space R^n on $T_{m_0}M$ by the formula $\theta(r_1, \ldots, r_n) = r_1 e_1 + \ldots + r_n e_n$. Let us consider a non-atonomous basis vector field $E(\theta^{-1} v(t))$

on the principal bundle of orthonormal frames $O(M)$ (see [2]). Apparently, this field is smooth with respect to $\sigma \in O(M)$, so that only one integral curve $\sigma(t)$, $\sigma(0) = \sigma$, passes through each point $\sigma \in O(M)$. The curve $\pi\,\theta(M)$ is the curve γ in question (π is the natural projection of $O(M)$ on M). Indeed, for any point t^* in the domain where γ is well defined, the vectors $\dot{\gamma}(t^*)$ and $v(t^*)$ are connected along γ by a parallel vector field $\theta(t)(\theta^{-1}v(t^*))$. Since M is complete, the fibres of $O(M)$ are compact, and parallel translation conserves the norm of a vector, the boundedness of $\|v(t)\|$ on I implies that γ is defined on the entire segment I. The theorem is proved.

The curve $\gamma(t)$ constructed by $v(t)$ will be denoted below by $Sv(t)$.

Remark. It can easily be seen that $Sv(t) = \delta^{-1}\int_0^t v(s)ds$, where δ is Cartan's development [2].

Let us consider the Banach space $C^0(I, T_{m_0}M)$ of continuous mappings of I into $T_{m_0}M$ and the Banach manifold $C^1(I, M)$ of C^1-smooth mappings of I into M. By virtue of Theorem 1.1, the operator $S: C^0(I, T_{m_0}M) \to C^1(I, M)$ is defined correctly. If M is a Euclidean space, Sv is an integral with a variable upper limit.

Apparently, S is a homeomorphism of $C^0(I, T_{m_0}M)$ onto its image $C^1_{m_0}(I, M)$ in $C^1(I, M)$, where the manifold $C^1_{m_0}(I, M)$ consists of C^1-curves γ such that $\gamma(0) = m_0$.

Among the other properties of S we shall need the following.

Theorem 1.2. Let U_k be a ball of radius k with centre at the zero of the space $C^0(I, T_{m_0}M)$. Then for curves $\gamma(t)$ belonging to the set SU_k the inequality $\|\dot{\gamma}(t)\| \leqslant k$ holds at each point $t \in I$.

This statement is obvious, since parallel translation conserves the norm of a vector.

Theorem 1.3. Let a point $m_1 \in M$ be not conjugate with m_0 along some geodesic of the Levi-Civita connection on M. Then for any geodesic $a(t)$ $(a(0) = m_0,\ a(1) = m_1)$, along which m_0 and m_1 are not conjugate, and for any number $k > 0$ there exists a number $L(m_0, m_1, k, a) > 0$ such that for $0 < t_1 < L(m_0, m_1, k, a)$ and for any curve $w(t) \in$

$\in U_k \subset C^0([0, t_1], T_{m_0}M)$ there exists, in a certain bounded neighbourhood of the vector $t_1^{-1}a(0) \in T_{m_0}M$, a unique vector $C_w \in T_{m_0}M$ continuously dependent on w, such that $S(w + C_w)(t_1) = m_1$.

Sketch of the proof. Apparently, for each $v(t) \equiv C \in T_{m_0}M$ we have $Sv(1) = \exp C$. Thus, according to the condition of the theorem, $S(\mathring{a}(0))(1) = m_1$ and S is the diffeomorphism of the neighbourhood of $\mathring{a}(0)$ in $T_{m_0}M$ onto the neighbourhood of point m_1 in M. Using the implicit function theorem, one can easily demonstrate that for some $\varepsilon > 0$ and any curve $\hat{w} \in U_\varepsilon \subset C^0([0, 1], T_{m_0}M)$ there exists, in a bounded neighbourhood of $\dot{a}(0)$ in $T_{m_0}M$, a unique vector $C_{\hat{w}}$ smoothly dependent on $\hat{w}$ and such that $S(\hat{w} + C_{\hat{w}})(1) = m_1$. Next, let t_1 be such that $t_1^{-1}\varepsilon > k$. For $w \in U_k \subset C^0([0, t_1], T_{m_0}M)$ we define $C_w \in$ $\in T_{m_0}M$ by the formula $C_w = t_1^{-1}C_{\hat{w}}$, where $\hat{w} \in U_\varepsilon \subset C^0([0, 1], T_{m_0}M)$, $\hat{w}(t) = t_1 \cdot w(t_1 \cdot t)$, $t \in [0, 1]$. Clearly, $S(\hat{w} + C_{\hat{w}})(t) = S(w + C_w)(t_1 \cdot t)$, i.e. $S(w + C_w)(t_1) = m_1$. As $L(m_0, m_1, k, a)$, we take the supremum of these t_1.

For a detailed proof of the theorem we refer the reader to [15].

It should be noted that in the case of a Euclidean M for ε in Theorem 1.3 we can take any number, that is the theorem holds for any t_1, $0 < t_1 < \infty$.

2. Let $\gamma(t)$ $(t \in I)$ be a C^1-curve in M and let $X(\gamma(t))$ be a continuous vector field along γ. Let $\Gamma X(\gamma(t))$ denote a curve in $T_{\gamma(0)}M$ obtained by parallel translation of the vectors $X(\gamma(t))$ along γ to the point $\gamma(0)$.

Lemma 1.1. (Compactness lemma). Let $\Xi \subset C^0(I, TM)$ be such that $\pi\Xi \subset C^1(I, M)$ ($\pi: TM \to M$ is the natural projection). If Ξ is relatively compact in $C^0(I, TM)$, then $\Gamma\Xi$ is relatively compact in $C^0(I, TM)$.

Sketch of the proof. Relative compactness of Ξ suggests relative compactness of $\pi\Xi$ in $C^0(I, M)$. The limiting point of $\pi\Xi$ need not necessarily be a C^1-curve; it can, however, easily be shown that this curve satisfies the Lipschitz condition (i.e. it is a Hölder curve of index 1). It is shown in [4] that a parallel translation coinciding with the limit of parallel translations is correctly defined along this curve. Thus, Γ transforms a convergent sequence into a convergent sequence.

Complete proof of the lemma can be found in [15] .

If $X(\gamma(t)) = \xi(t, \gamma(t))$, i.e. it is a restriction of a continuous vector field $\xi(t, m)$, $t \in I$, $m \in M$, onto γ, we shall write $\Gamma\gamma$ instead of $\Gamma\xi(t, \gamma(t))$. Thus, if the field $\xi(t, m)$ is fixed, we may consider the operator $\Gamma : C^1(I, M) \rightarrow C^0(I, TM)$, which is apparently continuous.

Let Ω_k be a set of curves in $C^1(I, M)$ such that at each point $t \in I$ the inequality $\|\dot{\gamma}(t)\| \leqslant k$ holds, where $k > 0$ is a real number, and such that the point set $\{\gamma(0) \mid \gamma \in \Omega_k\}$ is bounded in M.

<u>Theorem 1.4</u>. The set of curves $\Gamma\Omega_k$ is compact in $C^0(I, TM)$.

<u>Proof</u>. Clearly, Ω_k is compact in $C^0(I, TM)$. The continuity of the field $\xi(t, m)$ implies that the set of curves $\xi(t, \gamma(t))$, $\gamma \in \Omega_k$, is compact in $C^0(I, TM)$, and the statement follows from the compactness lemma.

<u>Corollary</u>. The operator Γ is locally compact.

<u>Proof</u>. For any $\gamma \in C^1(I, M)$ the continuous function $\|\dot{\gamma}(t)\|$ reaches its supremum K_γ on I. By the definition of a C^1-topology, for any γ' in a small C^1-neighbourhood of γ the inequality $\|\dot{\gamma}(t)\| < K + \varepsilon$ is valid.

3. Let us consider a continuous operator (superposition) $S \circ \Gamma$: $C^1_{m_0}(I, M) \rightarrow C^1_{m_0}(I, M)$.

<u>Theorem 1.5.</u> The curve γ is a fixed point of $S \circ \Gamma$, if and only if it is an integral curve of the field $\xi(t, m)$ with the initial condition $\gamma(0) = m_0$.

<u>Proof</u>. Let $\gamma(t)$ be an integral curve of the field $\xi(t, m)$, i.e. $\dot{\gamma}(t) = \xi(t, \gamma(t))$. It can easily be seen that in this case the operator Γ on γ is equal to S^{-1} and γ is a fixed point of $S \circ \Gamma$. Conversely, let γ be a fixed point of the operator $S \circ \Gamma$. By virtue of the construction of S and Γ, this means that if at each point $t \in I$ we take the vector $\xi(t, \ (t))$ and move it parallel to itself along γ to the point $\gamma(0) = m_0$ and then return this vector along the same curve to the point $\gamma(t)$, we obtain the vector $\dot{\gamma}(t)$. Hence, $\dot{\gamma}(t) = \xi(t, \gamma(t))$, $t \in I$. The theorem is proved.

Thus, $S \circ \Gamma$ is a direct analogue of Volterra's integral operator in the theory of ordinary differential equations in a Euclidean space. For some constructions with the operator $S \circ \Gamma$, see the survey [6] and references therein.

Let us consider a superposition $\Gamma \circ S$. It operates in the Banach space $C^0(I, T_{m_0}M)$ and is continuous. If $v = \Gamma \circ Sv$, then $Sv = S \circ \Gamma \circ Sv = S \circ \Gamma(Sv)$ is an integral curve of the field $\xi(t, m)$. Conversely, the equality $Sv = S \circ \Gamma(Sv)$ implies that $v = \Gamma \circ Sv$, since S is a one-to-one mapping.

Theorem 1.6. The operator $\Gamma \circ S$ is completely continuous.

Proof. Let U_k be a ball of radius k in $C^0(I, T_{m_0}M)$. Then by virtue of Theorem 1.2. $SU_k \subset \Omega_k$ and by Theorem 1.4 the set $\Gamma \circ SU_k$ is compact. The theorem is proved.

§ 2. Integral equations of geometric mechanics

We say that a mechanical system is defined if the following objects are given: a configuration space — a smooth finite-dimensional manifold M; kinetic energy $\mathbb{T}$ — a function on TM which is generated by the Riemannian metric $\langle \ , \ \rangle$ according to the formula $\mathbb{T}(X) = \frac{1}{2}\langle X, X\rangle$, $X \in TM$; a force field — a 1-form $\alpha(t, m, X)$ on M dependent, in general, on time t, the point $m \in M$, and the velocity vector $X \in T_mM$ (in other words, $\alpha(t, m, X)$ is a linear functional on the tangent space T_mM, which depends on t and $X \in T_mM$). In this section α is assumed continuous.

The concept of a force field can equivalently be formulated in terms of horizontal 1-forms on the phase space of the system — the tangent bundle TM [8,21].

Using the Riemannian metric $\langle \ , \ \rangle$, one can pass from the force 1-form $\alpha(t, m, X)$ to the corresponding vector field $A(t, m, X)$ by the relation $\langle A(t, m, X), Y\rangle = \alpha(t, m, X)(Y)$, which holds for each vector $Y \in T_mM$.

Remark. In sections to follow we shall consider a force field as a vector (many-valued, random, etc.) field; the corresponding 1-form can easily be obtained by the reader.

The trajectory $\gamma(t)$ of a mechanical system is described by a second-order differential equation, which is a geometrically invariant form of Newton's second law

$$\frac{D}{dt}\dot{\gamma}(t) = A(t, \gamma(t), \dot{\gamma}(t)) \tag{2.1}$$

Here $\frac{D}{dt}$ is the covariant derivative of the Levi-Civita connection of the metric $\langle \ , \ \rangle$ (the relevant notions of differential geometry can be found in [2,22]).

Remark. It should be noted that $\gamma(t)$ is a solution of (2.1) if and only if its derivative, i.e. the curve $(\gamma(t), \dot{\gamma}(t))$ in TM, is an integral curve of the special vector field (a second-order equation [24]) on TM

$$\xi_h(m, X) + \xi_v(t, m, X) \tag{2.2}$$

where $\xi_h(m, X) \in T_{(m,X)}TM$ is the geodesic pulverization of the Levi-Civita connection of the metric $\langle \ , \ \rangle$, $\xi_v(t, m, X)$ is the natural vertical lift of $A(t, m, X)$ to the point (m, X). Clearly, any second-order differential equation on TM can be represented in the form of (2.2), i.e. it can be written in an equivalent form (2.1).

Of interest in mechanics is the case where the initial-value problem for (2.1) has a unique solution. In this section we shall consider the general case of a continuous field A without assuming that the solution is unique.

In what follows we shall consider such mechanical systems that the trajectories of their inertial motion (in the absence of a force field) do not go to infinity for a finite time. In terms of Riemannian geometry, this means that the metric $\langle \ , \ \rangle$ is complete. In this case, the operator S is defined correctly (see Section 1), and Eq. (2.1) can be written in an equivalent integral form.

Theorem 2.1. Let $t \in I = [0, T]$, then the solution of Eq. (2.1) with the initial condition $\gamma(0) = m_0$, $\dot{\gamma}(0) = C \in T_{m_0}M$ (and only this solution) satisfies the integral equation

$$\gamma(t) = S\left(\int_0^t \Gamma A(\tau, \gamma(\tau), \dot{\gamma}(\tau))d\tau + C\right) \stackrel{def}{=} S\left(\int_0^t \Gamma \circ A\gamma \, d\tau + C\right) \tag{2.3}$$

where $\Gamma A(t, \gamma(t), \dot{\gamma}(t)) = \Gamma \circ A\gamma$ is a curve in $T_{m_0}M$ obtained by parallel translation of vectors of the field $A(t, \gamma(t), \dot{\gamma}(t))$ along

γ (see Section 1).

Proof. It is a simple matter to demonstrate that for $v \in C^0(I, T_{m_0}M)$ only the C^2-curve $\gamma(t) = S(\int_0^t v(\tau)d\tau + C)$ satisfies the conditions: $\gamma(0) = m_0$, $\dot{\gamma}(0) = C$, and the vector $\frac{D}{dt}\dot{\gamma}(t)$ at each point $t \in I$ is parallel along γ to the vector $v(t)$. Indeed, parallel translation of the vectors $\dot{\gamma}$ to some point $\gamma(t)$ produces a curve $\bar{\gamma}$ in the tangent space $T_{\gamma(t)}M$ such that by the definition of the covariant derivative [29] we have $\dot{\bar{\gamma}}(t) = \frac{D}{dt}\dot{\gamma}(t)$. Parallel translation of the curve $\bar{\gamma}$ along γ to the point m_0 generates the curve $\int_0^t v(\tau)d\tau + C$, i.e. the vector $\frac{D}{dt}\dot{\gamma}(t)$ is parallel along γ to the vector $v(t)$. Apparently, $\dot{\gamma}(t) = C$.

If by $v(t)$ we mean the curve $\Gamma \circ A\gamma$, equality (2.3) implies that for each $t \in I$ parallel translation of the vector $A(t, \gamma(t), \dot{\gamma}(t))$ along γ first from $\gamma(t)$ to $m_0 = \gamma(0)$ and then again to $\gamma(t)$ gives the vector $\frac{D}{dt}\dot{\gamma}(t)$. This completes the proof of the theorem (cf. proof of Theorem 1.5).

Let γ be a trajectory of a mechanical system, i.e. a solution of (2.1).

Definition 2.1. Following [27], we call the curve $v: I \to T_{\gamma(0)}M$ obtained by parallel translation of the velocity field $\dot{\gamma}(t)$ along γ to the point $\gamma(0)$ by the velocity hodograph of γ.

It can easily be seen that the equation of the hodograph constructed by the solution of (2.3) is of the form

$$v(t) = \int_0^t \Gamma \circ A \circ Sv(\tau)d\tau + C \tag{2.4}$$

where $\Gamma \circ A \circ Sv(\tau) = \Gamma A(\tau, Sv(\tau), \frac{d}{d\tau}Sv(\tau))$. Apparently, if v is a solution of (2.4), then Sv is a solution of (2.3) (i.e. the trajectory of a mechanical system).

Let $\int \Gamma \circ A \circ S_C$ denote the operator that maps $v \in C^0(I, T_{m_0}M)$ into $(\int_0^t \Gamma \circ A \circ Sv(\tau)d\tau + C) \in C^0(I, T_{m_0}M)$.

Theorem 2.2. The operator $\int \Gamma \circ A \circ S_C : C^0(I, T_{m_0}M) \to C^0(I, T_{m_0}M)$ is completely continuous.

Proof. The continuity of the operator follows from the continuity of S, A, and Γ . Let U_k be a ball of radius k with centre at the zero of the space $C^0(I, T_{m_0}M)$. Using Theorem 1.2, the continuity of the field A, and Lemma 1.1, we obtain that $\int \Gamma \circ A \circ S_C U_k$ is a compact (cf. proof of Theorem 1.6).

§ 3. A two-point boundary-value problem. Controllability of mechanical systems

Using integral-type operators with Riemannian parallel translation, we can analyse the solvability of a two-point boundary-value problem for second-order differential equations on a Riemannian manifold.

It is a well-known fact (see, e.g., [31]) that in the case of differential equations with a continuous bounded right-hand side in a vector space the two-point boundary-value problem always has a solution for each pair of points and any time interval. This is not, however, true for a non-linear space. Below we shall consider a mechanical system on a two-dimensional sphere with a smooth, atonomous, and velocity-independent (i.e. bounded) force field for which none of the trajectories emerging from the south pole reaches the north pole [15] . Nevertheless, for the points that are not conjugate along at least one geodesic of the Riemannian metric, which determines the kinetic energy of the system, the two-point boundary-value problem can be solved on some time interval for a continuous bounded force field [14,15] .

The same statement holds true for upper semicontinuous second-order differential inclusions with convex images [10] . Such solutions can be used to describe mechanical systems with control, as well as systems with discontinuous force fields (systems with switching, motion in several media, etc.). Thus, we have proved the controllability of a mechanical system for rather general assumptions in the case where the points of the configuration space are not conjugate along some geodesic.

1. First, we describe the example where the two-point boundary-value problem is unsolvable. Consider a mechanical system on a unit sphere S^2 embedded in R^3 with the force field $A(x, y, z) = (-y, x, 0)$, $(x, y, z) \in S^2$. Using the d'Alembert principle for the holonomic constraint $F(x, y, z) \equiv 1$, $F(x, y, z) = x^2 + y^2 + z^2$ (see, e.g., [32]), we obtain the system of equations in R^3 which describes trajectories

of the system

$$\ddot{x} = -y - 2\mathbb{T}x, \quad \ddot{y} = x - 2\mathbb{T}y, \quad \ddot{z} = 2\mathbb{T}z \tag{3.1}$$

where $\mathbb{T} = \frac{1}{2}(x^2 + y^2 + z^2)$ is kinetic energy.

Let N = (0, 0, 1) and S = (0, 0, -1) denote the north and south poles of the sphere, respectively. Let $\gamma(t) = (x(t), y(t), z(t))$ be the trajectory of a system emerging at a moment t_o from S with some initial velocity $v \neq 0$ (the contrary would mean $\gamma(t) \equiv S$). Clearly, $v \in T_S S^2$ is of the form $v = (X, Y, 0)$. Along $\gamma(t)$, the kinetic energy $\mathbb{T}$ increases monotonically unless γ reaches N or S. Indeed, using (3.1), we obtain $\dot{\mathbb{T}}(\gamma(t)) = -xy + yx$ (i.e. $\dot{\mathbb{T}}(\gamma(t_o)) = 0$), $\ddot{\mathbb{T}}(\gamma(t)) = x^2 + y^2 > 0$, and therefore $\dot{\mathbb{T}}(\gamma(t)) > 0$ for $\gamma(t) \neq N, S$. This implies that $\gamma(t) \neq N$ for all $t > t_o$, since $\mathbb{T}$ monotonically increases along γ from $\mathbb{T}(S) = 0$, but $\mathbb{T}(N) = 0$.

Consider the behaviour of the coordinate $z(t)$ of the trajectory $\gamma(t)$. Let t_1 be some other moment (after t_o) such that $\dot{z}(t_1) = 0$ (i.e. $z(t)$ increases on $[t_o, t_1]$). The last equation of system (3.1) shows that $z(t_1) > 0$ (consequently, $\ddot{z}(t_1) < 0$, i.e. $z(t_1)$ is a maximum of $z(t)$) and $z(t_1) < 1$, since $\mathbb{T}$ increases along $\gamma(t)$. Similarly, all the points $z(t_i)$ at which $\dot{z}(t_i) = 0$ satisfy the conditions $\operatorname{sign} z(t_i) = (-1)^{i+1}$ and $|z(t_i)| > |z(t_{i+1})| > 0$.

Thus, the trajectory $\gamma(t)$ oscillates about the equator of the sphere S^2, approaches it asymptotically, and never reaches the point N = = (0, 0, 1).

2\. Let $t \in R$, $m \in M$, $X \in T_m M$, where M is a Riemannian manifold with the metric $\langle\ ,\ \rangle$. Let also $\mathbb{A}(t, m, X)$ be a certain bounded convex set in $T_m M$. The correspondence $(t, m, X) \longrightarrow \mathbb{A}(t, m, X)$ defines a many-valued mapping $R \times TM \longrightarrow TM$ with convex images and the obvious property $\pi \mathbb{A}(t, m, X) = m$ ($\pi : TM \to M$ is the natural projection). Below we shall assume that $\mathbb{A}$ is upper-semicontinuous with respect to all the variables (for the fundamentals of the theory of many-valued mappings see, e.g., [5,9]).

Let us consider on M the second-order differential inclusion

$$\frac{D}{dt}\dot{\gamma}(t) \in \mathbb{A}(t, \gamma(t), \dot{\gamma}(t)) \tag{3.2}$$

Such inclusions describe mechanical systems with complicated force fields (the case of a flat configuration space is considered in [9]).

For example, $\mathbb{A}$ (t, m, X) may be the set of possible values of the control force at moment t at the point m of the configuration space for the velocity X. If the locally bounded force field A(t, m, X) of a mechanical system is discontinuous (say, for a system with switchings, for a system with dry friction, and for motion in several different media), one can pass from A(t, m, X) to a many-valued vector field $\mathbb{A}$(t, m, X). To this end, we put $\mathbb{A}$ (t, m, X) = = $\overline{co}$ Q(t, m, X), where Q(t, m, X) is the set of all limit points of sequences $A(t_k,m_k,X_k)$ for any sequence $(t_k, m_k, X_k) \to (t, m, X)$, $(t_k, m_k, X_k) \neq (t, m, X)$. It is shown in [10] that the many-valued vector field $\mathbb{A}$ thus constructed is upper semicontinuous with respect to all the variables and $\mathbb{A}$ = A if A is continuous.

A C^1-curve $\gamma(t)$ such that its derivative $\dot{\gamma}(t)$ is absolutely continuous and inclusion (3.2) holds for $\gamma(t)$ almost everywhere is called a solution of inclusion (3.2).

It should be noted that just like the transition from (2.1) to (2.2), one can pass from inclusion (3.2) to a first-order inclusion on TM. For upper semicontinuous first-order differential inclusions with convex images, the local solution existence theorem for the initial-value problem is valid in a finite-dimensional linear space (see, e.g., [9]). Thus, local solutions do exist for inclusion (3.2). From the point of view of mechanics, of interest is the case where the solution of a differential inclusion is locally unique. Some uniqueness theorems have been proved in [30] . In what follows we shall not assume that the solution is locally unique.

3. We now formulate and prove the main statement of this section.

Let M be a compact Riemannian manifold.

<u>Theorem 3.1</u>. Let a point $m_1 \in M$ be not conjugate with the point $m_0 \in M$ along some geodesic a(t) of the metric $\langle\ ,\ \rangle$ and let the field $\mathbb{A}$ (t, m, X) be uniformly bounded for all t, m, X. There exists a number $L(m_0, m_1, a)$ such that for any t_0, $0 < t_0 < L(m_0, m_1, a)$, inclusion (3.2) has a solution $\gamma(t)$ such that $\gamma(0) = m_0$ and $\gamma(t_0) = m_1$.

<u>Proof</u>. To study the global behaviour of the solutions of inclusion (3.2), we define an integral-type operator based on the constructions described in Sections 1 and 2.

Let $I = [0, T]$. Consider a many-valued vector field $\mathfrak{A}(t, \gamma(t), \dot{\gamma}(t))$ defined along the C^1-curve $\gamma(t) = Sw(t)$, $w \in C^0(I, T_{m_0}M)$, and carry out parallel translation of all the sets $\mathfrak{A}$ along γ to the point $m_0 = \gamma(0)$. Then for a fixed w we obtain a many-valued mapping of $\Gamma \circ Sw$ from the segment I to $T_{m_0}M$ with convex images. It is a simple matter to demonstrate, using the properties of parallel translation and upper semicontinuity of the field $\mathfrak{A}(t, m, X)$, that the mapping $\Gamma \circ S: C^0(I, T_{m_0}M) \times I \to T_{m_0}M$ is upper semicontinuous. Let us consider the set of all measurable sections $\mathcal{P}\Gamma \circ Sw$ of the many-valued mapping $\Gamma \circ Sw: I \to T_{m_0}M$ (according to [7], such sections do exist). Since the field $\mathfrak{A}$ is bounded, all the curves from $\mathcal{P}\Gamma \circ Sw$ are also bounded, i.e. integrable. Let us define a many-valued mapping $\int \mathcal{P}\Gamma \circ S$ with convex images in the Banach space $C^0(I, T_{m_0}M)$ by the formula

$$\int \mathcal{P}\Gamma \circ Sw = \left\{ \int_0^t v(\tau)d\tau \,\middle|\, v \in \mathcal{P}\Gamma \circ Sw \right\}$$

Lemma 3.1. The mapping $\int \mathcal{P}\Gamma \circ S$ is upper semicontinuous and maps bounded sets of the space $C^0(I, T_{m_0}M)$ into compact sets.

The proof of the lemma can be found in [10].

Let the many-valued vector field $\mathfrak{A}$ be bounded by a number $k > 0$. Clearly, for sufficiently small $t_1 > 0$ there holds the inequality $t_1 < L(m_0, m_1, kt_1, a)$, where $L(m_0, m_1, kt_1, a)$ is the number appearing in Theorem 1.3. We define the number $L(m_0, m_1, a)$ as the supremum of t_1 such that $t_1 < L(m_0, m_1, kt_1, a)$.

Let $t_0 < L(m_0, m_1, a)$. Without loss of generality, one can assume that the operator $\int \mathcal{P}\Gamma \circ S$ acts in $C^0([0, t_0], T_{m_0}M)$. Let us consider, on the ball $U_{kt_0} \subset C^0([0, t_0], T_{m_0}M)$, a many-valued upper semicontinuous compact mapping $\Xi w = \int \mathcal{P}\Gamma \circ S(w + C_w)$, where C_w is the vector from Theorem 1.3. Since parallel translation conserves the norm of a vector, one can easily see that Ξ maps U_{kt} into itself and, therefore, has a fixed point w_0 in U_{kt}, i.e. $w_0 \in \Xi w_0$ [5]. We now demonstrate that $\gamma = S(w_0 + C_{w_0})$ is the solution in question for (3.2). By construction, $\gamma(0) = m_0$, $\gamma(t_0) = m_1$, γ is a C^1-curve, $\dot{\gamma}$ is absolutely continuous. Since w_0 is a fixed point of Ξ, then $\dot{w}_0$ is a section of $\Gamma \circ S(w_0 + C_{w_0})$, i.e. the inclusion $w_0(t) \in \Gamma \circ S(w_0 + C_{w_0})(t)$ is valid at those points t at which $\dot{w}_0$ does exist. By construction and using the properties of the covariant derivative (cf. proof of Theorem 2.1), after parallel translation of $\dot{w}(t)$ and $\Gamma \circ S(w_0 + C_{w_0})(t)$ along γ to the point $\gamma(t)$, we obtain $\frac{D}{dt}\dot{\gamma}(t)$ and $\mathbb{A}(t, \gamma(t), \dot{\gamma}(t))$, respectively. Thus, $\frac{D}{dt}\dot{\gamma}(t) \in \mathbb{A}(t, \gamma(t), \dot{\gamma}(t))$, which proves the theorem.

We note that if the points m_0 and m_1 are not conjugate along several geodesics, any of these points can be used to prove the existence of a solution. Naturally, the numbers L and the solutions constructed by different geodesics will, in general, be distinct.

If the configuration space M (with the metric $\langle\ ,\ \rangle$) is a compact manifold of non-positive curvature, it follows from Theorem 3.1 and from the condition that the manifold is compact and conjugate points are absent, that there exists a number L such that for arbitrary points m_0 and m_1 and for any t_0, $0 < t_0 < L$, the two-point boundary-value problem has a solution.

If M is a flat manifold, the number L is equal to ∞ (see Theorem 1.3), and we obtain the classical result about the solvability of a two-point boundary-value problem for arbitrary points and any time interval.

Remark. By construction, $\frac{D}{dt}\dot{\gamma}(t)$ is a measurable vector field along the solution obtained. Thus, if $\mathbb{A}(t, m, X)$ is the set of possible values of the control force, Theorem 3.1 implies that there exists a measurable control that sends m_0 into m_1.

§ 4. Mechanical systems with random force fields

The purpose of this section is to construct and analyse stochastic differential equations that describe mechanical systems with random force fields (the classical "flat" case can be found, e.g., in [3]). Unlike the well-known methods (see, e.g., [25]), in this case the stochastic equation on a Riemannian manifold can be written in the natural, geometric-invariant form in terms of the operators S and Γ.

1. Let, as previously, M be a complete Riemannian manifold. Besides the continuous vector field A(t, m, X) (see Section 2), we consider

on M the field of linear operators $B(t,m,X)$ which is continuous with respect to all the variables (for $t \in I = [0, T]$, $m \in M$, and $X \in T_m M$, the operator $B(t,m,X)$ belongs to $L(T_m M)$, i.e. to the space of all linear mappings in $T_m M$). For $\gamma \in C^I(I,M)$, we define a curve $\Gamma B(t, \gamma(t)), \dot{\gamma}(t) = \Gamma \circ B\gamma \in L(T_{\gamma(0)}M)$ in analogy with the curve $\Gamma \circ A\gamma$ (see Section I and 2). Also, we put $\Gamma \circ B \circ Sv = \Gamma B(t, Sv(t), \frac{d}{dt} Sv(t))$. It can easily be seen that the mappings $\Gamma \circ B : C^I_{\gamma(0)}(I,M) \longrightarrow C^0(I, L(T_{\gamma(0)}M)$ and $\Gamma \circ B \circ S : C^0(I, T_{\gamma(0)}M) \longrightarrow C^0(I, L(T_{\gamma(0)}M))$ are continuous.

Let (Ω, F, P) be a probability space and F_t a non-decreasing sequence of σ-subalgebras of the σ-algebra F, i.e. $F_s \subset F_t \subset F$ for $0 \leqslant s \leqslant t \leqslant T$ (see [I2]).

Lemma 4.I. Let $v : I \times \Omega \longrightarrow T_{m_0}M$ be such that:

I) $P\{\omega \mid v(t, \omega) \in C^0(I, T_m M)\} = I$;

2) for each fixed $t \in I$, $v(t, \omega)$ is measurable with respect to F_t. Then for each fixed $t \in I$ the mappings $Sv(t, \cdot) : \Omega \longrightarrow M$, $\Gamma \circ A \circ Sv(t, \cdot) : \Omega \longrightarrow T_{m_0}M$, and $\Gamma \circ B \circ Sv(t, \cdot) : \Omega \longrightarrow L(T_{m_0}M)$ are measurable with respect to F_t. (The proof of this lemma can be found in [I9,20]).

We note that $P\{\omega \mid Sv(t, \omega) \in C^I_{m_0}(I,M)\} = I$ for v satisfying the conditions of Lemma 4.I.

Let us introduce notations: CL_2 is the set of mappings $v : I \times \Omega \longrightarrow T_{m_0}M$ satisfying the conditions of Lemma 4.I, and belonging to the space $C^0(I, L_2(\Omega, T_{m_0}M)$ of continuous images I in the space $L_2(\Omega, T_{m_0}M)$ of mappings which are aquare-integrable in $T_{m_0}M$; SCL_2 is the image of CL_2 in $C^I_{m_0}(I,M)$ under the mapping S.

Let the fields A and B linearly bounded at infinity, i.e. the following inequality holds:

$$\|A(t,m,X)\| + \|B(t,m,X)\| \leqslant K(1+\|X\|), \ K>0,$$

where the norms in the respective spaces are defined by the Riemannian metric. We say that A and B are natural if they are continuous with respect to all the variables and linearly bounded at infinity.

Consider a Wiener's process b in $T_{m_0}M$ subordinate to the flow of σ-algebras F_t (for the main definitions of the theory of stochastic processes see, e.g., [I2]).

Lemma 4.2. If $\gamma(t,\omega) \in SCL_2$ and A and B are natural, then
1) the Itô stochastic integral (see [11,12])

$$\int_0^t \Gamma B(\tau, \gamma(\tau,\omega), \frac{d}{d\tau}\gamma(\tau,\omega))db(\tau) \stackrel{def}{=} \int_0^t \Gamma\circ B\gamma db$$

is correctly defined;

2) the family of curves $\int_0^t \Gamma\circ A\gamma d\tau + \int_0^t \Gamma\circ B\gamma db$ belongs to CL_2.

2. Let in each tangent space T_mM there be defined a Wiener's process b subordinate to the flow of σ-algebras F_t. We consider the motion of a mechanical system under the action of a force field A(t, m, X) + + B(t, m, X)$\dot{b}$, where $\dot{b}$ is the Gaussian white noise and the linear operator B(t, m, X) describes the dependence of stochastic force perturbation on time, configuration, and velocity. The trajectory of this mechanical system with the initial condition $\gamma(0,\omega) \equiv m_0$, $\frac{d}{dt}\gamma(0,\omega) = v(\omega) \in L_2(\Omega, T_{m_0}M)$ is described by the stochastic equation (a generalization of the classical Langevin equation)

$$\gamma(t,\omega) = S\left(\int_0^t \Gamma\circ A\gamma d\tau + \int_0^t \Gamma\circ B\gamma db + v\right). \tag{4.1}$$

Remark. A Wiener's process subordinate to a flow of σ-algebras F_t is, in fact, described by the expectation and variance, so that in the derivation of Eq. (4.1) the dependence of the white noise on the point in the configuration space can be neglected.

It follows from Lemma 4.2 that the right-hand side of Eq. (4.1) is correctly defined for $\gamma(t,\omega) \in SCL_2$ and natural A and B.

If the motion of a mechanical system is described by Eq. (4.1), the velocity hodograph obeys the equation

$$v(t,\omega) = \int_0^t \Gamma\circ A\circ Sv d\tau + \int_0^t \Gamma\circ B\circ Sv db + v(\omega) \tag{4.2}$$

Theorem 4.1. If $v(t,\omega) \in CL_2$ and satisfies Eq. (4.2), then $Sv(t,\omega) \in SCL_2$ satisfies Eq. (4.1).

3. Let $\tilde{F}$ denote a σ-algebra of cylindrical sets on $\tilde{\Omega} = C^0(I, T_{m_0}M)$. Using the homeomorphism of S, we can define the corresponding σ-algebra on $\tilde{\tilde{\Omega}} = C^1_{m_0}(I, M)$.

Definition 4.1. Equation (4.1) is said to have a solution if there exist a stochastic process $\gamma(t,\omega)$ with sample paths almost surely in $C^1_{m_0}(I, M)$, which defines a measure $\widetilde{P}$ on $(\widetilde{\Omega}, \widetilde{F})$, and a Wiener's process on $(\widetilde{\Omega}, \widetilde{F}, \widetilde{P})$ in $T_{m_0}M$ such that Eq. (4.1) is satisfied and $\gamma(t,\omega) \in SCL_2$.

We note that the solution satisfying Definition (4.1) is weak in the sense of terminology of the theory of stochastic equations (see [11]).

Following [11] , we say that the solution of Eq. (4.1) is weakly unique if for any two solutions of Eq. (4.1) the corresponding measures on $(\widetilde{\Omega}, \widetilde{F})$ coincide.

Let us consider the case of a deterministic initial velocity $v(\omega) = C \in T_{m_0}M$.

Theorem 4.2. If A and B are natural, Eq. (4.1) has a solution.

Sketch of the proof. Consider the hodograph equation (4.2). This is a diffusion-type equation in a vector space. Using the properties of operators S and Γ and the condition that A and B are natural, one can easily demonstrate that the coefficients $\Gamma \circ A \circ S$ and $\Gamma \circ B \circ S$ of of Eq. (4.2) satisfy the weak solution existence theorem for such an equation [11, Ch. III, § 2] . Application of Theorem 4.1 completes the proof.

The next statesments are also proved by considering Eq. (4.2) and using the results of the classical theory.

Theorem 4.3. Let A(t,m,X) and B(t,m,X) be natural and let the operator B(t, m, X) be reversible for all t, m, X. Then the sufficient condition of weak uniqueness of the solution of Eq. (4.1) is that the solution of the equation

$$\gamma(t,\omega) = S\left(\int_0^t \Gamma \circ B_\gamma \, db\right) \tag{4.3}$$

is weakly unique.

Theorem 4.4. Let A(t, m, X) be natural and such that Eq. (2.1) has a unique solution for the initial-value problem. Let $B_\varepsilon(t, m, X)$ be continuous with respect to the variables ε, t, m, X (ε is a numerical parameter), be linearly bounded at infinity by the constant $K > 0$ for all ε, be identically zero at $\varepsilon = 0$, and tend uniformly to 0 on each compact in $I \times TM$ as $\varepsilon \longrightarrow 0$.

Let also the solution of the equation

$$\gamma(t,\omega) = S(\int_0^t \Gamma\circ A\gamma d\tau + \int_0^t \Gamma\circ B_\varepsilon db + C_\varepsilon) \tag{4.4}$$

be weakly unique for all ε and let $C_\varepsilon \longrightarrow C$ as $\varepsilon \longrightarrow 0$. Then for $\varepsilon \longrightarrow 0$ the measures on $C^1_{m_0}(I, M)$ corresponding to the solutions of Eq. (4.4) weakly converge to the measure confined on the unique solution of Eq. (2.1).

Example. For $B = \varepsilon E$ (E is an identity operator), Eq. (4.3) has, apparently, a unique solution. Thus, if $A(t, m, X)$ is natural, the solution of the equation

$$\gamma(t,\omega) = S(\int_0^t \Gamma\circ A\gamma d\tau + \varepsilon b + \varepsilon C_1 + C)$$

does exist and is weakly unique. For instance, if $A(t, m, X)$ is locally Lipschitz with respect to m and X, this equation satisfies Theorem 4.4.

4. Let us now discuss the existence of a strong solution to Eq. (4.1) and strong uniqueness of this solution.

If the coefficients $\Gamma\circ A\circ S$ and $\Gamma\circ B\circ S$ of hodograph equation (4.2) satisfy some Lipschitz condition [11], then for the initial velocity $v(\omega) \in L_2(\omega, T_{m_0}M)$ Eq. (4.1) has a strong solution which is strongly unique (see, e.g., [19]). These coefficients contain, however, the operators S and Γ, which are related to a parallel translation, i.e. the restricting condition is imposed on the entire system, rather than on the force field alone, and it is generally difficult to verify whether Eq. (4.1) has a strong solution. There are, nevertheless, mechanical systems for which strong solutions do exist.

(i) Reaction of the medium. $A(t, m, X) = \varphi(t, \|X\|)A'_m(X)$, $B(t, m, X) = \psi(t, \|X\|)B'_m(X)$, where $\varphi(t, \|X\|)$ and $\psi(t, \|X\|)$ are scalar functions; A' is the field of orthogonal automorphisms of tangent spaces which commutes with parallel translation along any curve; B' is the field of mappings of tangent spaces into their linear groups which also commutes with parallel translation. For instance, A' and B' can be equal to $\pm E$. In particular, the Ornstein-Uhlenbeck processes are described in this form.

(ii) Forces defined in a "fixed" coordinate system. The mappings $A(t): T_{m_0}M \longrightarrow T_{m_0}M$ and $B(t): T_{m_0}M \longrightarrow L(T_{m_0}M)$ are defined at point m_0;

the corresponding mappings at the point $Sv(t)$ are obtained by parallel translation of A and B along Sv to this point.

Mechanical meaning of parallel translation is discussed in Section 5.

Theorem 4.5. If in type (i) and type (ii) systems the coefficients A and B are natural and Lipschitz in $T_{m_0}M$ with respect to X, the corresponding equations (4.1) have a strong solution which is strongly unique.

Proof. In the cases considered above, Eq. (4.2) in the vector space $T_{m_0}M$ is of the form

$$v(t,\omega) = \int_0^t A(\tau, v(\tau,\omega))d\tau + \int B(\tau, v(\tau,\omega))db(\tau) + V(\omega)$$

According to the condition, this equation has a strong solution which is strongly unique [11] . Using Theorem 4.1, we complete the proof of the theorem.

Definition 4.2. The expectation of the process $\gamma(t,\omega) \in SCL_2$ is said to be the curve $S(Mv(t,\omega) \in C^1_{m_0}(I, M)$, where $Mv(t,\omega)$ is the expectation of the corresponding hodograph of the velocity $v(t,\omega)$. Similarly, the variance $\mathcal{D}(t,\gamma)$ of this process is said to be the scalar function $M(Mv(t,\omega) - v(t,\omega))^2$.

It should be noted that for a system involving reaction forces the expectation of the solution of stochastic equation (4.1) is equal to the solution of deterministic equation (2.1).

Theorem 4.6. Let Theorem 4.5 be valid. Then for $\varepsilon \to 0$ the solution of the stochastic equation

$$\gamma(t,\omega) = S(\int_0^t \Gamma \circ A_\gamma d\tau + \varepsilon \int_0^t \Gamma \circ B_\gamma db + \varepsilon V(\omega) + C) \tag{4.5}$$

tends, in the topology SCL_2, to the solution of deterministic equation (2.3), which is equivalent to (2.1); the expectation of the solution of Eq. (4.5) tends uniformly to the solution of Eq. (2.1).

That the solution of Eq. (4.5) tends to the solution of Eq. (2.1) in the topology SCL_2 means that the variance of the solution of (4.5) uniformly tends to zero.

§ 5. Mechanical systems with Riemannian parallel translation

In certain problems, where the global parallelization of the tangent bundle is absent, this parallelization can be replaced by Riemannian parallelism along a curve [16-18] . Such a supplantation is correct because the vector of pendulum oscillation direction (or the gyro-fixed frame in the tangent space) remains parallel along the trajectory if the motion takes place in a non-flat manifold. This mechanical interpretation of Riemannian parallel translation belongs to Radon (see [23]). Using a similar idea, Synge [27] introduced the hodograph concept for a generalized mechanical system.

Let us consider the motion of a mechanical system in a force field A under the action of a "control" force Φ which depends on time, configuration, and velocity; however, because of a delay typical of real situation, this force becomes effective only after some time interval h . In this case, the equation of motion takes the form

$$\frac{D}{dt}\dot{\gamma}(t) = A(t, \gamma(t), \dot{\gamma}(t)) + \| \Phi(t-h, \gamma(t-h), \dot{\gamma}(t-h)), \tag{5.1}$$

where the symbol $\|$ denotes Riemannian parallelism along the solution of this equation.

If we consider the motion of a system in a velocity field V with a delayed "control" velocity W, we obtain the equation

$$\dot{\gamma}(t) = V(t, \gamma(t)) + \| W(t-h, \gamma(t-h)). \tag{5.2}$$

If M is a Euclidean space, Eqs. (5.1) and (5.2) are rather simple differential equations with discrete constant delay [26] . On a curved manifold, however, these equations become much more complicated.

First, since parallel translation is, generally, defined along C^1-smooth curves and depends on the curve and its derivative, Eq. (5.2) is a neutral-type equation [26] .

Second, when Eq. (5.1) is reduced to the corresponding first-order equation on the tangent bundle TM, the equation thus obtained will be an equation with distributed delay. The reason for this fact is that in a curved manifold parallel translation of vertical vectors on TM in the standard Riemannian metric (induced by the metric on M) is not a lift of parallel translation on M.

Third, the first-order equation with distributed delay on TM, which corresponds to Eq. (5.1), is a neutral-type equation, since its right-hand side is not defined on an arbitrary continuous curve in TM and is not continuous in the C^0-topology.

Of special interest (on a curved manifold) are equations describing mechanical systems in which the control force or velocity depends only on time, i.e. the time behaviour of the control force or velocity is "programmed" in a "fixed" coordinate system

$$\frac{D}{dt}\dot{\gamma}(t) = A(t, \gamma(t), \dot{\gamma}(t)) + \| \Phi(t) \tag{5.1'}$$

$$\dot{\gamma}(t) = V(t, \gamma(t)) + \| W(t) \tag{5.2'}$$

where Φ and W are vector functions in the tangent space to M in the initial motion configuration m_0.

Importantly, Eqs. (5.1) and (5.2) can be reduced to Eqs. (5.1') and (5.2'), respectively. Indeed, let $\varphi \in C^1([-h, 0], M)$ be the initial condition for (5.1) and (5.2). We define $\Phi(\theta)$ and $W(\theta)$ on the segment $[-h, 0]$ in the tangent space $T_{\varphi(0)}M$ by parallel translation of the vectors Φ and W along φ to the point $m_0 = \varphi(0)$. Using the new time variable $t = h + \theta$, we obtain equations of the type of (5.1) and (5.2) whose solutions coincide with the solutions of the initial equations for $t < h$.

All these equations are correct, provided their solutions are C^1-smooth. Otherwise, one cannot speak about parallel translation along a solution.

<u>Theorem 5.1</u>. Let V(t, m) be a continuous vector field on M and W(t) a continuous curve in $T_{m_0}M$. Then,

1) equation (5.2') has a local C^1-smooth solution;
2) if V(t, m) is locally Lipschitz with respect to m for each t, this solution is unique.

<u>Sketch of the proof</u>. Let us fix an orthogonal basis $\mathcal{O}$ in $T_{m_0}M$ and consider on the bundle O(M) (see Section 1) a horizontal vector field $d\pi^{-1}V + E(\mathcal{O}^{-1}W(t))$, where $\pi : O(M) \to M$ is the natural projection, E is the basis vector field. The integral curve of this horizontal vector field, which originates in $\mathcal{O}$, is, obviously, projected into the solution in question (cf. proof of Theorem 1.1). To prove the uniqueness of the solution, suffice it to observe that $E(\mathcal{O}^{-1}W(t))$ is

smooth for each t.

Theorem 5.2. Let A satisfy the Carathéodory condition (see, e.g., [7]) and let $\Phi(t)$ be a measurable function in $T_{m_0}M$. Then for each initial condition $\dot{\gamma}(0) = C \in T_{m_0}M$:

1) equation (5.1') has a local C^1-smooth solution;

2) if A is locally Lipschitz for each fixed t, this solution is unique.

Sketch of the proof. Consider $O(M) \times R^n$ ($n = \dim M$). Let $\lambda : O(M) \times R^n \longrightarrow TM$ be the natural projection [2]. We note that $d\lambda : T(O(M) \times R^n) \rightarrow TTM$ maps isomorphically the connection of O(M) onto the connection of TM, and also maps TR^n onto vertical subspaces in TTM. Let a vector field on $O(M) \times R^n$ be defined as the sum of three vector fields: (a) $d\lambda^{-1} Z$ in the connection on O(M), where Z is a geodesic pulverization on TM; (b) $d\lambda^{-1} A^{\ell}$ on TR^n, where A^{ℓ} is the vertical lift of the field A in TM (the lift of the vector A(t, m, X) is calculated at the point $(m, X) \in TM$); (c) $d\lambda^{-1}(d\pi\, E(\mathcal{O}^{-1}\Phi(t)))^{\ell}$ on TR^n, where the superscript ℓ denotes, as before, vertical lift, $\mathcal{O}$ is a fixed orthonormal basis in $T_{m_0}M$, $E(\mathcal{O}^{-1}\Phi(t))$ is the basis vector field. The absolutely continuous integral curve of the field thus constructed (the relevant existence theorem can be found in [7]), which originates at $(\mathcal{O}, \mathcal{O}^{-1}C)$, gives, after applying λ and projecting onto M, the C^1-smooth solution in question. To prove that this solution is unique, note only that the vector fields (a) and (c) are, by construction, smooth on $O(M) \times R^n$ for each fixed t.

If M is a complete Riemannian manifold, some existence criteria for global solutions of Eqs. (5.1) and (5.2) can be derived. These are Whitner-type conditions on A and V. If these conditions are satisfied, a new complete Riemannian metric can be constructed on M in which the norm of the velocity vector to the solution is bounded on any finite interval [16,18]. This implies that the solution exists on the whole real line.

It should be noted that all the above statesments remain true if the component Φ in Eq. (5.1) depends on the acceleration $\frac{D}{dt}\dot{\gamma}(t-h)$ and W in Eq. (5.2), on the velocity $\dot{\gamma}(t-h)$. Naturally, in Eq. (5.1) we should require that the initial condition be C^2-smooth [16].

Papers [16,18] deal with the properties of shift operators along the trajectories of Eqs. (5.1) and (5.2). The methods described in [6] are applied to derive existence conditions for fixed points of these operators, i.e. for periodic solutions of equations with periodic right-hand sides.

The equations of mechanical systems (5.1') in which the control force Φ may belong to some set $\mathcal{F} \subset T_{m_0}M$ have been studied in [17]. Clearly, the set of possible values of the control force at points of a trajectory is obtained by parallel translation of $\mathcal{F}$ along the trajectory. If $\mathcal{F}$ contains some neighbourhood of the zero in $T_{m_0}M$, which means that the control force may either vanish or assume arbitrary small values along any direction, one can pass from Eq. (5.1') to a differential inclusion of the type of (3.2).

Using the results of [10] (Theorem 3.1 of the present paper), we have proved the existence of a control force $\Phi(t) \subset \mathcal{F}$ which is measurable with respect to t and sends system (5.1') from point m_0 to a point m_1 which is not conjugate with m_0 along some geodesic. We have also proved that the set of trajectories of the system connecting m_0 and m_1 is compact, which suggests the existence of optimal control.

REFERENCES

1. Arnold V.I. Mathematical methods of classical mechanics. Moscow, 1974 (in Russian).
2. Bishop R.L. and Crittenden R.G. Geometry of manifolds.- Academic Press, N.Y.-London, 1964.
3. Bolotin V.R. Random vibrations of mechanical systems. Moscow, 1979 (in Russian).
4. Borisov Yu.F. Parallel translation along Hölder curves in a Riemannian space.- Dokl. Akad. Nauk SSSR, 1971, 197, No.5 (in Russian).
5. Borisovich Yu.G., Gelman B.D., Myshkis A.D., and Obukhovsky V.V. Topological methods in the theory of fixed points of many-valued mappings.- Usp. Matem. Nauk, 1980, 35, No. 1 (in Russian).
6. Borisovič Ju. and Gliklih Ju.E. Fixed points of mappings of Banach manifolds and some applications.- Nonlinear analysis: Theory, Methods, and Applications, 1980, vol. 4, No. 1.
7. Warga J. Optimal control of differential and functional equations.- Acad. Press, N.Y.-London, 1972.
8. Vershik A.M. and Faddeev L.D. Lagrangian mechanics in invariant exposition.- in: Problems of theoretical physics. Leningrad, 1975, No. 2 (in Russian).
9. Gelig A.Kh., Leonov G.A., and Yakubovich V.A. Stability of nonlinear systems with a non-unique equilibrium state. Moscow, 1978 (in Russian).
10. Gelman B.D. and Gliklikh Yu.E. A two-point boundary-value problem in geometric mechanics with discontinuous forces.- Prikladnaya Matematika i Mekhanika, 1980, 43, No. 3 (in Russian).

11. Gikhman I.I. and Skorokhod A.V. The theory of random processes. Moscow, 1975, vol. 3 (in Russian).

12. Gikhman I.I. and Skorokhod A.V. Introduction to the theory of random processes. Moscow, 1977 (in Russian).

13. Gliklikh Yu.E. Integral operators on a manifold.- Trudy Matem. Fakulteta, Voronezh State Univ., 1971, No. 4 (in Russian).

14. Gliklikh Yu.E. On a certain generalization of the Hopf-theorem about geodesics.- Usp. Matem. Nauk, 1974, 29, No. 6 (in Russian).

15. Gliklikh Yu.E. A two-point boundary-value problem in geometric mechanics of systems with a bounded force field. Deposited at VINITI, June 6, 1977, No. 2217-77 (in Russian).

16. Gliklikh Yu.E. On a certain analogue of differential equations with discrete delay on a Riemannian manifold. Deposited at VINITI, March 20, 1980, No. 1089-80 (in Russian).

17. Gliklikh Yu.E. On systems of geometric mechanics with a delayed control force. Deposited at VINITI, October 14, 1980, No. 4384-80 (in Russian).

18. Gliklikh Yu.E. On equations with discrete delay on non-linear spaces.- in: Operator methods in non-linear analysis. Voronezh, 1982 (in Russian).

19. Gliklikh Yu.E. and Fedorenko I.V. On geometrization of one class of mechanical systems with random force perturbations. Deposited at VINITI, October 21, 1980, No. 4481-80 (in Russian).

20. Gliklikh Yu.E. and Fedorenko I.V. On equations of geometric mechanics with random force fields. in: Approximate methods of analysing differential equations and their applications. Kuibyshev, 1981, No. 7 (in Russian).

21. Godbillon C. Géometrie différentielle at mécanique analytique. Hermann, Paris, 1969.

22. Gromoll D., Klingenberg W., and Meyer W. Riemannsche Geometrie im grossen. Springer-Verlag, Berlin-Heidelberg-N.Y., 1968.

23. Klein F. Vorlesungen über höhere Geometrie (dritte Auflage bearbeitet und herausgegaben von W. Blaschke). Julius Springer, Berlin, 1926.

24. Lang S. Introduction to differential manifolds. Interscience, N.Y.-London, 1962.

25. McKean, Jr. H.P. Stochastic integrals.- Acad. Press, N.Y.-London, 1969.

26. Myshkis A.D. On certain problems in the theory of differential equations with deviated argument.- Usp. Matem. Nauk, 1977, 32, No. 2 (in Russian).

27. Synge J.L. Hodographs of general dynamical systems.- Trans. Roy. Soc. Canada, Ser. 3, 1926, vol. 25.

28. Synge J.L. Tensorial methods in dynamics. University of Toronto, 1936.

29. Sternberg S. Lectures on differential geometry. Prentice Hall, N.J., 1964.

30. Filippov A.F. Differential equations with discontinuous right-hand side.- Matem. sbornik, 1960, 51, No. 1 (in Russian).

31. Hartman P. Ordinary differential equations.- Wiley, N.Y.-London-Sydney, 1964.

32. Schmutzer E. Grundprinzipien der klassischen Mechanik und der klassischen Feldtheorie. VEB Deutscher Verlag der Wissenschaften, Berlin, 1973.

ON HAMILTONIAN SYSTEMS WITH DYNAMICAL SYMMETRIES

S.I. Pidkuiko and A.M. Stepin

Department of Mechanics and Mathematics,
Moscow State University, 119899, Moscow, USSR

New examples of completely integrable Hamiltonian systems, which have recently been discovered, are based on the Lax representation of the equations of motion. For a system with finitely many degrees of freedom, this approach was first suggested by H. Flashka [1] and J. Moser [2]. Some time later, A.S. Mishchenko and A.T. Fomenko [3] constructed series of left-invariant metrics on semisimple Lie groups for which geodesic flows are completely integrable. It is known [4] that the flow of geodesics on a Lie group relative to a left-invariant metric is factorized over the Euler system $x = [x, Ax]$. For the metrics constructed in [3], the integrability of the Euler systems on the orbits of adjoint action is based on the invariance of the system with respect to transformations of the form $x \mapsto x + \lambda a$.

A general scheme of constructing integrable Hamiltonian systems, which is based on a consistent use of the Lax representation, has been suggested by M. Adler and B. Kostant (see [5]). Complete integrability and the validity of the Lax representation of the equations of motion in this scheme is related to the decomposition of a Lie algebra into a linear sum of Lie subalgebras. Using precisely this method, M. Adler offered a group-theoretic explanation of the (L, A) pair suggested by J. Moser [2].

The existence of a finite-dimensional algebra of the integrals of motion enables one to reduce the system, i.e. to turn to a system with a lesser number of degrees of freedom. In this case, the dynamics of the reduced system may become more complicated, so that it is sometimes reasonable to use the inverse transition: namely, from the system considered to a system with a larger number of degrees of freedom. Such an extended system may have a sympler dynamics.

The description and development of the Adler-Kostant scheme through the reduction of a Hamiltonian system and application of this scheme to graded algebras is considered by A.G. Reyman and M.A. Semenov-Tian-Shansky [6,7].

It is also of interest to realize an integrable system as a factor-system of the flow of geodesics on a symmetric space. A relationship between integrable n-particle systems on a line and geodesic flows on symmetric spaces was established by M.A. Olshanetsky and A.M. Perelomov [8] . Then, D. Kazhdan, B. Kostant, and S. Sternberg [9] used the reduction method to describe this relationship for a system of n particles with the interaction potential $\sin^{-2}x$. I.V. Mikityuk has demonstrated that a system of n particles on a line with the interaction potential $\sinh^{-2}x$ can be obtained by the reduction of a geodesic flow in the co-tangent bundle of the space $M = GL(n, \mathbb{C})/U(n)$. He proved the existence of n independent integrals in involution for a geodesic flow in T^*M.

Some of the systems just mentioned are considered below in greater detail.

§ 1. Integrable systems with quasiperiodic motion

1. The action-angle variables (see [4]) are a useful tool for studying completely integrable systems with quasiperiodic motion (in particular, the problem of their non-degeneracy).

Let n independent functions $F_1, \ldots, F_n$ in involution be defined on a 2n-dimensional symplectic manifold (X, ω^2) and let there exist a compact level-manifold $M(c_1^o, \ldots, c_n^o) = \{F_1 = c_1^o, \ldots, F_n = c_n^o\}$ on which the forms dF_i, $i = 1, \ldots, n$, are linearly independent. Then there exists a neighborhood U of the manifold $M(c_1^o, \ldots, c_n^o)$ which is composed of level-manifolds $M(c_1, \ldots, c_n)$ and is diffeomorphic to the product $D^n \times T^n$ of an n-dimensional disk and an n-dimensional torus.

Let us introduce in U coordinates of the angle type. To this end, we consider the action S of the group $\mathbb{R}^n$ in U generated by commuting vector fields JdF_k, $k = 1, \ldots, n$; here $J: T^*X \to TX$ is the operator (of the symplectic structure) acting according to the formula $\omega^1(\eta) = \omega^2(\eta, J\omega^1)$, $\eta \in TX$. Fix a smooth section $x_o(c)$, $c \in D^n$, of the projection $U = D^n \times T^n \to D^n$ and denote by $\Gamma(c) \subset R^n$ a stationary subgroup of the point $x_o(c)$ with respect to the action S. There exists a smooth family of isomorphisms $r(c)$ of the group $\mathbb{R}^n$ onto itself such that $r(c)(\Gamma(c)) = \mathbb{Z}^n$. For $x \in M(c) \subset U$, there exists an element $t(x) \in \mathbb{R}^n$ such that $x = S_{t(x)}x_o(c)$.

We set

$$(\psi_1, \ldots, \psi_n) = r(c)(t(x)) \mod 1.$$

By virtue of the construction of the coordinates $\psi_1, \ldots, \psi_n$, it follows that on each torus M(c) the vector fields $\frac{\partial}{\partial \psi_i}$, i = = 1,2, ..., n, can be represented as linear combinations of the vector fields JdF_k with constant coefficients

$$\frac{\partial}{\partial \psi_i} = \sum_{k=1}^{n} a_{ik} \, JdF_k, \quad i = 1, \ldots, n. \tag{1}$$

On each torus, the variables $\psi_1, \ldots, \psi_n$ are angles. In U, however, these variables need not necessarily be in involution and are, generally, not angular variables in U.

Relations (1) imply that the variables $I_k(F_1, \ldots, F_n)$, k = 1, ...,n, which are canonically conjugate with the coordinates $\psi_1, \ldots, \psi_n$, must satisfy the conditions

$$dI_k = \sum a_{kl} \, dF_l, \quad k = 1, \ldots, n.$$

The functions I_k do exist, provided the forms $\sum a_{kl} dF_l$ are closed, i.e.

$$\frac{\partial a_{kl}}{\partial F_m} = \frac{\partial a_{km}}{\partial F_l} \tag{2}$$

To verify these conditions, we note that the matrix a_{kl} is inverse to the matrix of the Poisson brackets $\{F_k, \psi_l\}$. Furthermore, since the Poisson brackets $\{\psi_l, \psi_m\}$ are independent of the variables ψ_s and since the form ω^2 is closed, it follows that integrability conditions (2) are satisfied.

By construction, the variables I_k are in involution. However, we have already pointed out that the variables ψ_l, l = 1, ..., n, do not, in general, satisfy this property. This shortcoming can be eliminated by passing to the coordinates

$$\varphi_i = \psi_i + \omega_i(I_1, \ldots, I_n), \quad i = 1, \ldots, n.$$

The condition that there exist corrections ω_i for which φ_i are in involution is of the form

$$d\Big(\sum_{k,l} \{\psi_k, \psi_l\} \, dI_k \wedge dI_l\Big) = 0.$$

This condition does hold, since the form ω^2 is closed.

The variables $I_1, \dots, I_n$; $\varphi_1, \dots, \varphi_n$ thus constructed are called action-angle variables.

2. Using this procedure, one can define partial action-angle variables constructed by N.N. Nekhoroshev [10]. Indeed, let independent functions $F_1, \dots, F_{n+k}$ that are in involution with $F_1, \dots, F_k$ be defined on a 2n-dimensional symplectic manifold (X, ω^2). Suppose there exists a compact level-manifold $M(c_1^0, \dots, c_{n+k}^0) = \{F_1 = c_1^0, \dots, F_{n+k} = c_{n+k}^0\}$ on which the forms dF_i, $i = 1, 2, \dots, n+k$, are linearly independent. Then there exist a neighborhood U of the manifold $M(c_1^0, \dots, c_{n+k}^0)$, which is composed of level-manifolds $M(c_1, \dots, c_{n+k})$, and the coordinates $I_1, \dots, I_k$, $\varphi_1, \dots, \varphi_k$, p_j, q_j, $j = 1, \dots, n-k$, in U such that

$$\omega^2 = d\Big(\sum_{s=1}^{k} I_s\, d\varphi_s + \sum_{j=1}^{n-k} p_j\, dq_j\Big).$$

To prove this statement, one should introduce, using the above procedure, angle-type variables $\psi_1, \dots, \psi_k$, construct the conjugate actions $I_1(F_1, \dots, F_k), \dots, I_k(F_1, \dots, F_k)$, and "correct" the variables ψ_s:

$$\varphi_s = \psi_s + \omega_s(I_1, \dots, I_k), \quad s = 1, \dots, k.$$

To construct the coordinates p_j and q_j, one should use reduction. Let us consider a hypersurface $\Pi = \{\varphi_1 = 0, \dots, \varphi_k = 0\}$. The fields JdI_m are transversal to Π, i.e. there exists a projection on Π. The hypersurface Π is invariant with respect to commuting vector fields $Jd\varphi_m$. Thus, a group $\mathbb{R}^k$ acts on Π. Tangent spaces to the orbits of this action are the kernels of the restriction of the form ω^2 on Π. Let us factorize Π with respect to the action of $\mathbb{R}^k$. The image of the restriction of ω^2 on Π is correctly defined on a (2n-2k)-dimensional quotient space. This form is non-degenerate, and it remains to use the Darboux theorem.

3. The commonly used formulas for the action variables

$$I_k = \frac{1}{2\pi} \int_{\gamma_k} \sum p_i\, dq_i \tag{3}$$

where $\gamma_1, \dots, \gamma_n$ are basis cycles on the torus $M(c_1, \dots, c_n)$ continuously dependent on $(c_1, \dots, c_n)$ can be derived as follows. Since the form $\sum p_i dq_i - \sum I_k d\varphi_k$ is closed, the difference

$$\int_{\gamma_K} \sum p_i dq_i - \int_{\gamma_K} \sum I_K d\varphi_K$$

is constant. Hence, the function I_k coincides, to within a constant, with the integral

$$\frac{1}{2\pi} \int_{\gamma_K} \sum p_i dq_i \tag{4}$$

In a general case where the global canonical coordinates p_i, q_i do not exist, formulas (3) are reduced to

$$I_K = \frac{1}{2\pi} \int_{\gamma_K} \omega^1 ,$$

where ω^1 is an arbitrary form the differential of which is ω^2. Such a form always exists in a neighborhood of the n-dimensional torus $M(c_1, \ldots, c_n)$.

The definition of action variables by formulas (3) is rather effective in the following sense. If the curvatures of the hypersurfaces $F_k = c_k$ are uniformly bounded, the cycles γ_K can be chosen with an arbitrary accuracy given a priori. Also, integrals (4), as well as their derivatives with respect to p_i and q_i, can be calculated with an arbitrary accuracy. Restricting ourselves to the second-order derivatives, we can calculate $\det \frac{\partial^2 H}{\partial I_k \partial I_\ell}$ approximately. If the result exceeds, by absolute value, the calculation accuracy, we may infer that a quasiperiodic motion in a system with the Hamiltonian H is nondegenerate. An example is considered in Item 5 of this section.

4. As was already noted, semisimple Lie groups admit the left-invariant metrics for which geodesic flows are Liouville-integrable. It is interesting to find out whether quasiperiodic motions in the corresponding Euler systems are non-degenerate.

Let us consider Lie algebra SO(4) with the metric defined by the operator $\varphi_{a,b}\, x = ad_a^{-1} ad_b\, x$, where a and b are diagonal matrices with diagonal elements ia_j and ib_j. The system

$$\dot{x} = [x, \varphi_{a,b}\, x] \tag{5}$$

is Liouville-integrable on the orbits of general position of the adjoint representation. The corresponding integrals can be evaluated explicitly. To this end, we write E_{ij}, $i < j$, for the matrix in which only the elements at places (i,j) and (j,i) are non-zero and equal to 1 and -1, respectively. Let the coordinates in SO(4) be defined by the formula

$$X = \alpha E_{12} + \beta E_{13} + \gamma E_{14} + \delta E_{23} + \rho E_{24} + \varepsilon E_{34}.$$

The integrals of system (5) are of the form (see [3])

$$F_1(x) = (a_1+a_2)\alpha^2 + (a_1+a_3)\beta^2 + (a_1+a_4)\gamma^2 + (a_2+a_3)\delta^2 + (a_2+a_4)\rho^2 + (a_3+a_4)\varepsilon^2,$$

$$F_2(x) = (a_1^2+a_1a_2+a_2^2)\alpha^2 + (a_1^2+a_1a_3+a_3^2)\beta^2 + (a_1^2+a_1a_4+a_4^2)\gamma^2 + (a_2^2+a_2a_3+a_3^2)\delta^2 + (a_2^2+a_2a_4+a_4^2)\rho^2 + (a_3^2+a_3a_4+a_4^2)\varepsilon^2.$$

They are functionally independent and are in involution on the orbits of the adjoint representation: $\mathrm{Tr}x^2 = \text{const}$, $\det x = \text{const}$. The orbits of general position are manifolds diffeomorphic to $S^2 \times S^2$.

I.V. Mikityuk has demonstrated that for almost all the matrices a and b system (5) is non-degenerate on the orbits of general position. We now sketch the proof of this statement. There exist coordinates $x_1, y_1; x_2, y_2; x_3, y_3$ in which the orbits are described by the equations

$$x_1^2 + x_2^2 + x_3^2 = R_1^2, \quad y_1^2 + y_2^2 + y_3^2 = R_2^2.$$

As a, we take the matrix satisfying the conditions $a_1 = a_4$, $a_2 = a_3$, and $a_1 \neq a_2$. In this case, F_1 and F_2 are expressed in the new variables as

$$F_1 = (a_1-a_2)(x_3^2+y_3^2), \quad F_2 = 2(a_1-a_2)x_3y_3.$$

Let local coordinates $x_1, x_2; y_1, y_2$ be introduced on the orbit. In these coordinates the Kirillov form is written as

$$\omega^2 = (R_1^2 - x_1^2 - x_2^2)^{-1/2} dx_1 \wedge dx_2 - (R_2^2 - y_1^2 - y_2^2)^{-1/2} dy_1 \wedge dy_2.$$

The form ω^2 is exact, i.e. $\omega^2 = d\,\omega^1$, where

$$\omega^1 = \frac{(R_2^2 - y_1^2 - y_2^2)^{1/2} - R_2}{y_1^2 + y_2^2}(y_1 dy_2 - y_2 dy_1) - \frac{(R_1^2 - x_1^2 - x_2^2)^{1/2} - R_1}{x_1^2 + x_2^2}(x_1 dx_2 - x_2 dx_1).$$

The connected components of the level sets of the functions F_1 and F_2 are tori

$$\{(x_1, x_2, y_1, y_2) : x_1^2 + x_2^2 = r_1^2,\ y_1^2 + y_2^2 = r_2^2\}$$

and the basis cycles of these tori are of the form

$$\gamma_1(r_1) = \{(x_1, x_2, 0, 0) : x_1^2 + x_2^2 = r_1^2\},$$

$$\gamma_2(r_2) = \{(0, 0, y_1, y_2) : y_1^2 + y_2^2 = r_2^2\}.$$

The action variables I_1 and I_2 are calculated explicitly

$$I_k = \frac{1}{2\pi} \int\limits_{\gamma_k} \omega^1 = \frac{1}{2\pi} \left(\sqrt{R_k^2 - r_k^2} - R_k \right).$$

Straightforward verification shows that the Hessian $\frac{\partial^2 F_2}{\partial I^2}$ is non-degenerate. Thus, the system with the Hamiltonian F_2 is non-degenerate on the orbits of general position. A real-analytic dependence of the Hessian on a implies that system (5) is non-degenerate for almost all a.

5. Let us give another example of how the non-degeneracy of a quasi-periodic motion can be proved by using action variables.

Consider a system of n particles moving along a straight line due to pairwise interaction with the potential $a^2\wp(x)$. This system is known [11] to have n momentum-polynomial integrals F_k, $k = 1, 2, \ldots, n$, which are in involution. Separating the motion of the centre-of-mass, one obtains a system with quasiperiodic motion. A.S. Logachev has proved that such a system is non-degenerate.

The proof is based on the following consideration. The limit, as $a \to 0$, of a system with the Hamiltonian

$$H = \frac{1}{2} \sum p_i^2 + a^2 \sum_{i<j} \wp(x_i - x_j) \qquad (6)$$

is a billiard-type system (see Section 3) with the Hamiltonian

$$H^\circ = \frac{1}{2} \sum p_i^2 + \sum_{i<j,k} \delta(x_i - x_j + k\omega_1), \qquad (7)$$

in which the motion (after separation of the centre-of-mass) is quasiperiodic and non-degenerate.

For a system with generalized Hamiltonian (7), one can introduce the variables I_k^0, which are analogous to action variables; in this case, det $\frac{\partial^2 H^0}{\partial I_k^0 \partial I_l^0}$. It turns out that as $a \to 0$ the action variables I_k of the system with Hamiltonian (6) tend (together with the first- and seconder-order derivatives with respect to F_k) to the action variables of billiard system (7). Thus, $\frac{\partial^2 H}{\partial I_k \partial I_l} \to \frac{\partial^2 H^0}{\partial I_k^0 \partial I_l^0}$ for $a \to 0$, and non-degeneracy follows as a consequence.

We should also mention the paper by I.M. Krichever [12] in which angle-type variables were constructed for system (6).

§ 2. Hamiltonian systems with scattering

1. Consider a Hamiltonian system

$$\dot{q} = \frac{\partial H_0}{\partial p}, \quad \dot{p} = -\frac{\partial H_0}{\partial q} \tag{8}$$

with a non-compact phase space X. Suppose the trajectories of this system tend to infinity for $t \to \pm\infty$. If the function H on X is such that at infinity the difference $H - H_0$ decreases rather rapidly, each motion $f(t, x)$, $f(0, x) = x$, of the system

$$\dot{q} = \frac{\partial H}{\partial p}, \quad \dot{p} = -\frac{\partial H}{\partial q} \tag{9}$$

behaves asymptotically, for $t \to \pm\infty$, as a certain motion $f_0(t, x^{\pm})$, $f_0(0, x) = x$, of system (8). The mappings $\sigma_{\pm} : x \to x^{\pm}$ are canonical. Clearly, $\sigma_{\pm}(f(t, x)) = f_0(t, \sigma_{\pm}(x))$, i.e. the mappings $\sigma_{\pm}$ realize the conjugacy of perturbed system (9) and unperturbed system (8).

Consider the mappings $F: X \xrightarrow{\sigma_+} X \to X/\xi$, where ξ is a partition of X into the trajectories of system (8). This mapping is constant along the trajectories of system (9).

The mapping $\sigma = \sigma_+ \sigma_-^{-1}$ is called the scattering mapping of system (9). The direct problem in scattering theory is to establish existence conditions for the mapping σ and to study its properties. The inverse problem is to reconstruct the perturbation $H - H_0$ by the scattering mapping σ .

The perturbation $H - H_0$ depends, as a rule, only on configurational coordinates in the phase space $X = TM$. However, in some problems of differential geometry (see, e.g., [13]), the direct and inverse problems of scattering theory are treated under metric perturbation.

2. We now consider these problems for a system of n particles on a line with the Hamiltonian

$$H = \frac{1}{2}\sum p_i^2 + \sum_{i<j} U(x_i - x_j) \qquad (10)$$

Suppose the interaction potential U(x) falls rather rapidly at infinity (together with its derivative) and $U'(x) < 0$ for $x > 0$. Then the particles behave, asymptotically, as free ones, i.e.

$$x_k(t) - \xi_k^+ t - \eta_k^+ \to 0,$$

$$p_k(t) - \xi_k^+ \to 0$$

for $t \to +\infty$, $k = 1, \dots, n$. The quantities ξ_k^+ and η_k^+ , $k = 1, \dots, n$, are called asymptotic (for $t \to +\infty$) momenta and phases. The quantities ξ_k^- and η_k^- are defined in a similar fashion. In the variables ξ^+, η^+ (ξ^-, η^- , respectively), the motion becomes linear.

If the particles are numbered from left to right, i.e. $x_1 < \dots < x_n$, then $\xi_1^+ < \dots < \xi_n^+$ and $\xi_n^- < \dots < \xi_1^-$. In a general case, ξ_k^+ depend both on ξ_i^- and η_i^- , $i = 1, \dots, n$. There exist, however, potentials for which ξ_k^+ depend only on ξ_i^- . These are potentials of the form $U(x) = \frac{c^2}{sh^2 x}$ and $U(x) = \frac{c^2}{x^2}$. System (10) with such potentials has a complete set of momentum-polynomial integrals F_i, $i = 1, 2, \dots, n$, which are in involution. If in relations $F_k = \text{const}$ the time is let tend to infinity, we obtain that the asymptotic velocities $\xi_k^\pm$ satisfy the equations

$$\sum_{k=1}^{n} (\xi_k^\pm)^\ell = c_\ell, \quad \ell = 1, \dots, n,$$

which imply that the sets of asymptotic momenta $\{\xi_1^-, \dots, \xi_n^-\}$ and $\{\xi_1^+, \dots, \xi_n^+\}$ coincide. Taking into account the inequalities

$\xi_1^+ < \ldots < \xi_n^+$ and $\xi_n^- < \ldots < \xi_1^-$ we obtain $\xi_k^+ = \xi_{n-k+1}^-$.

That the scattering on momenta is of billiard type for the potentials $\sinh^{-2}x$ and x^{-2} is not accidental. We now demonstrate, using the method suggested by P.P. Kulish [14] , that scattering on momenta is precisely of this type whenever ξ_k^+ depends only on ξ_i^- .

Let, for simplicity, n = 3. As $\eta_1^- \to -\infty$, the evolution of the system can be described as follows: so long as the first particle is far on the left, the second and third particles experience scattering through momentum exchange; when the first particle approaches the second one, the third particle is far on the right and affects very little (for $\eta_1^- \to -\infty$) the scattering of the first and second particles, the latter acquiring the momentum of the first particle and approaching the third one. The first particle, which is now heavily lagged, influences little the scattering of the second and third particles. As a result of these collisions, the asymptotic momenta ξ_k^- undergo the permutation $(\xi_1^-, \xi_2^-, \xi_3^-) \to (\xi_3^-, \xi_2^-, \xi_1^-)$.

If scattering on momenta is of billiard type, then, as was shown by P.P. Kulish [14] , the scattering mapping can be calculated. Indeed, since the scattering mapping is canonical, $\sum d\xi_k^- \wedge d\eta_k^- = \sum d\xi_k^+ \wedge d\eta_k^+$. Taking into account that $\xi_k^+ = \eta_{n-k+1}^-$, we have $d\left(\sum_k (\eta_k^- - \eta_{n-k+1}^+) d\xi_k^-\right) = 0$. Consequently, there exists a function $G(\xi_1^-, \ldots, \xi_n^-)$ which determines the phase shift $\delta_k = \eta_{n-k+1}^+ - \eta_k^-$ by the formula

$$\delta_k = \frac{\partial G}{\partial \xi_k^-} .$$

Thus, the scattering mapping is of triangular form

$$\xi_k^+ = \xi_{n-k+1}^- ,$$

$$\eta_k^+ = \eta_{n-k+1}^- + \delta_{n-k+1}(\xi_1^-, \ldots, \xi_n^-) .$$

The problem is, therefore, reduced to finding the function G. Since this function does not depend η_i^- , the scattering mapping σ is the superposition of scatterings due to pairwise collisions, and the function G is of the form $\sum_{k<l} g(\xi_k^- - \xi_l^-)$. Hence,

$$\delta_k = \sum_{l>k} g'(\xi_k^- - \xi_l^-) - \sum_{l<k} g'(\xi_l^- - \xi_k^-) . \tag{11}$$

The function $\delta(\xi) = g'(\xi)$ is the phase shift in a two-particle problem and can be expressed in terms of the potential [15].

3. Let us show how to obtain the necessary condition for the potential U, which is satisfied for system (9) with Hamiltonian (10) and a triangular scattering mapping.

We note, first of all, that the motion (symmetric about the origin) in a system of n particles on a straight line may be considered as the motion of a system of $\left[\frac{n}{2}\right]$ particles with the Hamiltonian

$$H = \frac{1}{2}\sum p_i^2 + \sum_{j<k} U(x_j - x_k) + c\sum U(x_i) + \frac{1}{2}\sum U(2x_i) + \sum_{j<k} U(x_j + x_k), \qquad (12)$$

where c = 1 for odd n and c = 0 for even n.

Hamiltonians of this form were suggested by M.A. Olshanetsky and A.M. Perelomov [16], proceeding from group-theoretic considerations. In this paper, they arise according to the reduction scheme.

In a particular case (n = 3) we obtain that the restriction of the phase flow of a three-particle system on the set of configurations that are symmetric about the origin coincides with the phase flow of a system with one degree of freedom and the potential

$$U(x) + \frac{1}{2}U(2x).$$

In accordance with what has just been said, we calculate scattering for the system under consideration. On the one hand, the phase shift of the third particle is $\delta_{2U(x)+U(2x)}(\xi)$ (here δ_U denotes the phase shift of a particle moving in the field of the potential U). On the other hand, by virtue of Eq. (11), the phase shift of the third particle is

$$\delta_U(\xi) + \delta_U(2\xi)$$

It is a simple matter to verify that $\delta_U(c\xi) = \delta_{c^2U}(\xi)$, $c > 0$.

Thus, the necessary condition for scattering to be of billiard type is

$$\delta_{U(x)}(\xi) + \delta_{4U(x)}(\xi) = \delta_{2U(x)+U(2x)}(\xi). \qquad (13)$$

The left-hand side of this equation can be written as the phase shift $\delta_{\bar{U}}(\xi)$ corresponding to a certain potential $\bar{U}$. To do this, we use the integral representation of the phase shift $\delta_{U}(\xi)$ obtained by J. Moser

$$\delta(\xi) = V(\xi^2) + \int_0^{\xi^2} \left[\left(1-\frac{s}{\xi^2}\right)^{-1/2} - 1\right] dV(s) \tag{14}$$

where V is a function inverse to the function 4U. Let $4\bar{U}$ denote a function which is inverse to V(y) + V(y/4). Then Eq. (13) takes the form

$$\delta_{\bar{U}}(\xi) = \delta_{2U(x)+U(2x)}(\xi) \tag{15}$$

If the inverse scattering problem for the system in question admitted a unique solution, the necessary condition of billiard scattering would be the equality of $\bar{U}(x)$ and 2U(x) + U(2x).

As was shown by J. Moser [15] , the potential is determined almost uniquely by the phase shift. To be more exact, let $\delta(y)$ be a continuously differentiable function on $(0, y_1]$ satisfying the condition

$$\int_0^{y_1} y^{\gamma} \, |d\delta(y)| < \infty, \quad \gamma \in (0, 1/2].$$

Then the function $V(y) = \frac{1}{\pi}\int_0^y \frac{\delta(t)dt}{\sqrt{t(y-t)}}$ is a solution of Eq. (14), and the general solution of Eq. (14) in the class of functions satisfying

$$\int_0^{y_1} y^{\theta} \, |dV(y)| < \infty, \quad \theta \in (0,1), \tag{16}$$

differs from the solution V(y) by $cy^{-1/2}$, where c is a constant.

Thus, from (15) we obtain the following necessary condition of billiard scattering in a system with the interaction potential U: the function

$$V\left(\frac{y}{2}\right) + V(2y) + \frac{c}{\sqrt{y}}$$

is inverse to the function $y = \frac{1}{2} U(x) + U(\frac{x}{2})$, provided V satisfies condition (16).

Here is a simple consequence: if in a system with the potential U(x) scattering is of billiard type, $U(x) \sim cx^{\alpha}$ for $x \to 0$, and $\int^{\infty} U^{\theta} dx < \infty$ for $\theta \in (0, 1)$, then $\alpha = 2$.

A similar statement holds for the asymptotic behavior of $U(x) \sim cx^{\alpha}$ for $x \to \infty$.

4. Using the necessary condition of Item 3, A.Yu. Plakhov has proved that the meromorphic potential U, for which an n-particle system on a straight line has billiard scattering is either $\frac{a\lambda^2}{sh^2 \lambda x}$ or $\frac{a}{x^2}$, $a > 0$.

Some time earlier, Ya.G. Sinai proved the following statement: if the potential $U(x) = \frac{1}{x^2 P(x^2)}$, where P is a polynomial, is characterized by billiard scattering, then P = const.

The result of A.Yu. Plakhov can, seemingly, be extended to an arbitrary analytic potential. As is shown in [17], the potential U satisfies an algebraic differential equation and is, therefore, analytic, provided the system has an additional momentum-polynomial integral. This shows, in particular, that there exist integrable systems that do not admit momentum-polynomial integrals.

In the class of smooth potentials, the description of potentials with a triangular scattering mapping is a much more complicated problem. All what has been established up to now is that the set of such potentials is nowhere dense.

A.Yu. Plakhov considered the problem of determining potentials with billiard scattering in the class of exponentially decaying functions. He has proved that among potentials of the form $U(x) = \exp(-kx + a(x))$, $k > 0$, only the potential $a^2 \sinh^{-2}(\frac{kx}{2})$ has billiard scattering under the following conditions: $a'(x) \to 0$ for $x \to \infty$, $a''(x) \to 0$ for $x \to \infty$, $a(x) \to \infty$ for $x \to +0$, and U does not vanish in some neighborhood of the origin.

5. Scattering in a system with an analytic interaction potential is related to the Stokes phenomenon, i.e. to a jumpwise change in the asymptotic expansions for the solution in the transition across cer-

tain lines in the complex plane. If the phase variables depend on time algebraically, a distinction of asymptotic solutions for $t \to +\infty$ and $t \to -\infty$ corresponds to the transition (when moving along the real time axis from $-\infty$ to $+\infty$) from one sheet of the Riemannian surface of the phase space to another sheet [18] . A.Yu. Plakhov and A.M. Stepin took this fact into account while determining potentials for which the phase variables depend on time algebraically.

First we demonstrate, following J. Moser (see [19]), that the potential $U(x) = x^{-2}$ does possess this property. We set

$$L_{jk}(x,p) = p_j \delta_{jk} + \frac{i}{x_j - x_k}$$

$$B_{jk}(x,p) = i\delta_{jk} \sum_{k \neq j} \frac{1}{(x_j - x_k)^2} + \frac{i}{(x_j - x_k)^2}$$

It is a simple matter to verify that

$$\dot{L} = [B, L], \tag{17}$$

where $\dot{L}$ is the derivative of the matrix L along the trajectory of the system

$$\dot{x} = \frac{\partial H}{\partial p}, \ \dot{p} = -\frac{\partial H}{\partial x}, \ H = \frac{1}{2} \sum p_i^2 + \sum_{i<j} \frac{1}{(x_i - x_j)^2} . \tag{18}$$

Let a matrix function U(t) satisfy the equation $\dot{U}(t) = B(x(t),$ $p(t))U(t)$, where (x(t), p(t)) is a solution of system (18). It follows from (17) that the matrix $U^{-1}(t)L(x(t), p(t))U(t)$ is constant. We may assume that $U = E + O(1)$ for $t \to +\infty$. Introduce a matrix $M(x, p, t) = K(x) + tL(x, p)$, where K(x) is a diagonal matrix with diagonal elements $x_1, \ldots, x_n$. It turns out that the matrix $U^{-1}MU$ is constant along the trajectory (x(t), p(t)). Calculating $U^{-1}MU$ for $t = 0$ and $t \to +\infty$, we derive the relation

$$U^{-1} K(x) U = L^t(\xi^+, \eta^+). \tag{19}$$

Since the motion is linear in the variables ξ^+, η^+ , it follows from (19) that the functions $x_i(t)$, $i = 1, \ldots, n$, depend on t algebraically.

We now assume that an analytic potential U possesses this property. Let σ denote the scattering mapping for a system with the potential U. Since an algebraic function has a finite number of branches, there exists a positive integer $l = l(x^-, p^-)$ such that

$$\sigma^{l}(x^-, p^-) = (x^-, p^-)$$

It can be shown that all the numbers $l(x^-, p^-)$ are uniformly bounded. Thus, there exists a positive integer m such that σ^m = id. Choosing x^- in such a manner that the mapping σ is sufficiently well approximated by the superposition of pairwise collisions, we obtain the following property of the mapping σ^2 :

$$\sigma^2(x^-, p^-) \simeq (x^- + \check{\Delta}(p^-) + \Delta(\check{p}^-), p^-),$$

where $\Delta(p^-) = (\delta_1(p^-), \ldots, \delta_n(p^-))$ and $(p_1, \ldots, p_n)^\vee = (p_n, \ldots, p_1)$.

Applying this property to the mapping σ^{2m} = id, we find

$$\check{\Delta}(p) + \Delta(\check{p}) = 0$$

Since $\Delta(\check{p}) = \check{\Delta}(p)$, then $\Delta(p) \equiv 0$. Fixing $p_2 - p_j$, $j > 2$, and setting $p_1 - p_2 = p$, we have $\delta(p)$ = const. Taking into account that $\Delta(p) \equiv 0$, we finally obtain $\delta(p) \equiv 0$, whence it follows that $U(x) = \frac{a^2}{x^2}$, $a \neq 0$.

This result enables one to offer examples of infinite-sheet analytic functions that have, nevertheless, finitely many branching points of finite order. Consider, for instance, the interaction potential $U(x) = \frac{1}{x^4}$. The function $t(x) = t_0 + \int_{x_0}^{x} \frac{dx}{\sqrt{h - x^{-4}}}$ is not algebraic. It has four singular points $x_k = h^{-4} i^k$, $k = 0,1,2,3$. They all are branching points of order 2.

§ 3. Integrable billiard systems

1. Let D be a domain in R^n with a piecewise-smooth boundary ∂D. The flow T_t in the space $TD = D \times R^n$ of tangent vectors in D, which acts according to the relation

$$T_t(x, p) = (x + tp, p), \quad x \in D, \; p \in \mathbb{R}^n$$

is called a billiard in D. If the support x of a tangent vector $(x, p) \in TD$ meets the boundary ∂D, it is instantaneously reflected from ∂D according to the law: the angle of incidence is equal to the angle of reflection, i.e. the tangent component of the vector p is conserved, while the normal component changes the sign. If the support of the vector (x, p) meets a kink of the boundary, its further motion is not defined.

The phase space of a billiard in D is said to be a quotient space $\overline{D} \times R^n/\sim$ with respect to the equivalence relation: $(x, p) \sim (x, p')$ if $p - p' = 2(n, p)n$, where n is the normal to ∂D at point x.

A billiard system in D is called regularizable if the flow can be extended by continuity to the entire phase space. In this case, by a regularized phase space we mean the phase space with the following additional point identification: $\lim T_t x = \lim T_t x$.

A billiard system is called completely integrable if it is regularizable and if there exists a partition of the regularized phase space which is fixed relative to the flow T_t almost all elements of which are tori in the induced topology, the flow T_t being quasiperiodic on each torus.

It turns out [20] that a completely integrable billiard in a polyhedron is closely related to a root system.

Theorem

1. A billiard in a polyhedron is regularizable if and only if the polyhedron is an affine Weyl chamber of a certain root system.

2. A billiard in a polyhedron is completely integrable if and only if it is regularizable.

Complete integrability of the billiard in the affine Weyl chamber K stems from the following consideration. An affine Weyl group W, which acts in R^n, contains a subgroup Π of parallel translations along n independent directions. Since $K = R^n/W$, the torus R^n/Π acts in the phase space of the billiard in K, the orbits of this action being invariant tori of the billiard in K.

The proof of the converse statement is based on the following theorem due to Stiefel: a finite group generated by *reflections* in R^n is a Weyl group if it preserves a lattice. Since the billiard in the polyhedron is regularizable, the space R^n can be paved with the images of this polyhedron under successive reflections about the faces. This fact and Stiefel's theorem imply that the polyhedron in question is an affine Weyl chamber of a certain root system.

In connection with the second statement we should note that a regularizable billiard may not be completely integrable (the corresponding domain is not, of course, a polyhedron).

Billiard systems frequently occur as limiting cases of smooth dynamical systems. For example, billiard systems in affine Weyl chambers of root systems of series A_n, B_n, C_n, D_n, and BC_n may be interpreted as limiting ones for Hamiltonian systems with the Hamiltonian suggested in [16] . Another example is a billiard in an ellipse. Complete integrability of this billiard is related to complete integrability of a geodesic flow on an ellipsoid the limiting case of which is a billiard in an ellipse.

2. A billiard in an ellipse has a momentum-quadratic integral, which does not depend on energy. The following question arises in this connection: what are plane domains (with piecewise-smooth boundaries) in which the billiard has a momentum-polynomial energy-independent integral? It should be noted that the case of integrals that depend analytically on the momentum is reduced to the case of momentum-polynomial integrals. Indeed, the homogeneous components of the Taylor expansion of an analytic integral are also integrals of the system. As in the case of smooth Hamiltonian systems, the existence of a complete set of integrals in involution for a billiard in a bounded domain leads to complete integrability of this billiard system. Thus, billiards in plane domains that possess an additional integral represent new examples of integrable systems.

Consider a plane domain bounded by a piecewise-smooth curve defined by the equation

$$f(x, y) = 0$$

We assume that the billiard in this domain has a momentum-polynomial integral $F(x, y, p, q)$ independent of the Hamiltonian $H = \frac{1}{2}(p^2 + q^2) + \delta(f(x, y))$. ($F$ may be considered as a homogeneous polynomial in p and q.) The condition $\{F, H\} = 0$ inside the domain implies that the coefficients of the integral F are polynomials of x and y. The condition that F is constant under the reflection from the domain boundary is written as

$$F(x, y, p, q) = F(x, y, -ap - bq, -bp + aq), \qquad (20)$$

where

$$a = \frac{f_x^2 - f_y^2}{f_x^2 + f_y^2}, \quad b = \frac{2f_x f_y}{f_x^2 + f_y^2}.$$

Equating the coefficients at $p^k q^l$ in Eq. (20), we obtain a system of equations for the function $f(x, y)$, which yields the following necessary condition for the integral F to exist.

If a billiard in a domain bounded by the curve $f(x, y) = 0$ has an additional integral of odd degree, which is a polynomial in p and q, the curve $f(x, y) = 0$ satisfies the equation

$$\frac{dy}{dx} = \frac{P(x, y)}{Q(x, y)}, \tag{21}$$

where P and Q are polynomials of x and y and are linear combinations of the coefficients of the integral F.

If the degree of the integral F is even, the curve $f(x, y) = 0$ satisfies the equation

$$\left(\frac{dy}{dx}\right)^2 + \frac{P(x,y)}{Q(x,y)}\frac{dy}{dx} - 1 = 0, \tag{22}$$

where P and Q are linear combinations of the coefficients of the integral F.

Using the explicit form of the polynomials P and Q, one can show that the curve $f(x, y) = 0$ is algebraic.

This approach to constructing integrable billiards offers new examples of integrable systems even for momentum-quadratic integrals. Indeed, Eq. (22) is in this case of the form

$$y'^2 + 2y' \frac{a_{20} - a_{02}}{a_{11}} - 1 = 0, \tag{23}$$

where $a_{20}p^2 + a_{11}pq + a_{02}q^2$ is an integral of the system. For $a_{20} = y$, $a_{11} = x$, and $a_{02} = 0$, the function $x^2 + y^2 = (c^2x^2 - y^2)^2 D$ is an integral of Eq. (23). Thus, in the domain bounded by the curves

$$y = \frac{D + x^2 - C^4x^4}{\pm 2C^2x^2}$$

the billiard has the additional integral $yp_1^2 - xp_1p_2$.

We note here that F. Lund has obtained the following result: the equation KdV and related equations can be realized as geodesic flows on quadrics in a Hilbert space. This result gives rise to the following questions: do integrable geodesic flows exist on algebraic surfaces of order higher than two?; do integrable billiards exist in domains bounded by algebraic curves of order higher than two? The example just considered gives an answer to the second question.

3. The examples presented above show how billiard systems can be obtained from smooth dynamical systems. It appears that the reverse procedure is also possible: namely, smooth dynamical systems can be obtained from billiard systems. For example, the limiting case of a billiard in a convex domain is a geodesic flow on the boundary of this domain.

An interesting example of a Hamiltonian system, which is obtained as a limit of a billiard system, has been constructed by A.Yu. Plakhov. Consider a system of three particles on a straight line which interact in the billiard fashion. The coordinates of these particles are denoted by x_1, y, and x_2, $x_1 \leqslant y \leqslant x_2$, and their masses, by M, μ, and M, respectively. If the velocity of the middle particle v is let tending to ∞ in such a manner that μv^2 = const, the heavy particles x_1 and x_2 will move as in a Hamiltonian system with the Hamiltonian

$$H = \frac{1}{2}(p_1^2 + p_2^2) - \frac{c}{(x_1 - x_2)^2} .$$

Here is the sketch of the proof. Let at the initial moment the points x_1, y, and x_2 have velocities v_1, v, and v_2, respectively. Conservation of momentum and energy implies that at the moment when the particle x collides with the particle x_2

$$\mu v + v_2 = \mu \bar{v} + \bar{v}_2 ,$$

$$\mu v^2 + v_2^2 = \mu \bar{v}^2 + \bar{v}_2^2 ,$$

where v and v_2 are the velocities of particles y and x_2 after collision. We assume that $v \gg v_i$, $i = 1,2$, and that $\mu v \ll 1$. Then,

$$\bar{v} = -v + 2v_2 + o(1) ,$$

$$\bar{v}_2 = v_2 + 2\mu v + o(\mu v) .$$

The moment of the second collision of particles y and x_2 is given by

$$\Delta t = \frac{x_2 - x_1}{v - 2v_2 + o(1)} + \frac{x_2 - x_1}{v - 2v_2 + 2v_1 + o(1)} = \frac{2(x_2 - x_1)}{v} + o\left(\frac{1}{v}\right).$$

Hence,

$$\frac{\Delta v_2}{\Delta t} = \frac{\mu v^2}{x} + o\left(\frac{1}{v}\right), \quad x = x_2 - x_1 .$$

Letting v tend to ∞ in such a way that μv^2 = const, we can express the constant on the right-hand side in terms of the initial data. We have $\Delta v = 2v_1 - 2v_2 + O(1)$, whence

$$\frac{\Delta v}{\Delta t} = - \frac{v_1 - v_2}{x_1 - x_2} v + O\left(\frac{1}{v}\right).$$

In the limit we obtain

$$\frac{\sqrt{\mu}\, \dot{v}}{\sqrt{\mu}\, v} = - \frac{v_1 - v_2}{x_1 - x_2} = - \frac{\dot{x}_1 - \dot{x}_2}{x_1 - x_2} .$$

Integration yields $\sqrt{\mu}\, v(x_1 - x_2)$ = const, i.e. $\mu v^2 = \dfrac{C_1}{(x_1 - x_2)^2}$.

Thus, the equation of motion of the heavy particle takes the form

$$\ddot{x}_2 = - \frac{C_1}{(x_1 - x_2)^3} .$$

REFERENCES

1. Flashka H. On the Toda lattice. 1.- Phys. Rev., 1974, B9.
2. Moser J. Three integrable Hamiltonian systems connected with isospectral deformations.- Adv. Math., 1975, 16, No. 2.
3. Mishchenko A.S. and Fomenko A.T. Integrability of Euler's equations on semisimple Lie algebras.- in: Trudy seminara po vektornomu i tenzornomu analizu. Moscow, 1977, No. 19 (in Russian).
4. Arnold V.I. Mathematical methods of classical mechanics.- Moscow, 1974 (in Russian).
5. Adler M. On a trace functional for formal pseudodifferential operators and symplectic structures of the Korteweg-de Vries equations.- Invent. Math., 1979, 50, No. 3.
6. Reyman A.G. and Semenov-Tian-Shansky M.A. Reduction of Hamiltonian systems and graded Lie algebras.- Usp. Mat. Nauk, 1979, 34, No. 4 (in Russian).
7. Reyman A.G. and Semenov-Tian-Shansky M.A. Reduction of Hamiltonian systems, affine Lie algebras, and Lax equations.- Invent. Math., 1979, 54.

8. Olshanetsky M.A. and Perelomov A.M. Explicit solutions of certain completely integrable Hamiltonian systems.- Funktsionalny analiz i yego prilozheniya, 1977, 11, No. 1 (in Russian).

9. Kazhdan D., Kostant B., and Sternberg S. Hamiltonian group action and dynamical systems of Calogero type.- Comm. Pure Appl. Math., 1978, 31, No. 4.

10. Nekhoroshev N.N. Action-angle variables and their generalizations.- Trudy Moskovskogo Matem. Obshchestva, Moscow, 1972, 26 (in Russian).

11. Pidkuiko S.I. On a certain integrable problem of n bodies.- Usp. Mat. Nauk, 1978, 33, No. 3 (in Russian).

12. Krichever I.M. Elliptic solutions of the Kadomtsev-Petviashvili equation and integrable systems of particles.- Funktsionalny analiz i yego prilozheniya, 1980, 14, No. 4.

13. Gluck H. and Singer D. Scattering problems in differential geometry.- Lect. Notes Math., 1977, 597.

14. Kulish P.P. Factorization of classical and quantum S-matrices and conservation laws.- Teor. i Matem. Fiz., 1976, 26, No. 2 (in Russian).

15. Moser J. The scattering problem for some particle systems on the line.- Lect. Notes Math., 1977, 597.

16. Olshanetsky M.A. and Perelomov A.M. Completely integrable Hamiltonian systems connected with semisimple Lie algebras.- Invent. Math., 1976, 37, No. 2.

17. Pidkuiko S.I. and Stepin A.M. Polynomial integrals of Hamiltonian systems.- Dokl. Akad. Nauk SSSR, 1978, 239, No. 1 (in Russian).

18. Stepin A.M. Polynomial integrals of Hamiltonian systems.- Asterisque, 1978, 51.

19. Airault H., McKean H.P., and Moser J. Rational and elliptic solutions of Korteweg-de Vries equation and a related many-body problem.- Comm. Pure Appl. Math., 1977, 30.

20. Pidkuiko S.I. Completely integrable billiard-type systems.- Usp. Mat. Nauk, 1977, 32, No. 1 (in Russian).

ON THE THEORY OF GENERALIZED CONDENSING PERTURBATIONS OF CONTINUOUS MAPPINGS

V.G. Zvyagin

Department of Mathematics, Voronezh State University
394693, Voronezh, USSR

Condensing vector fields constitute one of the well-known classes of mappings for which the degree theory has been constructed. For this class of mappings, the degree theory is developed in many papers (see, e.g., [1-5]).

The class of condensing vector fields may be looked upon as a class of condensing perturbations of an identity mapping. Hetzer [6] extended the degree theory to linear Fredholm mappings plus a condensing mapping with a constant, i.e. to the class of condensing perturbations of a linear Fredholm mapping of non-negative index. In paper [7] Hetzer used the degree theory to prove the existence of periodic solutions of neutral-type differential equations; in paper [8] , to prove the existence of weak solutions for a non-linear elliptic boundary-value problem; and in paper [9] , to solve the elliptic boundary-value problem with a resonance under the assumptions that are a generalization of the Landesman-Laser conditions [10] . In paper [11] the Fredholm alternative is generalized to the case of linear Fredholm mappings plus a condensing mapping with a certain constant, and application is considered to neutral-type periodic solutions and to a boundary-value problem for an ordinary second-order differential equation. Some applications of Hetzer's degree theory are discussed in [12,13] .

Dmitrienko and Zvyagin [14] considered mappings of the form: an arbitrary mapping A plus a continuous mapping which is condensing relative to A. The problem of homotopic classification for these mappings can be reduced to a similar problem for completely continuous perturbations of an arbitrary mapping. It should be noted that the latter problem was analysed by Hoim and Spaier [15] . In paper [14] a degree theory has been constructed for mappings of the form: A plus a condensing mapping relative to A in the case where A is a proper Fredholm mapping of non-negative index.

§ 1. Homotopic classification

1. Fundamental concepts. Let M be an arbitrary bounded subset of a Banach space, and let $\overline{co}(M)$ stand for a convex closure of this subset. The following concept is well known (see, e.g., [4]).

Definition 1.1. A non-negative function ψ defined on bounded subsets $\{M\}$ of a Banach space is called a measure of non-compactness if it satisfies the conditions:

(1) $\psi(\overline{co}\, M) = \psi(M)$;

(2) $M_1 \subseteq M_2$ implies $\psi(M_1) \leq \psi(M_2)$;

(3) the set M is compact if and only if $\psi(M) = 0$.

he most familiar measures of non-compactness also satisfy the following conditions:

(4) $\psi(M \cup N) = \max\{\psi(M), \psi(N)\}$;

(5) $\psi(\alpha M + N) \leq |\alpha|\psi(M) + |\beta|\psi(N)$, where $\alpha M + \beta N = \{x : x = \alpha u + \beta v, u \in M, v \in N\}$, in which case $\psi(\alpha M) = |\alpha|\psi(M)$, where $\alpha M = \{x : x = \alpha u, u \in M\}$.

In what follows we shall assume that the measures of non-compactness used below satisfy conditions (1)-(5). As an important example of the measure of non-compactness, we shall consider Kuratowski's measure of non-compactness.

Definition 1.2. Kuratowski's measure of non-compactness $\gamma(M)$ is the infimum of those $\alpha > 0$ for which M can be covered with finitely many sets of diameter α .

Kuratowski's measure of non-compactness satisfies conditions (1)-(5).

Let D be an open bounded domain in a Banach space E. As usual, $\overline{D}$ is the closure of D, and ∂D is its boundary. Let $A:\overline{D} \to F$ be an arbitrary mapping of $\overline{D}$ into a Banach space F, and let $f:\overline{D} \to F$ be a bounded mapping (i.e. the image of a bounded set under the mapping f is a bounded subset of the space F).

The following notion has been suggested in [14] .

Definition 1.3. A bounded mapping $f:\overline{D} \to F$ is called A-condensing if $\psi[f(M)] < \psi[A(M)]$ for any set $M \subseteq \overline{D}$ such that $\psi[f(M)] \neq 0$.

In this definition $\psi[f(M)]$ and $\psi[A(M)]$ are the measures of non-compactness for the sets $f(M)$ and $A(M)$, respectively.

Definition 1.4. A non-empty set T F is called A-fundamental for the mapping f:D F if

(1) $f(A^{-1}(T)) \subseteq T$;

(2) $Ax \in \overline{CO}\,[f(x) \cup T]$ implies $Ax \in T$.

This concept is a generalization of the concept of a fundamental set, which is successfully used in the theory of condensing vector fields.

2. The existence of an A-fundamental set. Let us consider the mapping $(A - f):\bar{D} \to F$ satisfying the condition

$$(A - f)(\partial D) \subseteq F \setminus 0 \tag{1.1}$$

where 0 is the zero of the space F. The mapping A - f with condition (1.1) is briefly written as $(A - f):(\bar{D}, \partial D) \to (F, F \setminus 0)$ and called a mapping of pairs. In this subsection we shall assume that A is an arbitrary mapping and f is an A-compact mapping with respect to some measure of non-compactness ψ.

Let $Q = (A - f)^{-1}(0)$. Condition (1.1) means that $Q \subseteq D = \bar{D} \setminus \partial D$.

Remark 1.1. Hereinafter, if not stated otherwise, we do not exclude the case $Q = \phi$, where ϕ is an empty set. In this case we imply that f(Q) is also an empty set.

Proposition 1.1. The set f(Q) is a compact.

Proof. Assume the converse, i.e. f(Q) is not a compact. Then $\psi[f(Q)] \neq 0$ and $\psi[f(Q)] < \psi[A(Q)]$. But $f(Q) = A(Q)$, whence

$$\psi(f(Q)) < \psi[A(Q)] = \psi[f(Q)]$$

This contradiction proves the proposition.

The following statement is a key point for the consideration to follow.

Theorem 1.1. There exists a convex, closed compact T which is A-fundamental for the mapping f and contains f(Q).

Proof. There exists at least one convex, closed A-fundamental set for the mapping f which contains f(Q). As such a set, we can take, for example, $\overline{CO}\{f(\bar{D}) \cup A(\bar{D})\}$. Consider the intersection $T = \bigcap_{\alpha} B_{\alpha}$ of all convex, closed, A-fundamental sets for the mapping f, which contain f(Q). Apparently, T is a convex, closed, A-fundamental set for f which contains f(Q). It remains to prove that T is a compact.

Introduce a set $\mathcal{P}(T) = \overline{co}\,[f(A^{-1}(T))]$ and verify that it is an A-fundamental set for the mapping f.

Since T is an A-fundamental set, $f(A^{-1}(T)) \subseteq T$ and, therefore, $\mathcal{P}(T) \subseteq T$ because T is convex and closed. In this case,

$$f[A^{-1}(\mathcal{P}(T))] \subseteq f[A^{-1}(T)] \subseteq \mathcal{P}(T)$$

The first property of an A-fundamental set is proved.

Let now

$$A(x) \in \overline{co}\,[f(x), \mathcal{P}(T)]. \tag{1.2}$$

Using the inclusion $\mathcal{P}(T) \subseteq T$, we obtain $A(x) \in \overline{co}(f(x), T)$ and also $A(x) \in T$, since T is an A-fundamental set. Thus, $x \in A^{-1}(T)$ and $f(x) \in f(A^{-1}(T)) \subseteq \mathcal{P}(T)$, so it follows from (1.2) that $A(x) \in$ $\in \overline{co}(\mathcal{P}(T)) = \mathcal{P}(T)$. The second property of an A-fundamental set is also proved.

Let us verify the inclusion $f(Q) \subseteq \mathcal{P}(T)$. Indeed, $A(Q) = f(Q) \subseteq T$. Hence, $Q \subseteq A^{-1}(T)$ and $f(Q) \subseteq f(A^{-1}(T)) \subseteq \mathcal{P}(T)$. Since the set T with given properties is minimal, we obtain $\mathcal{P}(T) = T$. This relation will now be used to prove that T is compact. Assume the converse, i.e. that T is non-compact or, in other words, that $\psi(T) > 0$. Then

$$\psi[T] = \psi[\mathcal{P}(T)] = \psi[\overline{co}\,(f(A^{-1}(T)))] =$$
$$= \psi[f(A^{-1}(T))] < \psi[A(A^{-1}(T))] = \psi[T \cap A(\bar{D})] \tag{1.3}$$

In these manipulations we have used the non-compactness of $f(A^{-1}(T))$, because otherwise $\mathcal{P}(T) = \overline{co}\, f(A^{-1}(T))$ is compact and, therefore, T is also compact. Thus, in (1.3) we obtain $\psi(T) < \psi[T \cap A(\bar{D})]$, which contradicts the second property of the measure of non-compactness. Hence, T is compact, which proves the theorem.

Remark 1.2. The proof of Theorem 1.1 shows that T admits such additional conditions that are conserved under the operation of intersection of two sets and under the action of the operator $\mathcal{P}$. For example, if A and f are odd mappings, an A-fundamental set T can be chosen symmetric. Furthermore, in Theorem 1.1 we can require that the inclusion R   T be valid for an arbitrary compact. In this case, we put $\mathcal{P}(T) = \overline{co}\,[f(A^{-1}(T)) \cup R]$, all other reasoning being the same.

3. Homotopic classification. Let f_o, $f_1:\bar{D} \to F$ be two continuous A-condensing mappings.

Definition 1.5. The mappings $(A-f_o)$, $(A-f_1):(\bar{D}, \partial D) \to (F, F \setminus 0)$ are called J-homotopic if there exists a continuous mapping $f:\bar{D} \times [0, 1] \to F$, called an A-condensing homotopy, such that:

(1) $f(x, 0) = f_o(x)$, $f(x, 1) = f_1(x)$, $x \in \bar{D}$;

(2) $\psi[f(M \times [0, 1])] < \psi[A(M)]$ for any set $M \subseteq \bar{D}$ for which $\psi[f(M \times [0, 1])] \neq 0$;

(3) $A(x) - f(x, t) \neq 0$ for all $x \in \partial D$ and $t \in [0, 1]$.

Let $[\bar{D}, F, A]_y$ denote the set of homotopic classes of mappings of the form $(A - f):(\bar{D}, \partial D) \to (F, F \setminus 0)$, where f is a continuous A-condensing mapping, and let k_o, $k_1 : \bar{D} \to F$ be two completely continuous mappings.

Definition 1.6. The mappings $(A - k_o)$, $(A - k_1) : (\bar{D}, \partial D) \to (F, F \setminus 0)$ are called C-homotopic if there exists a completely continuous mapping $k:\bar{D} \times [0, 1] \to F$ such that:

(1) $k(x, 0) = k_o(x)$, $k(x, 1) = k_1(x)$, $x \in \bar{D}$;

(2) $A(x) - k(x, t) \neq 0$ for all $x \in \partial D$, $t \in [0, 1]$.

Let $[\bar{D}, F, A]_c$ stand for the set of homotopic classes of mappings of the form $(A - k):(\bar{D}, \partial D) \to (F, F \setminus 0)$, where k is a completely continuous mapping.

Since each completely continuous mapping is condensing, the natural mapping

$$i_* : [\bar{D}, F, A]_c \to [\bar{D}, F, A]_y$$

is valid.

Theorem 1.2 (on the homotopic classification). The mapping i_* is a bijection, i.e. a one-to-one mapping.

Proof. We now demonstrate that i_* is surjective. Let $\alpha \in [D, F, A]_y$ be an arbitrary homotopic class and let $(A - f):(\bar{D}, \partial D) \to (F, F \setminus 0)$ be some representative of this class. Also, let T be a convex, closed, A-fundamental compact for the mapping f which contains $f(Q)$, where $Q = (A - f)^{-1}(0)$. As is noted in Remark 1.1, the case $Q = \phi$ is not excluded either). By virtue of Theorem 1.1, the compact T does exist.

Let $\rho : F \to T$ denote the retraction of the space F onto the compact T. Such a retraction exists, according to Dugunji's theorem. Let us define a completely continuous mapping $k(x) = \rho(f(x))$, $x \in \overline{D}$, and demonstrate that

$$(A-f)^{-1}(0) = (A-k)^{-1}(0). \tag{1.4}$$

Let $x \in (A - f)^{-1}(0) = Q$. Then $f(x) \in f(Q) \subseteq T$ and $f(x) = \rho f(x)$, i.e. $x \in (A-k)^{-1}(0)$. Thus, we have proved the inclusion $(A - f)^{-1}(0) \subseteq$ $\subseteq (A - k)^{-1}(0)$. Let now $x \in (A - k)^{-1}(0)$, i.e. $A(x) = k(x) = \rho f(x) \in$ $\in T$. Then $x \in A^{-1}(T)$ and $f(x) \in f(A^{-1}(T)) \subseteq T$, i.e. $f(x) = \rho f(x) =$ $= k(x)$ and $x \in (A - f)^{-1}(0)$. Hence, the inverse inclusion $(A - k)^{-1}(0) \subseteq (A - f)^{-1}(0)$ is also valid, so that Eq. (1.4) is proved. It should be noted that if $(A - f)^{-1}(0) = \emptyset$ (an empty set), then $(A - k)^{-1}(0) = \emptyset$.

From (1.4) it follows that $(A - k)(\partial D) \subseteq F \setminus 0$, i.e. $(A - k):$ $(\overline{D}, \partial D) \to (F, F \setminus 0)$ is a mapping of pairs. Let β denote a homotopic class of the mapping $(A - k)$ in the set $[D, F, A]_c$. We now demonstrate that the mapping $(A - k)$ is J-homotopic to the mapping $(A - f)$. To this end, we consider the homotopy $g(x, t) = tk(x) + (1-t)f(x)$ and verify whether it satisfies the conditions of Definition 1.5:

(1) $g(x, 0) = f(x)$, $g(x, 1) = k(x)$, $x \in \overline{D}$;

(2) $\psi[g(M \times [0, 1])] = \psi[\overline{co}(k(M) \cup f(M))] = \psi[f(M)] <$ $< \psi[A(M)]$ for any set $M \subseteq \overline{D}$ such that $\psi[g(M \times [0, 1])] \neq 0$.

(3) let

$$Ax_0 - g(x_0, t_0) = 0, \quad t_0 \in [0,1], \quad x_0 \in \partial D. \tag{1.5}$$

Then $A(x_0) = g(x_0, t_0) \in \overline{co}(f(x), T)$. Since the set T is A-fundamental, $A(x_0) \in T$, whence $x_0 \in A^{-1}(T)$ and $f(x_0) \in f(A^{-1}(T)) \subseteq T$. Then $\rho f(x_0) = f(x_0)$ and Eq. (1.5 takes the form $A(x_0) - f(x_0) = 0$, whence $x_0 \in D = \overline{D} \setminus \partial D$. Thus, $A(x) - g(x, t) \neq 0$ for all $x \in \partial D$ and $t \in [0, 1]$.

Consequently, the mappings $(A - k)$ and $(A - f)$ are J-homotopic and, therefore, $i_*(\beta) = \alpha$, i.e. i_* is surjective.

We now prove that the mapping i_* is injective. Suppose $\beta_1, \beta_2 \in$ $\in [D, F, A]_c$ are two homotopic classes such that $i_*(\beta_1) = i_*(\beta_2) =$

$= \alpha \in [D, F, A]_y$. Let $(A - k_1)$ and $(A - k_2)$ be representatives of the classes β_1 and β_2, respectively. Then $(A - k_1)$ and $(A - k_2)$ are J-homotopic, i.e. there exists an A-condensing homotopy $f(x, t)$ that connects k_1 and k_2. According to Theorem 1.1 and Remark 1.2, there exists a convex, closed, A-fundamental compact T_1 for the mapping $f(x, t)$ which contains $f(Q_1) \cup k_1(D) \cup k_2(D)$, where $Q_1 = \{(x, t): Ax = g(x, t),\ x \in \bar{D},\ t \in [0, 1]\}$. Let $\rho_1 : F \quad T_1$ be a retraction. Then $A - \rho_1 f$ is a condensing C-homotopy between the mappings $(A - k_1)$ and $(A - k_2)$. Indeed,

(1) $A(x) - \rho_1 f(x, 0) = A(x) - \rho_1 k_1(x) = A(x) - k_1(x), \quad x \in \bar{D}$

$A(x) - \rho_1 f(x, 1) = A(x) - \rho_1 k_2(x) = A(x) - k_2(x), \quad x \in \bar{D}$

(here we have used the conditions $k_i(\bar{D}) \subseteq T_1$, $i = 1,2$, and the property of ρ_1);

(2) as in the proof of surjectivity, one can demonstrate that

$$(A - f)^{-1}(0) = (A - \rho_1 f)^{-1}(0),$$

which implies the fulfillment of the second condition of Definition 1.6. Thus, the mappings $(A - k_1)$ and $(A - k_2)$ are C-homotopic and, therefore, $\beta_1 = \beta_2$. Hence, we have proved the injectivity and Theorem 1.2.

§ 2. The degree of $\Phi_n C^1$J-mappings

Theorem 1.2 enables one to introduce the concept of degree for classes of mappings which are A-condensing perturbations of the familiar classes of mappings. We shall demonstrate this for the class of $\Phi_n C^1$J-mappings. First, we recall some useful definitions [18].

Definition 2.1. A continuous mapping $A : D \quad F$ of a closed domain $\bar{D} \subseteq E$ into F is called a C^1-smooth Fredholm mapping of index n ($\Phi_n C^1$-mapping, in brief) if the contraction $A|_{\text{int } D}$ (int D is the interior of $\bar{D}$) belongs to the class C^1 and for each $x \in D$ the Fréchet derivative $DA(x)$ is a linear Fredholm operator of index n. The set of $\Phi_n C^1$-mappings $A : \bar{D} \to F$ is denoted by the symbol $\Phi_n C^1(\bar{D}, F)$.

Let $A : \bar{D} \to F$ be a $\Phi_n C^1$-mapping and let $k : \bar{D} \to F$ be a completely continuous mapping. The mapping of the form $A - k$ is called a $\Phi_n C^1$BH-mapping. The set of $\Phi_n C^1$BH-mappings $(A - k) : \bar{D} \to F$ is denoted by $\Phi_n C^1 BH(\bar{D}, F)$.

Let $(A - k) \in \Phi_n C^1 BH(D, F)$ and let A be a proper mapping (a mapping $A:\bar{D} \to F$ is called proper if the inverse image $A^{-1}(K)$ is a compact for any compact $K \subseteq F$). Furthermore, if $(A - k)(\partial D) \subseteq F \setminus 0$ and $n \geq 0$, there is defined, for the mapping $(A - k)$, the degree $\deg_2(A-k, \bar{D}, 0)$ with respect to the point $0 \in F$ [19]. We now use this fact to define the degree of a new class of mappings of Banach spaces.

Let $\Phi_n C^1 Y(D, F)$ denote the set of mappings of the form $(A - f)$: $\bar{D} \to F$, where A is a $\Phi_n C^1$-mapping, f is a continuous A-condensing mapping. Let also $(A - f) \in \Phi_n C^1 J(\bar{D}, F)$, where A is a proper mapping, $n \geq 0$, and $(A - f)(\partial D) \subseteq F \setminus 0$. We denote by $\alpha = [A - f] \in \in [\bar{D}, F, A]_J$ a homotopic class of the mapping $(A - f)$. According to Theorem 1.2, there exists a unique homotopic class $\beta \in [\bar{D}, F, A]_c$ such that $i_*(\beta) = \alpha$. Let $(A - k)$ be any representative of the class β. Then $(A - k) \in \Phi_n C^1 BH(\bar{D}, F)$, A is a proper mapping, and $(A - k)(\partial D) \subseteq F \setminus 0$, i.e. there is valid the degree $\deg_2(A-k), \bar{D}, 0)$ with respect to the point $0 \in F$.

Let $\deg_2(A-f, \bar{D}, 0)$ -- the degree of the mapping $(A - f):(\bar{D}, \partial D) \to \to (F, F \setminus 0)$ with respect to the point 0 -- be defined by the relation

$$\deg_2(A-f, \bar{D}, 0) = \deg_2(A-k, \bar{D}, 0),$$

where $(A - k)$ is any representative of the class $i_*^{-1}[A - f]$.

The ordinary properties of the degree are valid for $\deg_2(A-f, \bar{D}, 0)$. Here we shall consider the most important of them.

<u>Theorem 2.1</u>. If the degree $\deg_2(A-f, D, 0)$ is defined for the mapping $(A - f):D \quad F$ and if the degree is not zero, the equation

$$A(x) - f(x) = 0 \tag{2.1}$$

has a solution in D.

<u>Proof</u>. Assume the converse, i.e. that Eq. (2.1) does not have a solution. Then, according to (1.4), the equation $A(x) - \rho f(x) = 0$ does not have a solution either (here ρ is a retraction on a certain compact, convex, closed, A-fundamental set T). Hence, $\deg_2(A-\rho f, \bar{D}, 0) = = 0$. But $(A - \rho f) \in i_*^{-1}[A - f]$, so that $\deg_2(A - f, \bar{D}, 0) = 0$. The contradiction proves the theorem.

<u>Definition 2.2</u>. Two $\Phi_n C^1 J$-mappings (A_0-f_0), $(A_1-f_1):(\bar{D}, \partial D) \to \to (F, F \setminus 0)$ are called homotopic if there exists a proper C^1-mapping $\Phi: \bar{D} \times [0, 1] \to F$ such that:

(1) $\Phi(\cdot, 0) = A_0$, $\Phi(\cdot, 1) = A_1$;

(2) for each $t \in [0, 1]$ $\Phi(\cdot, t) \in \Phi_n C^1(\bar{D}, F)$ and there exists a Φ-condensing mapping $f: \bar{D} \times [0, 1] \to F$ such that

(3) $f(\cdot, 0) = f_0$, $f(\cdot, 1) = f_1$, and

(4) $(\Phi - f)(x, t) \neq 0$, $x \in \partial D$, $t \in [0, 1]$.

Theorem 2.2. If two $\Phi_n C^1 J$-mappings $(A_0 - f_0)$, $(A_1 - f_1): (\bar{D}, \partial D) \to (F, F \setminus 0)$ $(n \geq 0)$ are homotopic, then

$$deg_2(A_0 - f_0, \bar{D}, 0) = deg_2(A_1 - f_1, \bar{D}, 0).$$

Proof. Let ρ denote the retraction on a compact, convex, closed, A-fundamental set containing $f(\Phi - f)^{-1}(0)$. By definition

$$deg_2(A_0 - f_0, \bar{D}, 0) = deg_2(A_0 - \rho f_0, \bar{D}, 0) \tag{2.2}$$

and

$$deg_2(A_1 - f_1, \bar{D}, 0) = deg_2(A_1 - \rho f_1, \bar{D}, 0), \tag{2.3}$$

But the $\Phi_n C^1 BH$-mappings $(A_0 - \rho f_0)$ and $(A_1 - \rho f_1)$ are connected by the homotopy $\Phi - \rho f$ satisfying the condition $\Phi(x, t) - \rho f(x,t) \neq \neq 0$ for $x \in \partial D$ and $t \in [0, 1]$, because $(\Phi - \rho f)^{-1}(0) = (\Phi - f)^{-1}(0)$. Hence,

$$deg_2(A_0 - \rho f_0, \bar{D}, 0) = deg_2(A_1 - \rho f_1, \bar{D}, 0). \tag{2.4}$$

The proof of the theorem follows from (2.2), (2.3), and (2.4).

Then we have the following analogue of the Borsuk theorem on the oddness of the degree of an odd mapping.

Theorem 2.3. Let $\bar{D} \subseteq E$ be a symmetric domain containing the zero of of the space E, let $A: \bar{D} \to F$ be a proper, odd $\Phi_0 C^1$-mapping, and let $f: \bar{D} \to F$ be an odd, continuous, A-condensing mapping. Then whenever $(A-f)(x) \neq 0$, $x \in \partial D$, the degree $deg_2(A - f_1, \bar{D}, 0)$ is equal to unity.

Proof. Consider a convex, closed, A-fundamental compact T containing $f[(A-f)^{-1}(0)]$ and satisfying an additional symmetry condition (see Remark 1.2). Let ρ be a retraction on T. Then the mapping $\rho_1(x) = \frac{1}{2}[\rho(x) - \rho(-x)]$ is an odd retraction on T. By definition, $deg_2(A-f, \bar{D}, 0) = deg_2(A - \rho_1 f_1, \bar{D}, 0)$. The statement follows from the fact that the equality $deg_2(A - \rho_1 f_1, \bar{D}, 0) = 1$ holds for the odd $\Phi_0 C^1 BH$-mapping $A - \rho_1 f$.

Remark 2.1. We can also consider the mappings of the form (S-f): $(\bar{D}, \partial D) \to (F, F \setminus 0)$, where S is a demicontinuous bounded operator satisfying the condition α) of I.V. Skrypnik [16] and f is a continuous S-condensing operator. As before, choosing some representative S-k (k is a completely continuous mapping) of the homotopic class $i_*^{-1}[S-f]$, we set $\deg(S-f, \bar{D}, 0) = \deg(S-k, \bar{D}, 0)$. The degree $\deg(S-k, \bar{D}, 0)$ has been constructed in [16].

§ 3. Solvability theorems

Here we formulate several existence theorems for the solutions of the equation

$$A(x) + f(x) = 0, \quad x \in E,$$

where A is a $\Phi_n C^r$-mapping, which is proper on each bounded closed set, f is an A-compact mapping.

Definition 3.1. The mapping A:E F is called α -homogeneous if $A(t, x) = t^\alpha A(x)$ for $t \geq 0$ and any $x \in E$.

Theorem 3.1. Let $A: E \to F$ be an α -homogeneous $\Phi_0 C^r$-mapping ($r \geq 1$), which is proper on each bounded closed subset of the space E. Let the equation

$$A(x) = 0, \quad x \in E \tag{3.1}$$

have a unique solution $x = 0$, and let the degree $\deg_2(A, B_1, 0)$ be non-zero (here B_1 is a unit ball of the space E). Then if the condition

$$\lim_{\|x\| \to \infty} \frac{\|f(x)\|}{\|x\|^\alpha} = 0$$

is satisfied for an A-condensing mapping $f: E \to F$, the equation

$$A(x) + f(x) = h \tag{3.2}$$

has a solution for any $h \in F$.

Proof. We now demonstrate that the homotopy $\Phi(x, t) = Ax + tf(x) - th$, $t \in [0, 1]$ does not vanish on a sufficiently large sphere of the space E. Assume the converse. Then there exists a sequence $x_n \in E$ such that $\|x_n\| \to \infty$ and the equality

$$A(x_n) - t_n f(x_n) - t_n h = 0 \tag{3.3}$$

holds true, where $t_n \in [0, 1]$ are certain numbers. Dividing the two sides of (3.3) by $\|x_n\|^\alpha$, we obtain

$$A\left(\frac{x_n}{\|x_n\|}\right) = t_n \frac{h}{\|x_n\|^\alpha} - t_n \frac{f(x_n)}{\|x_n\|^\alpha}. \qquad (3.4)$$

Let us set $y_n = \frac{x_n}{\|x_n\|}$. Then y_n belongs to the unit sphere S_1 of the space E. It follows from (3.4) that $\|A(y_n)\| \to 0$ for $n \to \infty$. Since the contraction $A|_{S_1}$ is a proper mapping, there exists a convergent subsequence $y_{nk} \to y_o$, where $\|y_o\| = 1$. Clearly, $A(y_o) = 0$. We have come to contradiction.

Thus, there exists a sphere $S_r \subseteq E$ for which $\Phi(x, t) \neq 0$, $x \in S_r$, $t \in [0, 1]$. Let B_r denote the ball bounded by the sphere S_r. Since Eq. (3.1) does not have solutions in the set $B_r \setminus B_1$, then $\deg_2(A, B_r, 0) = \deg_2(A, B_1, 0) \neq 0$. And since the mapping $A + f - h$ is homotopic to the mapping A in a given class of mappings, then, according to Theorem 2.2, $\deg_2(A + f - h, B_r, 0) \neq 0$. By virtue of Theorem 2.1, Eq. (3.2) does have a solution in the ball B_r. The theorem is proved.

If A is a linear operator (which is 1-homogeneous), by analogy with the case of compact perturbations (see [20]), we can prove the following statement.

Theorem 3.2. Let $A: E \to F$ be a bounded linear operator such that:

(1) the image Im A is a closed subspace;

(2) E_1 = Ker A -- the kernel of A -- has a closed complement E_2 in E.

Then for any A-condensing operator $f: E \to F$ such that

(3) $f(E) \subseteq \text{Im } A$;

(4) $f(x) = o(\|x\|)$ for $\|x\| \to \infty$

there is valid

$$\text{Im}(A + f) = \text{Im } A.$$

Proof. Represent the space E as a direct sum $E = E_1 \oplus E_2$. Then $A: E_2 \to \text{Im } A$ is a linear isomorphism and I is a homogeneous mapping. We shall assume that $x = x_1 + x_2$, where $x_i \in E_i$, $i = 1,2$. In view of this fact, for any fixed point $x_1 \in E$ we have the equation

$$Ax_2 + f(x_1 + x_2) = y, \quad y \in \text{Im } A. \qquad (3.5)$$

According to Theorem 3.1, this equation can be solved for any point $y \in \operatorname{Im} A$. The inverse inclusion $\operatorname{Im}(A + f) \subseteq \operatorname{Im} A$ follows from condition (3). The theorem is proved.

Theorem 3.2 implies that there exists (for $\|x_1\| \to \infty$) a continuous branch of solutions to Eq. (3.5), $x_1 + x_2(x_1)$, which tends to infinity.

Thus, we can infer some information on the global behaviour of solutions of the equations. Here is another statement of this type.

<u>Theorem 3.3</u>. Let $A: E \to F$ be a proper, odd $\Phi_n C^r$-mapping $(r \geq 1)$ of positive index n, and let $f: E \to F$ be an odd A-condensing mapping. Then for each $\rho > 0$ the equation

$$A(x) + f(x) = 0 \tag{3.6}$$

does have a solution with the norm $\|x\| = \rho$.

<u>Proof</u>. Assume the converse, i.e. that there exists a sphere $S_\rho = \{x \in E: \|x\| = \rho\}$ such that

$$(A + f)(x) \neq 0, \quad x \in S_\rho$$

Let $i: F \to F \times R^n$ be a natural linear embedding. Then the mapping $(iA): E \to F \times R^n$ is a proper odd $\Phi_0 C^r$-mapping, $(if): E \to F \times R^n$ is an odd (iA)-condensing mapping, and

$$i(A + f)(x) \neq 0, \quad x \in S_\rho .$$

By Theorem 2.3, $\deg_2(i(A + f), B_\rho, 0) = 1$, where $B_\rho = \{x \in E: \|x\| = \rho$ is a ball bounded by the sphere S_ρ. However, since the equation $i(A + f)(x) = y$ is not solvable in the ball B_ρ for arbitrarily small (with respect to the norm) points y of the form $y = (0, y_1)$, $y_1 \in R^n$, then $\deg_2(i(A + f), B_\rho, 0) = 0$. We have come to contradiction.

Consequently, on each sphere S_ρ there exists $x \in S_\rho$, which is a solution of Eq. (3.6). The theorem is proved.

For odd completely continuous perturbations of a linear Fredholm operator a similar result was proved by Rabinowitz [21]. An analogous statement is also valid for other classes of mappings (see, e.g., [22]).

Hetzer [9] extended the result of Landesman and Laser [10] to the class of condensing perturbations of linear Fredholm mappings. Let us formulate the generalization of the Landesman-Laser theorem to the class of mappings studied here.

Consider a mapping $(A + f)$, where $A:E \to F$ is a linear Fredholm operator of index $n \geq 0$ and $f:E \to F$ is an A-condensing mapping. Let $E = E_p \oplus E^p$ be a direct decomposition of the Banach space E, where $E_\rho = \operatorname{Ker} A$, and let $x = x_1 + x_2$ be the corresponding decomposition of the element $x \in E$. We shall take that the norm $\|x\|$ in E is defined as $\|x\| = \max\{\|x_1\|, \|x_2\|\}$, where $\|x_1\|$ and $\|x_2\|$ are the norms of the corresponding projections in the spaces E_ρ and E^ρ.

Let P stand for the operator of the linear projection of the Banach space F onto $F_q = \operatorname{Coker} A$. The operator $A|_{E^\rho} : E^\rho \to \operatorname{Im} A$ is reversible. We set $\alpha = \|(A|_{E^\rho})^{-1}\|$.

We recall that a linear operator B is called an asymptotic derivative (or derivative at infinity) of a mapping f if the representation $f(x) = Bx + \omega(x)$ is valid, where $\omega(x)$ satisfies $\lim \|\omega(x)\|/\|x\| = 0$.

In the theorem to follow we shall assume that an A-condensing mapping $f:E \to F$ has an asymptotic derivative B. Introduce the notation $\beta = \alpha \|I - P\| \|B\|$.

<u>Theorem 3.4</u>. Let there exist constants $M > 0$ and $N > 0$ for which $Pf(x_1 + x_2) \neq 0$ whenever $\|x_2\| < \rho_1$, $\rho_1 > N$, and $\|x_1\| \geq M\rho_1$. Let

$$\beta \max\{1, M\} < 1.$$

Then if the stable homotopic class η_r of the mapping $Pf|_{S_r^\rho} : S_r^{\rho-1} \to F_q \setminus 0$, where $S_r^{\rho-1} \subseteq E$ is a sphere of radius r and $r \geq M\rho_1$, is non-zero, the equation

$$A(x) + f(x) = 0$$

has a solution $x \in E$.

The proof of this theorem goes exactly in the same way as for the case of completely continuous perturbations of a linear Fredholm operator (see [18]).

§ 4. <u>The boundary-value problem for a certain class of non-linear partial different al equations</u>

In this section the Dirichlet problem for a certain class of nonlinear partial differential equations is interpreted as the operator equation

$$A(u) - f(u) = h, \tag{4.1}$$

where A is a proper Fredholm mapping of zero index and f is an A-condensing mapping. We shall also formulate an existence theorem for the solution of the Dirichlet problem for the corresponding equation.

Thus, we consider the Dirichlet problem for the equation

$$\sum_{|\alpha|=2m} a_\alpha(x, u, \ldots, D^\gamma u) D^\alpha u = h(x) + b(x, u, \ldots, D^{2m} u), \quad x \in \Omega \tag{4.2}$$

where

$$D_n^j u|_\Gamma = h_j(x'), \quad x' \in \partial\Omega, \; j = 0, \ldots, m-1,$$

and $D_n^i u|_\Gamma$ is the derivative of order j of the function $u(x)$ along the outward normal n to the boundary Γ ; Ω is a bounded domain in Euclidean space R^n with sufficiently smooth boundary Γ ; $W_p^{(2m)}(\Omega)$ and $W_p^{(2m-j-\frac{1}{p})}(\Gamma)$ (j = 0, ..., m-1) are the Sobolev-Slobodetsky spaces of real-valued functions ($m \geq 1$ is an integer); $a_\alpha(x, \xi_j)$ and $b(x, \eta_{2m})$ are real-valued functions.

As in [23], we shall assume that the following conditions are satisfied:

(1) the functions $a_\alpha(x, \xi_j)$ are continuous with respect to the variables (x, ξ_j) $(x \in \bar{\Omega}, -\infty < \xi_j < +\infty)$; $|j| < 2m - n/p$ if $2mp > n$, and the coefficients a_α do not depend on $u, \ldots, D^\gamma u$ if $2mp \leq n$;

(2) there exist the derivatives $\frac{\partial a_\alpha}{\partial \xi_\gamma}$ which are continuous functions of the variables (x, ξ_γ).

In this case, the operator

$$A(u) = \Big(\sum_{|\alpha|=2m} a_\alpha(x, u, \ldots, D^\gamma u) D^\alpha u, \; u|_\Gamma, D_n^1 u|_\Gamma, \ldots, D_n^{m-1} u|_\Gamma \Big)$$

acts from the space $W_p^{(2m)}(\Omega)$ to the space $\mathcal{L}_p(\Omega) \times \prod_{j=0}^{m-1} W_p^{(2m-j-\frac{1}{p})}(\Gamma)$ and is continuously differentiable, i.e. a C^1-smooth operator.

We shall also assume that for each function $u(x)$ from $W_p^{(2m)}(\Omega)$

$$\sum_{|\alpha|=2m} a_\alpha(x, u, \ldots, D^\gamma u)\,\xi^\alpha \geq c(u)\,|\xi|^{2m}, \quad c(u) > 0, \tag{4.3}$$

for any vector $\xi \in R^n$ and any point $x \in \bar{\Omega}$ (after, possibly, correcting $u(x)$ on a set of measure zero). Condition (4.3) is called the ellipticity condition for the corresponding quasilinear differential operator.

For each function $u(x) \in W_p^{(2m)}(\Omega)$ we define a linear operator $A_u(V)$ which also acts from $W_p^{(2m)}(\Omega)$ to $\mathcal{L}_p(\Omega) \times \prod_{j=0}^{m-1} W_p^{(2m-j-\frac{1}{p})}(\Gamma)$

$$A_u(V) = \Big(\sum_{|\alpha|=2m} a_\alpha(x, u, \ldots, D^\gamma u)\, D^\alpha u(x),\ V|_\Gamma,\ D_n^1 V|_\Gamma, \ldots, D_n^{m-1} V|_\Gamma\Big) \tag{4.4}$$

On the basis of these assumptions, it was shown in [23] that a linear operator is a zero-index Fredholm operator for any function $u(x) \in W_p^{(2m)}(\Omega)$.

Introduce the following notation: $\|V\|_{i,p}$ is the norm of an element V in the space $W_p^i(\Omega)$; $\|W\|_F$ is the norm of an element W in the space $F = \mathcal{L}_p(\Omega) \times \prod_{j=0}^{m-1} W_p^{(2m-j-\frac{1}{p})}(\Gamma)$. By S(i) we denote the number of multi-indices α with $|\alpha| \leq i$ for a positive integer i.

Suppose there exists a constant k, $0 < k < \infty$, such that

$$\|V\|_{2m,p} \leq k\,[\,\|A_u(V)\|_\Gamma + \|V\|_{0,p}\,] \tag{4.5}$$

for any $u \in W_p^{(2m)}(\Omega)$. Inequality (4.5) is an a priori estimate for solutions of the boundary-value Dirichlet problem with the parameter corresponding to mapping (4.4).

As for the function $b(x, \eta_0, \ldots, \eta_{2m})$, we shall assume that it satisfies the Caratheodori condition, i.e. it is continuous with respect to $\eta_0, \eta_1, \ldots, \eta_{2m}$ for almost all x and is x-measurable for all $\eta_0, \eta_1, \ldots, \eta_{2m}$ in the domain $(x, \eta_0, \ldots, \eta_{2m}) \in \bar{\Omega} \times R^{S(2m)}$.
We shall also assume that the inequality

$$|b(x, \eta_0, \ldots, \eta_{2m})| \leq \varphi(x) + C\Big(\sum_{|\alpha|<2m} |\eta_\alpha|^{r_\alpha} + \sum_{|\beta|=2m} |\eta_\beta|^{r_\beta}\Big)$$

holds true, where $\varphi(x) \in \mathcal{L}_p(\Omega)$; $r_\alpha < \frac{n}{n-(2m-|\alpha|)p}$ for $n > (2m -$

$- |\alpha|)p$ and r_α are arbitrary non-negative numbers for $n \le (2m - |\alpha|)p$; $r_\beta \le 1$ for $n > 2m$ and r_β are arbitrary non-negative numbers for $n \le 2m$; C is a positive constant.

It follows from these assumptions that the operator f defined by the relation

$$f(u)(x) = (b(x, u(x), \dots, D^{2m}u(x), 0, 0, \dots, 0)$$

acts from $W_p^{(2m)}(\Omega)$ to $F = \mathcal{L}_p(\Omega) \times \prod_{j=0}^{m-1} W_p^{(2m-j-\frac{1}{p})}(\Gamma)$

and is continuous.

Finally, we shall take that for some number $\ell \in [0, k^{-1})$ the inequality

$$|b(x, z, y_1) - b(x, z, y_2)| \le \ell \sqrt[p]{\sum_i (y_1^i - y_2^i)^p} \qquad (4.6)$$

is valid for almost all $x \in \bar{\Omega}$, for all $z \in R^{S(2m-1)}$, and for all $y_1, y_2 \in R^{S(2m)-S(2m-1)}$, where $y_j = (y_j^1, y_j^2, \dots y_j^{S(2m)-S(2m-1)})$, $j = 1, 2$.

A priori estimate (4.5) and inequality (4.6) imply that the operator f is A-condensing relative to the Kuratowski measure of non-compactness j(M) on any bounded subset of the space W_p ( ). We now prove this fact.

Let $D \subseteq W_p^{(2m)}(\Omega)$ be an arbitrary bounded subset. Verify the inequality

$$j(A(M)) \geq k^{-1} j(M) \qquad (4.7)$$

for any non-compact $M \subseteq D$. Assume the converse, i.e. that $j(A(M)) = d_1 < k^{-1} j(M) = k^{-1} d$, where $d = j(M) > 0$. We write $2\mu = k^{-1}d - d_1$. Then $\mu > 0$. For the set A(M) we shall choose the covering $\{D_i'\}_{i=1}^q$ with the diameter $D_i' < d_1 + \mu$, $i = 1, \dots, q$. Let us put $D_i = A^{-1}(D_i') \cap M$. Since the embedding of $W_p^{(2m-1)}(\Omega)$ in $W_p^{(2m)}(\Omega)$ is compact, for any $\delta > 0$ and for a given partition $\{D_i\}_{i=1}^q$ there exists a finite subpartition $\{D_{ij}\}$ of the set M by open subsets D_{ij} such that $\|u - v\|_{2m-1,p} < \delta$ for $u, v \in D_{ij}$. Then, for $u, v \in D_{ij}$ we have

$$\| A(u) - A(v) \|_F =$$

$$\| \sum_{|\alpha|=2m} a_\alpha(x,u,\dots,D^\gamma u) D^\alpha u - \sum_{|\alpha|=2m} a_\alpha(x,u,\dots,D^\gamma u) D^\alpha v +$$

$$+ \sum_{|\alpha|=2m} a_\alpha(x,u,\dots,D^\gamma u) D^\alpha v - \sum_{|\alpha|=2m} a_\alpha(x,v,\dots,D^\gamma v) D^\alpha v,$$

$$(u-v)|_\Gamma , D_n^1(u-v)|_\Gamma , \dots , D_n^{m-1}(u-v)|_\Gamma \|_F \geq$$

$$K^{-1} \|u-v\|_{2m,p} - \eta(\delta) \|v\|_{2m,p}$$

where $\eta(\delta) \longrightarrow 0$ for $\delta \longrightarrow 0$. Hence,

$$\|u-v\|_{2m,p} \leq K(d_1 + \mu + \eta(\delta) \|v\|_{2m,p})$$

for $u,v \in D_{ij}$. It means that $j(M) \leq K(d_I + \mu) = K(K^{-I}d - \mu) <$ $< d = j(M)$, and it follows that $j(M) < j(M)$, i.e. a contradiction. Thus (4.7) is proved.

Show that $j(f(M) \leq lj(M)$, where $l \in [0,K^{-I}]$ for each non-compact set $M \subseteq D$. Let, as above, $j(M) = d > 0$. Then for every $\varepsilon > 0$ there exists a finite number of open sets $D_i \subseteq M$ with diameter diam $D_i < d + \varepsilon$ such that $\bigcup_i D_i = M$. As above, for every $\delta > 0$ with respect to the given partition $\{D_i\}$ there exists a finite sub-partition $\{D_{ij}\}$ of M, such that $\|u - v\|_{2m-I,p} \leq$ $\leq \delta$ and $\|u - v\|_{2m,p} < d + \varepsilon$ for all $u,v \in D_{ij}$. Then for $u,v \in D_{ij}$

$$\| f(u) - f(v) \|_F \leq$$

$$\| b(x,u,\dots,D^{2m-1}u, D^{2m}u) - b(x,v,\dots,D^{2m-1}v, D^{2m}u) \|_{\mathcal{L}_p(\Omega)} +$$

$$+ \| b(x,v,\dots,D^{2m-1}v, D^{2m}u) - b(x,v,\dots,D^{2m-1}v, D^{2m}v \|_{\mathcal{L}_p(\Omega)} \leq$$

$$\leq \eta(\delta) + \left(\int_\Omega [b(x,v,\dots,D^{2m-1}v, D^{2m}u) - b(x,v,\dots,D^{2m-1}v, D^{2m}u)]^p dx \right)^{\frac{1}{p}} \leq$$

$$\leq \eta(\delta) + l \| D^{2m}(u-v) \|_{0,p} \leq$$

$$\leq \eta(\delta) + l \| u-v \|_{2m,p} \leq$$

$$\leq \eta(\delta) + 1(d+\varepsilon),$$

where $\eta(\delta) \to 0$ for $\delta \to 0$. Hence, $j(f(M)) \leq 1d = 1\, j(M)$.

Thus, let $M \subseteq \overline{D}$ be a set such that $j(A(M) \neq 0$. But in this case $j(M) \neq 0$, so that $j(A(M)) \geq k^{-1} j(M) > 1\, j(M) \geq j(f(M)$, i.e. $j(A(M)) > j(f(M))$. Consequently, f is an A-condensing mapping.

To apply the degree theory outlined in Section 2, we should verify that the operator A is proper.

Proposition 4.I. The mapping $A: W_p^{(2m)}(\Omega) \to \mathcal{L}_p(\Omega) \times \prod_{j=0}^{m-1} W_p^{(2m-j-\frac{1}{p})}(\Gamma)$ is proper on bounded, closed subsets of the space $W_p^{(2m)}(\Omega)$.

Proof. Let $\overline{D}$ be a bounded, closed subset of the space $W_p^{(2m)}(\Omega)$. It is required to demonstrate that $A^{-1}(K) \subseteq \overline{D}$ is compact for any compact K in $\mathcal{L}_p(\Omega) \times \prod_{j=0}^{m-1} W_p^{(2m-j-\frac{1}{p})}(\Gamma)$. To this end, it is sufficient to consider the following situation. Let $H^n = (h^n, h_o^n, \ldots, h_{m-1}^n)$ converge in $\mathcal{L}_p(\Omega) \times \prod_{j=0}^{m-1} W_p^{(2m-j-\frac{1}{p})}(\Gamma)$ to the element $Y = (y, y_o, \ldots, y_{m-1})$ as $n \to \infty$, let $A(u_n) = H^n$, and let u_n belong to $\overline{D}$. We now prove that in this case u_n contains a convergent subsequence.

Since the embedding of $W_p^{(2m)}(\Omega)$ in $W_p^{(2m-1)}(\Omega)$ is compact and $\overline{D}$ is bounded, the sequence $\{u_n\}_{n=1}^{\infty}$ contains a subsequence $\{u_{n_i}\}$, which is convergent in $W_p^{(2m-1)}(\Omega)$, i.e. there exists $u_0 \in W_p^{(2m-1)}(\Omega)$ such that

$$\|u_{n_i} - u_0\|_{2m-1,p} \to 0, \quad i \to \infty. \tag{4.8}$$

Consider the operator $A_{u_o}(u)$. It is a Fredholm operator of zero index and, therefore, it is proper on the bounded closed set $\bar{D}$. From condition (4.8) and from the fact that $A(u_n) = Au_n(u_n)$ converges to Y for $n \to \infty$ it follows that $A_{u_0}(u_n)$ also converges to Y for $n \to \infty$. Hence, $\{u_n\}$ does contain the convergent subsequence $\{u_{n_i}\}$. The proposition is proved.

Thus, under the above conditions the Dirichlet problem for Eq. (4.2) is equivalent to the solvability of the equation $A(u) - f(u) = h$, where A is a proper Fredholm mapping of zero index, f is an A-compact mapping.

In conclusion, we shall the solvability theorem for the Dirichlet problem for Eq. (4.2).

<u>Theorem 4.1</u>. Let the conditions formulated above for Eq. (4.2) be satisfied, and let the following conditions be also satisfied:

(1) there exists a constant $C_1 > 0$ such that

$$\|u\|_{2m,p} \le C_1 \|A_t(u)\|_F \,, \quad t \in [0,1], \tag{4.9}$$

where

$$A_t(u) = \Big(\sum_{|\alpha|=2m} a_\alpha(x, tu, \dots, tD^j u) D^\alpha u,\ u|_\Gamma, \dots, D_n^{m-1} u|_\Gamma\Big),$$

(2) there exists a constant $C_2 > 0$ such that

$$|f(x, u, \dots, D^{2m}u)| \le C_2 \Big(1 + \sum_{|\alpha| \le 2m} |D^\alpha u|\Big)^j$$

where $j < 1$. Then the Dirichlet problem for Eq. (4.2) has a solution.

<u>Proof</u>. Consider the homotopy $\Phi(u, t) = A_t(u) - tf(u)$. This homotopy does not vanish on a sufficiently large sphere. Clearly, the degree of the mapping A_0 on the corresponding ball is equal to unity. Then the theorem follows from the properties of the degree (Theorems 2.1 and 2.2).

Some other theorems on the solvability of the Dirichlet problem for Eq. (4.2) can be obtained by using the operator theorems for the existence of solutions (see Section 3).

<u>Remark 4.1</u>. Similarly, one may consider other boundary-value problems for Eq. (4.2) (including those with non-linear boundary conditions). The order $|j|$ of the derivatives $D^j u$ appearing in the coefficients $a_\alpha(x, u, \dots, D^j u)$ of Eq. (4.2) can be raised to $2m-1$ by

increasing p. To raise the order of $|j|$ with more smooth functions $a_\alpha(x, u, \ldots, D^j u)$, $h(x)$, f, and h_i, $i = 0, \ldots, m-1$, one should consider $A-f$ as a mapping acting from $W_p^{(2m+S)}(\Omega)$ to the space $W_p^{(S)}(\Omega) \times \prod_{j=0}^{m-1} W_p^{(2m+s-j-\frac{1}{p})}(\Gamma)$.

REFERENCES

1. Borisovich Yu.G. and Sapronov Yu.I. On a topological theory of condensing operators.- Dokl. Akad. Nauk SSSR, 1968, 183, No. 1 (in Russian).
2. Sapronov Yu.I. On homotopic classification of condensing mappings.- Trudy Matem. Fakulteta, Voronezh State University, 1972, No. 6 (in Russian).
3. Sadovsky B.N. Limiting-compact and condensing operators.- Usp. Matem. Nauk, 1972, 27, No. 1 (in Russian).
4. Krasnoselsky M.A. and Zabreiko P.P. Geometric methods of non-linear analysis, Moscow, 1975 (in Russian).
5. Sadovsky B.N., Akhmerov R.R., Kamensky M.I., and Potapov A.S. Condensing operators.- In: Itogi Nauki i Tekhniki, Moscow, 1980, vol. 18 (in Russian).
6. Hetzer G. Some remarks on Φ_+ operators and on the coincidence degree for a fredholm equation with noncompact nonlinear perturbations.- Ann. Soc. Sci. Bruxelles, 1975, vol. 89.
7. Hetzer G. Some applications of the coincidence degree for set-contractions to functional differential equations of neutral type.- Comment. Math. Univ. Carolinae, 1975, vol. 16.
8. Hetzer G. On the existence of weak solutions for some quasilinear elliptic variational boundary-value problems at resonance.- Comment. Math. Univ. Carolinae, 1976, vol. 17.
9. Hetzer G. A note on a paper of Howard Shaw concerning a nonlinear elliptic boundary value problem.- J. Different. Equat., 1979, 32, No. 2.
10. Landesman F. and Laser A. Non-linear perturbations of linear elliptic boundary value problems at reasonance.- J. Math. Mech., 1970, 19.
11. Hetzer G. and Stallbohm V. Eine Existenzaussage für asymptotisch lineare Störungen eines Fredholmoperators mit Index 0.- Manuscripta Math., 1977, 21.
12. Fitzpatrick P.M. Existence results for equations involving noncompact perturbations of Fredholm mappings with applications to differential equations.- J. Math. Annal. Appl., 1978, 6.
13. Hetzer G. Alternativ- und Veringungsprobleme bei Koinzidenzgleichungen mit Anwendungen auf Randwertprobleme bei neutralen Funktionaldifferential- und elliptischen Differentialgleichungen.- Bonn, 1980.
14. Dmitrienko V.T. and Zvyagin V.G. Homotopic classification of generalized condensing perturbations of mappings.- Proc. 4-th Tiraspol Symp. General Topology and its Applications. Kishinev, 1979 (in Russian).

15. Hoim P. and Spaier E. Compact perturbations and degree theory.- Lect. Notes Math., 1970, vol. 168.

16, Skrypnik I.V. Non-linear elliptic equations of higher order. Kiev, 1973 (in Russian).

17. Zvyagin V.G. On a certain topological method of studying boundary-value problems,which are non-linear with respect to a higher-order derivative. Proc. Conf. Non-linear Problems in Mathematical Physics, Kiev, 1980 (in Russian).

18. Borisovich Yu.G., Zvyagin V.G., and Sapronov Yu.I. Non-linear Fredholm mappings and Leray-Schauder theory.- Usp. Matem. Nauk, 1977, 32, No. 4 (in Russian).

19. Zvyagin V.G. On the existence of a continuous branch of eigenfunctions for a non-linear elliptic boundary-value problem.- Dif. Uravneniya, 1977, 13, No. 8 (in Russian).

20. Nirenberg L. Topics in nonlinear functional analysis. Courant Institute, New York, 1974.

21. Rabinowitz P.H. A note on nonlinear elliptic equations. Indiana Univ. Math. J., 1972, 22.

22. Babin A.V. On the properties of quasilinear elliptic mappings and non-linear elliptic boundary-value problems.- Vestnik MGU. Ser. Matem., Mekh., 1975, No. 5 (in Russian).

23. Pokhozhaev S.I. On non-linear operators with a weakly closed range and on quasilinear elliptic equations.- Matem. sbornik, 1969, No. 78 (in Russian).

ON SOLVABILITY OF NON-LINEAR EQUATIONS WITH FREDHOLM OPERATORS

Yu.G.Borisovich

Voronezh University
394693, Voronezh, USSR

Methods of nonlinear functional analysis, in particular various fixed point principles, are widely applied to investigation of nonlinear problems. Smooth and homotopic topology gives a lot of methods for the theory of nonlinear operator equations. At first one must point out versions of Leray-Schauder theory arose after 40th years of this century and the theory of nonlinear Fredholm mappings arised at the end of 60th years. Theory of games and optimisation lead to the investigation of multivalued mappings and their fixed points. For them corresponding topological principles are also constructed.

The present article deals with modern tendencies in above mentioned branches of nonlinear functional analysis and gives a survey of some new results.

I. Fredholm operators and continuation in parameter.

Let F_λ be a family of nonlinear operators, depending on a parameter $\lambda \in I = [0,I]$. Consider an equation

$$F_\lambda (x) = \theta \qquad (I)$$

in certain functional space. There is a standard method for constructing of a solution $x(\lambda)$ namely to extend in any possible way the known solution $x(0)$ to small λ and more as far as possible sometimes to $\lambda = I$.

Thus Poincaré's method of small parameter is applied to construct solutions in celestial mechanics and theory of oscilations.

Some problems need to use essentially "non-small" values of the parameter. We must point out another Poicaré's method arised in existence problem for closed geodesics on convex analytic surfaces. Here (I) is a differential equation of geodesics, and λ corresponds to a family of convex surfaces joining the surface under consideration with the standard sphere.

The original Poincaré's work on the geodesics has some lacunes (see [41]), this problem has been later studied by calculus of variations in the large (M. Morse, L.A. Lyusternik, A.G. Shnirel'man). But the mentioned method has usefull applications to convex closed sets : both in Weil problem namely a realization of ds^2 = Edu + 2Fdudv + Gdv^2 as an intrinsic regular metric of closed convex surface in R^3 and in the problem if convex surface is defined by its metric (A.D. Aleksandrov, A.V. Pogorelov, etc). The mentioned problems are close to the solvability question for nonlinear elliptic boundary value problems, for which in 1904 S.N.Bernshtein gave a continuation in a parameter principle of Dirichlet problem's solution in the class of analytic function.

H.Poincaré has formulated "parity conservation principle" for a number of solutions (I) with respect to λ . It is natural to describe this principle in terms of Fredholm operator theory. A brief sketch of notions of this theory should be given.

Let M and N be $C^{r \geqslant I}$ smooth Banach manifolds modeled on Banach spaces E and F. C^r smooth mapping f:M$\rightarrow$N is called a Fredholm one (or Φ-mapping) if Frechet differential $D_x f: E \rightarrow F_x, x \in$ M has finite dimensional kernel and cokernel; dim Ker-dim Coker (called analytical index ind$D_x f$ of the differential $D_x f$) does not depend on x$\in$M (for example if M is connected), this common value is called an index ind f of the mapping f. The class of the Fredholm mappings is denoted by Φ.

Let $L \subset N$ be a submanifold in N and f $\pitchfork$ L (f is transversal to L). The following property of Φ-mappings is essential for degree theory and continuation problem.

Proposition I. If f $\pitchfork$ L, full inverse image $f^{-I}(L)$ is either empty set or a submanifold in M such that

$$T_x(f^{-I}(L)) = (D_x f)^{-I}(T_{f(x)}L).$$

If dimL $< \infty$, $f^{-I}(L) \neq \emptyset$, f $\pitchfork$ L, then dim $f^{-I}(L)$ = dimL + ind f; in particular if L = y (a point in N) we have dim $f^{-I}(L)$ = ind f. A set $R_f = \{y \ N : f \pitchfork y\}$ is called a set of regular values of f, $S_f = N \setminus R_f$ is a set of singular (critical) values.

The mapping f is called proper (σ-proper) if for every compact $K \subset N$ the set $f^{-I}(K)$ is compact (is a union of not greater than countable number of compact sets). There is the following generalization of Sard's theorem.

Theorem I (Smale, Queenn) Let Fredholm C^r mapping f : M$\rightarrow$N be such that : I) it is σ-proper, 2) r $>$ max (0, ind f). Then R_f is a residual set.

Proposition I is true also for manifold M with boundary ∂M if both f $\pitchfork$ L and $f_{|\partial M} \pitchfork$ L. Here $f^{-I}(L)$ can have boundary $\partial f^{-I}(L) = \partial M \cap f^{-I}(L)$.

Let X,Y be compact manifolds and $X \cup Y = \partial Z$, where Z is a manifold with boundary. Then X and Y are called bordant. Classes of bordant manifolds $[X^s]$ of dimension s is a group $\mathfrak{N}^s$ of nonoriented bordisms, their direct sum $\oplus\sum_{s\geq 0} \mathfrak{N}^s$ is a ring $\mathfrak{N}$ (calculated by R.Thom).

Let every operator $F_\lambda : E_I \to E_2$ in (I) be Fredholm with the index 0; the family $\{F_\lambda\}$ be induced by a proper C^I Fredholm operator $F : E_I \times I \to E_2$ with the index I; E_I, E_2 be Banach spaces. Then the following proposition holds.

Proposition 2. (Parity conservation principle). If θ is a regular value of F, the number of solutions (I) has the same parity for all $\lambda \in I$.

This proposition follows from the bordism of 0-manifolds $F^{-I}_{\lambda=0}(\theta)$ and $F^{-I}_{\lambda=I}(\theta)$ that form boundary of I-manifold $F^{-I}(\theta)$.

If θ is critical we can obtain only the following proposition.

Proposition 3. (Caccioppoli principle of continuation in parameter). If F is C^2- smooth and $F_{\lambda=0}$ has continuous inverse, there exists a solution of (I) for every $\lambda \in$ I.

Really, according to Smale's theorem F has an arbitrary close to θ regular for F value y_o. One should consider an equation F (x) = y_o instead of (I), where $y_o \to \theta$.

It should be pointed out that this principle was not well grounded by Caccioppoli (another principle was correctly proved by him, namely the principle of completely invertibility, see the survey [I2]) it was formely formulated by L.A.Lyusternik [25] in terms of traversing of a singularity.

2. Topological characteristics.

Denote by $\Phi_q C^r$ the class of Fredholm C^r-mappings with index $q \geq 0$. Geometrical constructions of sect. I are connected with introduction of topological characteristic for proper $\Phi_q C^r$ mappings $f : M \to N$. For a point $y_* \in N$, $r > q+1$, according to Smale define $\deg_2(f,M,y_*) = f^{-I}(Z_*)$ where the right side is an element of the group $\mathfrak{N}^q$, $Z_* = y_*$ if y_* is regular, Z_* is close to y_* and regular if y_* is critic. If M has the boundary $\partial M \neq \emptyset$ the definition is correct if f is proper and continuous, $f \in \Phi_q C^{q+2}$ on $M \setminus \partial M$ and $f(\partial M) \not\ni y$.

Special case of Banach spaces: $M = \bar{\Omega}$ is a closed domain in E_I, $N = E_2$, $y_* = \theta \in E_2$ and $\deg_2(f, \bar{\Omega}, \theta)$ is well defined. If also q = 0, then $\mathfrak{N}^0 = Z_2$ and $\deg_2$ may be 0 or I (degree mod 2). Elworthy, Tromba [40] have introduced smooth structures on Banach manifolds of

Fredholm type (Φ -structures) and have constructed degree deg $(f,\bar{\Omega},\theta)$ with values in the group Z if q = 0.

The interest to characteristics of deg type is connected with their properties, which are important for applications:

I) if $\deg(f,\bar{\Omega},\theta) \neq 0$ then the equation f(x) = 0 has a solution in Ω;

2) deg is constant under homotopies.

Such characteristics are needed also for more general classes of mappings and the investigation of them is being continued.

In present article we consider nonlinear mappings

$$f : \bar{\Omega} \longrightarrow E_2, \; f = A - g \qquad (2)$$

where A, g : $\bar{\Omega} \longrightarrow E_2$ are mapping of closed bounded domain of a Banach space E_I into a Banach space E_2. A is proper, has closed graph, g is continuous with some additional properties (see below).

For f = A, where $A \in \Phi_q C^r$ and proper ($q \geqslant 0$, $r > q + I$), Smale [28], Elwothy and Tromba [40] have defined $\deg_2(f,\bar{\Omega},\theta)$ in the class of non-oriented bordisms $\mathfrak{n}^q$, if $\theta \bar{\in} f(\partial\bar{\Omega})$. Yu.I.Sapronov [I2] has reduced the value of r to r = I for q = 0. Articles [I2] , [26] include a more complicated case q > 0. For applications to boundary value problems it is important to study mappings of the type f = A - g where $A \in \Phi_q C^r$, g is a compact (non-smooth) operator. In this case the topological characteristic for $q \geqslant 0$, $r > q + I$ is defined in [23] . In [26] the value of r is reduced to r = I (see also [I2]).

In [II] the topological characteristic is constructed if the operator A is Fredholm only on $f^{-I}(\theta)$. In [8] g may be maltivalued mapping with convex images. We must note the first papers [34,39] devoted to the case where A is linear Fredholm and g is compact.

V.G.Zvyagin and V.T.Dmitrienko in [27] have introduced a new idea in a principle of compact contraction (see [I6,I4]). They have considered an operator f = A-g with non-compact g subordinated to A. Combining ideas of Fredholm and condensing mapping they have constructed the topological characteristic $\deg_2$ for f = A-g where A is proper and belongs to $\Phi_q C^r$, $q \geqslant 0$, $r > q+I$, and g is condensing with respect to A.

In [I3] we have introduced a notion of distinguishing mapping $\chi : 2^X \longrightarrow K$ from the space of closed subsets of a topological space X into a cone K of non-negative elements in real linear topological space, and also a notion of a concordant with χ mapping $F : U \longrightarrow X$, where U is open domain $\subset X$. F may be either multivalued or singlevalued. The concordance means that $\chi(\overline{FA}) = 0$ if $\chi(A) = 0$ and $\chi(\overline{FA}) \not\geqslant \chi(A)$ if $\chi(A) \neq 0$ for every set $A \subset U$. The principle of compact contraction has the form : if F is concordant with χ, there is a set Φ such that $\Phi \in \operatorname{Ker}\chi$, $U_\Phi = \Phi \cap U \neq \emptyset$, $F : U_\Phi \to \Phi$ (i.e. Φ is invariant under F) and $\Phi \supset R$, where R is any given set included in $\operatorname{Ker}\chi$.

If X = E is a Banach space, a family $\{\Phi\}$ consists of all convex closed sets, χ is a measure of noncompactness (Kuratowski's or Hausdorf's), then we obtain the principle of compact contraction for condensing operators $F : U \longrightarrow E$ not assuming F to be continuous or singlevalued. For multivalued maps we can formulate it in the following manner :

Theorem 2. Let $Y \in E$ be convex bounded damain, $f : Y \longrightarrow E$ - multivalued mapping, condensing with respect to Kuratowski's or Hausdorf's measure of noncompactness (i.e. $\chi(f(M) < \chi(M)$, $\forall M \subset Y, \chi(M) \neq 0$) and such that f(M) is compact for every compact M. Then there exists a fundamental compact set T such that $T \cap Y \neq \emptyset$, $T \supset T_0 = \operatorname{Fix} f$(the set of fixed points $x \in f(x)$), $T \supset R$ (given compact set).

A convex closed invariant set T is called the fundamental with respect to f if $X_0 \in \overline{CO}\,(f(X_0), T)$ necessary and sufficient leads to $X_0 \in T$.

The proof of theorem 2 is in analogy with [I3] , and the essential part of it is a construction of minimal fundamental set T^* as an intersection of all fundamental sets with mentioned properties. Then we prove the equality $\overline{CO}\,(f(T^* \cap Y) \cup R) = T^*$ and obtain that T^* is compact.

Let us continue the investigation of the mapping (2). Consider new possibly multivalued mapping :

$$I - g \circ A^{-I} : A(\bar{\Omega}) \longrightarrow E_2 \qquad (3)$$

Here $A(\bar{\Omega})$ is closed in E_2 because of A is closed. Also $(A-g)^{-I}(\theta) = A^{-I}T_0$, where $T_0 = \operatorname{Fix}(g \circ A^{-I})$.

Definition. The mapping g is called A-condensing if $g \circ A^{-I}$ is condensing with respect to Kuratowski's or Hausdorf's measure of noncompactness.

In articles [I6,I4] there is another definition equivalent to this.

Lemma. For the mapping $g \circ A^{-I}$ all conditions of theorem I are fulfilled.

The mapping (3) has a fundamental compact set $T \supset T_0$, $T \cap A(\bar{\Omega}) \neq \emptyset$. Consider a retraction $\rho_T : E_2 \to T$ and a compact multivalued vector field

$$I - (\rho_T \circ g) \circ (A^{-I}) : A(\bar{\Omega}) \to E_2, \qquad (3')$$

which is correspondant to a mapping

$$A - \rho_T \circ g : \Omega \longrightarrow E_2$$

with compact operator $\rho_T \circ g$.

Thus for every mapping (2) with A-condensing mapping g we have find the mapping (2') with compact mapping $\rho_T \circ g$. Properties of the correspondance (2)$\longmapsto$(2') are included in the following fundamental theorem.

Theorem 3 ([I6]) Let $A,g,k:\Omega \longrightarrow E_2$ be respectively proper, A-condensing, completely continuous . Let $[A - g]_\gamma$, $[A - k]_c$ be homotopy classes sets of the vector fields A - g, A - k not equal to zero on ∂M . Then the natural injection $\mathbb{I} : [A - k]_c \longrightarrow [A - g]_\gamma$ is a bijection.

If additionaly $A \in \Phi_q C^r$, $q \geqslant 0$, $r > q + I$, it is possible to construct the topological characteristic for the vector field A - g with A-condensing mapping g by an equality

$$\deg_2 (A - g, \overline{\Omega}, \theta) = \deg_2 (A - k, \Omega, \theta) \qquad (4)$$

The right-hand side of (4) is defined in [II] .

If A = I , I -condensing mapping is ordinary condensing one, then the corresponding analogy of theorem 3 (bijection principle) and the definition of the characteristic via (4) coincide with the same propositions of Yu.I.Sapronov's article [27] .

In this article Yu.I.Sapronov has finished the development of the idea of [I3] , where relative rotation of vector field in an invariant set T is defined but without an assumption of the fundamentality.

Yu.I.Sapronov's construction is now generally accepted such that sometimes it is described without the author's name as in [2,38] . It is applied to the theory of multivalued condensing mappings [9,I0] and to generalizing of the defenition (4) for the case where g is an upper semi-continuous A-condensing multivalued mapping with convex images. The theorem 2 unifies all of this cases: if g is upper semicontinuos A-condensing multivalued mapping with compact images, then $g \circ A^{-I}$ as before is under conditions of theorem 2, and we can apply the previous construction. Here the vector field $A - \rho \circ g$ has compact multivalued mapping $\rho \circ g$ with convex images, thus the degree is defined in [II] . The bijection principle (theorem 3) is also true.

The case when the multivalued mapping g is acyclic or generalized acyclic is more complicated. It seems that the bijection principle is not true here (at best the usual formula is not true), and the investigation of differential geometric properties of the Fredholm operator A should be done. The following construction is given by B.D.Gel'man and N.Benkafadar.

Let E_I, E_2 be Banach spaces, U a bounded open domain in E_I, $A: U \to E_2$ continuous mapping such that $A/_U$ belongs to the class $\Phi_o C^I$.

Denote the set of compact subsets of E_2 by $K(E_2)$. Upper semicontinuous mapping $g: U \to K(E_2)$ is called generalized acyclic if there exist a topological space Z and continuous mappings $t: Z \to \overline{U}$, $r: Z \to E_2$ such that the following conditions hold :

I) t is surjective and proper; 2) $t^{-I}(x)$ is an acyclic set for every $x \in \overline{U}$; 3) $r \circ t^{-I}(x) \subset g(x)$ for every $x \in U$.

A multivalued mapping $F = A-g$ is called a multivalued vector field with main Fredholm part A.

Let $g(\overline{U})$ be relatively compact in E_2 and $A(x) \overline{\in} g(x)$ for every $x \in \partial U$. It is possible to construct a topological invariant namely a local degree Deg $(A-g, U, \theta)$ for F, where Deg $(A-g, \overline{U}, \theta)$ is a set of natural numbers. If $g = 0$, this invariant has the same absolute value as usual degree of Fredholm mapping. In general case it has the following properties of local degree :

I) if Deg $(A-g, \overline{U}, \theta) \neq \{0\}$ (a set consists of the only point zero), the inclusion $A(x) \in g(x)$ has a solution in U ;

2) if $F(x,\lambda) = A(x) - g(x,\lambda)$ is a nondegenerate homotopy, $F_o(x) = F(x,0), F_I(x) = F(x,I)$, then Deg $(F_o, \overline{U}, \theta) =$ Deg $(F_I, \overline{U}, \theta) \neq \emptyset$;

3) if the set $g(x)$ is acyclic for every $x \in \overline{U}$, the set Deg $(A-g, \overline{U}, \theta)$ consists of the only number.

The construction of the local degree is made by the following manner.

In analogy with the corresponding construction of [I2] it is possible to construct a finite dimensional approximation $g_I = \pi \circ g$ of the multivalued mapping g, oriented finite dimensional $M^n \subset E_I$ and a finite dimensional space $E_2^n \subset E_2$, such that $F_I = A - \pi \circ g : M^n \to E_2^n$ and as before F_I is generalized acyclic multivalued mapping. The local degree Deg $(F_I, \overline{U}, \theta)$ is obtained as a corresponding generalization of Dold's index. It is shown that the degree does not depend on the construction.

Quite recently so called "semi-differentiable" Fredholm mapping have been introduced [24] . It is a generalization of Fredholm properties to non-smooth case. The notion is a natural enlargement of almost strong differentiable Fredholm mappings investigated by E.Ya. Antonovskii [35,36] . The following definitions and a theorem are given by M.N.Kreǐn.

A mapping $f: \Omega \to E_2$ of a domain $\Omega \subset E_I$ (a Banach space) into a Banach space E_2 is called semidifferentiable if for every $x_o \in \Omega$ a linear bounded operator $\tilde{f}(x_o) \in L(E_I, E_2)$ is defined and for every ε there exist a neighbourhood $\vartheta(x_o, \varepsilon) \subset \Omega$ and a finite codimensional subspace $E_{x_o,\varepsilon} \subset E_I$ such that for $(x,y) \subset \vartheta(x_o,\varepsilon) \times \vartheta(x_o, \varepsilon)$, $x-y \in E_{x_o,\varepsilon}$ the inequality $\| f(x) - f(y) - f(x_o)(x-y)\| < \varepsilon \|x-y\|$ holds. The operator $f(x_o)$ is called a semiderivative of f in X_o.

If the semiderivative of f is a Fredholm operator in each point x_o, the operator f is called semidifferentiable Fredholm.

Theorem 4. If $f : \Omega \longrightarrow E$ is a proper semidifferentiable Fredholm mapping, then for every $y \in f(\Omega)$ there sxists a finite dimensionall ball $D_y \ni y$ such that $f^{-I}(D_y)$ is a topological manifold of the dimension dim D_y + ind f.

This theorem allows to make use of finite dimensional reduction for constructing of topological degree for f [24] ; it is defined as a degree of the mapping

$$f|_{f^{-I}(D_y)} : f^{-I}(D_y) \longrightarrow D_y.$$

For mappings of the type $f + k$, where f is proper semidifferentiable Fredholm and k is continuous compact, it is also possible to define the degree. Replace k by finite dimensional mapping k_ε which is ε- close to k such that $f + k_\varepsilon$ is semidifferentiable, and define the degree of $f + k$ to be equal to the degree of $f + k_\varepsilon$. For sufficiently small ε the mappings k_ε are finite dimensional homotopic to each other, therefore the degree does not depend on the choice of sufficiently close to k finite dimensional approximations.

It is possible to construct the degree for some other perturbations of semidifferentiable Fredholm mapping making use of the same methods as for perturbations of smooth Fredholm mapping:

a) k is multivalued upper semicontinuous compact mapping with convex images. In [8] a compact singlevalued ε- approximation k_ε for k is constructed. The degree of $f + k_\varepsilon$ is called the degree of $f + k$. It also does not depend on the choice of k_ε sufficiently close to k.

b) k is f-condensing with respect to certain measure of noncompactness (it means that the mapping $k \circ f^{-I}$ is condensing). Because of the imbedding of homotopy classes $[f + k]_y \supset [f + k]_c$ is a bijection for every proper continuous mapping f (here y means a class of condensing mappings, c means a class of continuous compact mappings [I6]), it is possible to define the degree of $f + k$ via the degree of f plus compact mapping of the $f + k$ homotopy class.

Homotopic stability of the characteristic $\deg_2$ makes possible the extention of Poincaré's continuation principle for corresponding classes of operators $F_\lambda : \overline{\Omega} \longrightarrow E$, depending on the parameter λ , $0 \leqslant \lambda \leqslant I$. That is why the constructing of algorithms for calculating of deg_2 is an actual problem (for some partial cases it is calculated in [I2]). Probably effective results of N.M.Bliznyakov on the calculation of index in a singular point of planar vector field will be usefull here (see [4] for an algebraic algorithm). Note also estimations of index obtained by methods of singularities theory [I,3I-33,4-6].

For investigation of equivariant Fredholm mappings it is possible to use rather general results of T.N.Shcholokova on index calculation for equivariant mappings of spheres [37] .

3. Applications to existence problems in differential equations.

A very interesting example of applications of A-condensing mappings to solvability of certain Dirichlet problem is given by V.G. Zvyagin (see his article in this book). It should be pointed out that the right hand side of the equation includes a nonbounded nonlinearity in the highest derivatives, such that Holder's constant of the nonlinearity is subordinated to the constant of the a priory eatimation of correspondant linear problem.

Leray-Schauder's method of partial inversion of quasilinear part is not adequate here to obtain this theorem because the complete continuity is lost.

V.T.Dmitrienko has given another interesting application of A-condensing vector fields to ordinary differential equatins, which are not resolved with respect to the highest derivatives. We shall describe one of the results.

Consider a second order differential equation

$$A(t,x,\dot{x},\ddot{x},) = f(t,x,\dot{x},\ddot{x}), \quad t \in [0,\omega],$$

(here $A,f : [0,\omega] \times R^n \times R^n \times R^n \longrightarrow R^n$ are continuous, A is C^I - smooth with respect to vector variables) and Picard's boundary value problem

$$x(0) = C_o,\ x(\omega) = C_I;\ C_o,\ C_I \in R^n.$$

Suppose that for every (t,u,v,w) the following conditions hold:

(A_I) the derivative $A_w'(t,u,v,w)$ is invertible;

(A_2) $|A(t,u,v,w) - A(t,u,v,w_o)| > |f(t,u,v,w) - f(t,u,v,w_o)|$ for all $w_o \in R^n$;

(A_3) $|A(t,u,v,w)| \geqslant |w|^m - \sum_{i=o}^{m-I} a_i (|u,v,w|^i),\ a_i > 0,$

$$|f(t,u,v,w)| \leqslant \sum_{i=0}^{m-I} \alpha_i \, |(u,v,w)|^{\,i} + b_m |w|^{\,m}, \alpha_i > 0,\ b_m < I.$$

Theorem 5. Suppose that the finite dimensional degree $\deg(\varphi, \Omega, \theta)$ is well defined and not equal to zero on domaines $\Omega = (|y| < R)$ with sufficiently big R, where $\varphi(y) = A(0, C_o, \frac{CI - C_o}{\omega}, y)$. Also let the conditions (A_I) - (A_3) hold. Then the boundary value problem has a solution.

The scheme of the proof is the same as for V.G.Zvyagin's theorem. But at first we need to rewrite the boundary conditions in the form

$$x(0) = C_o,$$

$$\dot{x}(0) = \frac{C_I - C_o}{\omega} - \frac{I}{\omega} \int_0^{\omega} \int_0^{n} \ddot{x}(s)\, ds du$$

and to introduce operators A,f : $C^2([0,\omega], R^n) \to C^o([0,\omega], R^n)$ by the equalities

$$[A(x)]\,(t) = A\,(t,x,\dot{x},\ddot{x}),$$

$$[f(x)]\,(t) = f\,(t,x,\dot{x},\ddot{x})$$

The characteristic which is introduced in [II] for the mappings of the type $f + k$, where f is Fredholm on the compact $(f+k)^{-I}(\theta)$ and k is completely continuous, makes possible to improve the existence theorem for Monge-Ampere boundary value problem:

$$z_{xx} z_{yy} - z_{xy} = \varphi(x,y,z,p,q),$$

$$z|_{\Gamma} = g(s)$$

This problem was considered in [3] in Hölder functions space with the condition $\varphi'_z \geqslant 0$.

Suppose, that $g(s) \in C^{n+2,\lambda}(\Gamma)$, $\varphi \in C^{n,\lambda}(G)$, $G = \overline{\Omega} \times R^3$, Ω is a circle in R^2, replace the condition $\varphi'_z \geqslant 0$ by the condition $\varphi \geqslant C(\|z\|_I) > 0$ with a continuous function C, and take the majorant of the form

$$\varphi(x,y,z,p,q) \leqslant \Phi(x,y) f(p^2 + q^2) \text{ with } \|z\| \leqslant m = \max g(s),$$

where $f \in C^o$, $f \in C^{n+2,\lambda}$ such that the inequality

$$\sup_{\Omega} \Phi(x,y) \cdot \operatorname{mes} \Omega < \int_{-\infty}^{\infty}\int_{-\infty}^{\infty} [\sup f(\xi^2 + \eta^2)]^{-I} dpdq$$

holds if $(\xi - p)^2 + (\eta - q)^2 \leqslant M_H(\gamma)$. Then the existence theorem is obtained in the class $C^{n+2,\lambda-\varepsilon}$ for certain ε, The abstract scheme for this theorem is described in [I7]. In Sobolev spaces the analogy of this generalization is given in [29] .

4. Local investigation of Fredholm mappings.

Let $f : (E_I, \theta) \longrightarrow (E_2, \theta)$ be $\Phi_q C^r$-mapping. The structure of the set $f^{-I}(\theta) \cap U(\theta)$ for sufficiently small domain $U(\theta)$ is an important question for applications. The case $q > 0$, $\dim \operatorname{Ker} D_f(0) > q$ is of special interest: it includes the bifurcation problem for equation $f(x, \lambda) = 0$ with respect to parameter $\lambda \in R^q$ in a neighbourhood of (0,0) if f(x,0) is of index 0. This problem is investigated by a lot of authors via the use of ideas from the theory of singularities. Yu.I.Sapronov and V.R.Zachepa have introduced a notion of correct bifurcation, which means that the singular solution is isolated and carpeting functional exists. It is shown that the germ of the solution set is of conoid type in a point of correct bifurcation, that the topological type of the conoid is stable under correct deformations of the equation. The notion of finite definiteness of a Fredholm equation in a singular point is introduced. It means that the topological type of the germ does not change when the function is replaced by its correspondant Taylor approximation.

The following theorem is a necessary and sufficient condition for r-definiteness.

Theorem 6. The equation $f(x) = \theta$ is r-definite in the point θ if and only if θ is an isolated singularity of the equation $f_I(x) = \theta$ for arbitrary $f_I \in j_0^r(f)$, where $j_0^r(f)$ is r-jet of the mapping f.

V.R.Zachepa has applied this theory to the investigation of Karman's equation of flexible plates

$$\Delta^2 \omega = \lambda [f, \omega] + [\omega, f], \quad \Delta^2 f = [\omega, \omega] \text{ in } \Omega$$

$$\omega = \omega_x = \omega_y = 0, \quad f = f_x = f_y = 0 \text{ in } \partial\Omega$$

where Ω is a bounded domain in R^2. This problem is reduced to an operator equation in a pair of spaces $\overset{\circ}{W}{}_2^4(\Omega)$, $L_2(\Omega)$:

$$B_u = \lambda A_u + C(u) \tag{5}$$

where B is linear Fredholm operator with index zero, A is linear, C is quadratic. Let dim KerB = n. Consider the bifurcation equation for (5) and 3 first members of its Taylor series, which form the equation

$$\mu^T \xi + \omega^3(\xi) = 0 \tag{$*$}$$

V.R.Zachepa has proved that the existence of two simple solutions among 2n solutions of (*) leads to the existence of a pair of Karman's equation solutions which are stable with respect to perturbations of the order not less than 4 in u and not less than 2 in μ ; that Karman's equation is 3-definite if and only if for $\mu \neq 0$ the linear operator $T + \frac{\partial \omega^3(\xi(\mu))}{\partial \xi}$ is not degenerate on the solutions $\xi = \xi(\mu)$ of (*).

Yu.N.Zavarovskii and Yu.I,Sapronov [20,2I] have investigated nonlinear Kirchhof equations describing the equlibrium form of thin elastic pivot with fixed ends in 3-space with loading along the middle line. The matrix form of the equation is written so:

$$M \frac{d\varkappa}{ds} + \frac{d\varkappa}{ds} M + [\varkappa^2, M] - \lambda[\varepsilon, f^{-I} \varepsilon f] = 0$$

where $\varkappa = f^{-I} \frac{df}{ds}$, f(s) is an orthogonal matrix describing the natural 3-frame of main stresses, s is a parameter of the middle line, λ is a parameter (effort). The left hand side of Kirchhof equation is a variational derivative of the energy functional $w(f) = \frac{1}{2}\langle M\varkappa + \varkappa M, \varkappa\rangle - \lambda\langle \varepsilon, f^{-I}\varepsilon f\rangle$. It is shown that under correspondant conditions on M the equations has a two-dimensional degeneration. A special matrix-algebraic technic is constructed. It makes possible to calculate the normal form of the bifurcation equation and of key function. The normal form of key function is proved to be Legendre's. This results make possible to calculate the asymptotics for small solutions of Kirchhof equation.

The qualitative behaviour of bifurcation equation solutions (for Kirchhof equation) is defined by the extremals of the following family of smooth functionals:

$$x^4 + Nx^2y^2 + y^4 + \mu(x^2 + y^2),$$

if $N^2 \neq 4$.

REFERENCES

I. Arnol'd V.I. Index of singular point of vector field, inequalities of Petrovskiĭ-Oleinik and mixed Hodge structures (Russian). - Funct.anal. appl., I978, vol.I2, No2.

2. Ahmerov R.R.,Kamenskiĭ M.I.,Potapov A.C.,Sadovskiĭ B.N. Condensing operators (Russian). - In: Itogi nauki i tehniki. Mat.analis, VINITI, Moscow, I980, vol.I8.

3. Bakel'man I.Ya. Geometrical methods of elliptic equations solutions (Russian). Moscow, I965.

4. Bliznyakov N.M. Calculation and eatimation of index of singular point of planar vector field (Russian). Dep. VINITI, I979, N304I-79.

5. Bliznyakov N.M. On estimations of topological index of singular point of vector field (Russian). - Dep. VINITI, I979, N589-79.

6. Bliznyakov N.M. On estimates of rotation of vector field on algebraic manifolds (Russian). - Funct. anal. appl., I979, vol.I3, No2

7. Borisovich Yu.G. Topology and nonlinear functional analisys (Russian). - Uspehi mat. nauk, I979, vol. 34, No6.

8. Borisovich Yu.G. On the theory of topological degree for Fredholm mappings perturbed by multivalued operators (Russian). - Dep. VINITI, I980, N5026-80.

9. Borisovich Yu.G., Gel'man B.D., Myshkis A.D., Obuchovskiǐ V.V. Topological methods in fixed points theory of multivalued mappings (Russian). - Uspehi mat. nauk, I980, vol. 35, NoI.

IO. Borisovich Yu.G., Gel'man B.D., Myshkis A.D., Obuchovskiǐ V.V. Multivalued mappings (Russian). - In: Itogi nauki i tehniki. Mat. analis, VINITI, Moscow, I982, vol. I9.

II. Borisovich Yu.G., Zvyagin V.G. On certain topological principle for solvability of equations with Fredholm operators (Russian). - Doklady AN Ukr. SSR, I976, Ser.A, No3.

I2. Borisovich Yu.G., Zvyagin V.G., Sapronov Yu.I. Nonlinear Fredholm mappings and Leray - Schauder theory (Russian). - Uspehi mat. nauk, I977, vol. 32, No4.

I3. Borisovich Yu.G., Sapronov Yu.I. On topological theory of compact contracting mappings (Russian). - Trudy seminara po funkc. analizu, Voronezh, I969, NoI2.

I4. Dmitrienko V.T. Homotopic classification of certain class of multivalued mappings (Russian). - Dep. VINITI, I980, No209I-80.

I5. Dmitrienko V.T. On properness of certain class of differential operators (Russian). - In: Uravneniya na mnogoobraziyah, Voronezh, I982.

I6. Dmitrienko V.T., Zvyagin V.G. Homotopic classification of generalized condensing perturbations of mappings (Russian). - 4th Tiraspol symposium on general topology, Kishinev, I979.

I7. Zakalyukin B.M. Algebraic calculability of index of singular point of vector field (Russian). - Funkct. anal. appl., I972, vol. 6, No I.

I8. Zachepa V.R. Finite-definite equations (Russian). - Dep. VINITI, I980, No36I5-80.

I9. Zachepa V.R., Sapronov Yu.I. Regular bifurcations and regular deformations of nonlinear Fredholm equations (Russian). - Dep. VINITI, I980, No36I7-80.

20. Zavarovskiĭ Yu.N., Sapronov Yu.I. Normal form of key function in the problem of critical loadings of elastic pivots (Russian). - Dep. VINITI, 1981, No4185-81.
21. Zavarovskiĭ Yu.N., Sapronov Yu.I. Two-dimensional degenerations in the problem of critical loadings of elastic pivots (Russian). - Dep. VINITI, 1981, No2602-81.
22. Zabreĭko P.P., Krasnosel'skiĭ M.A., Strygin V.V. On principles of invariance of rotation (Russian). - Izvestiya vysch. uch. zavedenii - matematika, 1972, vol. 5(120).
23. Zvyagin V.G. The existence of a continuous branch of eigenfunctions for nonlinear elliptic boundary value problem (Russian). - Differential equations, 1977, No8.
24. Kreĭn M.N. Semidifferentiable mappings and their topological properties (Russian). - Dep. VINITI, 1979, No2177-79.
25. Lyusternik L.A. Some problems of nonlinear functional analysis (Russian). - Uspehi mat. nauk, 1956, vol. II, No6.
26. Ratiner N.M. On the degree theory of Fredholm mappings with non-negative index (Russian). - Dep. VINITI, 1981, No1493-81.
27. Sapronov Yu.I. On homotopic classification of condensing mappings (Russian). - Trudy matematicheskogo fakulteta, Voronezh University, 1972, No6.
28. Smale S. An infinite dimensional version of Sard's theorem. - Amer. J. Math., 1965, vol. 87.
29. Skrypnik I.V., Shishkov A.E. On solvability of Dirichlet problem for Monge-Ampère equations (Russian). - Doklady AN Ukr.SSR, 1978, Ser.A, No3.
30. Skrypnik I.V. Nonlinear elliptic equations of the highest order (Russian). Kiev, 1973.
31. Hovanskiĭ A.G. Index of polynomial vector field (Russian). - Funct anal. appl., 1979, vol. 13, No1.
32. Hovanskiĭ A.G. On certain class of transcedent simultaneous equations (Russian). - Doklady AN SSSR, 1980, Vol. 255, No4.
33. Hovanskiĭ A.G. Newton's polygon and the index of vector field (Russian). - Uspehi mat. nauk, 1981, vol. 36, No4.
34. Shvarc A.S. Homotopy topology of Banach spaces (Russian). - Doklady AN SSSR, 1964, t. 154.
35. Antonovskiĭ E.Ya. Some topological and analytical properties of almost strong differentiable mappings of Hilbert spaces (Russian). - Uspehi mat. nauk, 1977, vol. 32, No5.
36. Antonovskiĭ E.Ya. Investigation of non-smooth mappings of Hilbert spaces via methods of smooth analysis (Russian). - Uspehi mat.

nauk, 1980, vol. 35, No3.

37. Shchelokova T.N. To the problem of degree calculation for equivariant mapping (Russian). - Sibirskiĭ mat. zhurnal, 1978, vol. 19, No2.

38. Ewert G. Homotopical properties and the topological degree for γ-contraction vector fields. - Bull. Acad. Polon. Sci., 1980, vol. 28, No5-6.

39. Nirenberg L. Generalized degree and nonlinear problems. - In: Contributions to Nonlinear Functional Analysis. New York-London, 1971.

40. Elworthy K.D., Tromba A.J. Degree theory on Banach manifolds. - Proc. Symp. Pure Math, A.M.S. v. 18, 1970.

41. Anosov D.V. Translation editor's notes to the book; Klingenberg W. Lectures on closed geodesics (Russian edition). Moscow, 1982. pp.370-372.

ON CERTAIN PROPERTIES OF EXTREMALS IN VARIATIONAL PROBLEMS

A.T. Fomenko

Department of Mechanics and Mathematics,
Moscow State University, 119899, Moscow, USSR

This survey is a brief report of new results obtained by the author and his pupils in 1980-1981 through the activities of the seminar "Modern methods in geometry" headed by the author at the Department of Mechanics and Mathematics of Moscow State University. The results presented in the paper are mainly concerned with geometric properties of extremals in multidimensional variational problems for the functional of Riemannian volume and for the Dirichlet functional.

§1. Globally minimal surfaces in Riemannian manifolds and geometric properties of the volume of these surfaces

Let X^k be a globally minimal surface in a Riemannian manifold M^n. A question arises: what are geometric properties of the surrounding manifold that determine the volume of a minimal surface, $\mathrm{vol}_k X$, where k is the dimension of X? Let M be an orientable, smooth, compact, connected manifold with the boundary $\partial M = M_1 \cup M_2$, where M_i are connected orientable manifolds. Sometimes, we assume that the boundary M_1 is empty. Let $f: M \to R$ be a Morse's function on M the critical points of which do not belong to the manifold boundary; it is assumed that $0 \leqslant f \leqslant 1$, $f|_{M_1} = 0$, $f|_{M_2} = 1$. Suppose the critical points of the function are not local minima or maxima (provided the boundaries are both non-empty). If the boundary M_1 is empty, we shall assume that the level set $\{f = 0\}$ consists of one point, which is an absolute, non-degenerate minimum of the function. Consider a family of level hypersurface $\{f = r\}$ that foliate the manifold M. Let there be defined, on the manifold, a vector field v such that all its singular points are non-degenerate and do not belong to the manifold boundary. Suppose that: (a) $v(f) > 0$ on the complement to the singular points of v; (b) the set of critical points of the function f is contained in the set of singular points of v; (c) the indices of all the singular points of v are equal to neither zero nor n. As such a field, we can, for example, take the field grad f. The field satisfying conditions (a)-(c) is called f-monotonic.

Let $\partial_1 X = X \cap M_1$, $\partial_2 X = X \cap M_2$, $\partial X = X \cap \partial M$, and let $A = A^{k-1} \subset \subset M_2$ be a fixed (k-1)-dimensional "contour", i.e. a (k-1)-dimensional compact. We write H(A) for the group of (k-1)-dimensional (co)homologies with coefficients in G. Let L be an arbitrary non-empty fixed subgroup (subset) in the (co)homology group H(A). Consider the class O(L) of all compacts $X \subset M$ such that: (1) $\dim X = k$, $\Psi_k(x, X) \geqslant 1$ for any point $x \in X \setminus \partial X$ (here Ψ_k denotes the standard function of spherical density of the subset X in the manifold M); (2) $\partial_2 X \subset A$; (3) the embedding $i: A \longrightarrow \partial_2 X \rightarrow X/\partial_1 X$ implies that $i_* L = 0$, where $i_*: H_{k-1}(A) \rightarrow H_{k-1}(X/\partial_1 X)$, or $L \subset \operatorname{Im} i^*$, where $i^*: H^{k-1}(X/\partial_1 X) \longrightarrow \longrightarrow H^{k-1}(A)$.

Let the class O(L) be non-empty. If $d_k = \inf \operatorname{vol}_k X$, it follows from the general existance theorem (see [1]) that there exists a globally minimal surface $X_o \in O(L)$ such that $\operatorname{vol}_k X_o = d_k$.

Let the surface $X_o/\partial_1 X_o$ pass through a singular point $* = \pi M_1$ in the quotient space M/M_1, where $\pi: M \rightarrow M/M_1$ is factorization, i.e. $\partial_1 X_o \neq \emptyset$. Let us construct a real-valued function, which is called the function of volume of a given globally minimal surface. We set, by definition, $\Psi(X_o, f, r) = \operatorname{vol}_k(X_o \cap \{f \leqslant r\})$. Clearly, this function does not decrease with respect to r. The following problem is of interest for applications: find an exact lower estimate for the function Ψ in terms of the Riemannian metric of the surrounding manifold, this estimate being universal, i.e. independent of the topology of the globally minimal surface X_o. When saying "exact", we mean that the estimate in question must become an equality for sufficiently rich series of particular triplets (M, f, X).

Given an f-monotonic field. Let γ denote its integral trajectories on M. Since the field is f-monotonic, almost all the trajectories originating at the boundary M_1 reach the upper boundary M_2. Consider the field $-v$ and draw an integral trajectory $\gamma(\tau)$ from each point x on the level surface $\{f = r\}$. Two cases are possible: (a) the trajectory terminates on M_1; (b) the trajectory terminates at some singular point of the vector field. The set of separatrix trajectories (case (b)) has a zero measure. Let H be an (n-1)-dimensional hyperplane orthogonal to a vector v at a point x and let Π be an arbitrary (k-1)-dimensional plane in this hyperplane. Consider the exponential mapping (along geodesics) $\exp_x : T_xM \to M$, and let $S_\varepsilon = \exp_x D_\varepsilon^{k-1}$, where $D_\varepsilon^{k-1} \subset \Pi$ is a ball of small radius ε with centre at point 0 in the plane Π ; $x = \exp_x(0)$. Then S_ε is a (k-1)-dimensional ball of radius ε with centre at x. From each point of this ball we draw an integral trajectory of the field -v and extend it until it either meets the lower boundary of the manifold or terminates at a singular point of the vector field. The set of all these trajectories forms a tube C_ε , which is a polyhedral complex of dimension k. Furthermore, this tube is a k-dimensional submanifold with boundary which is smooth almost everywhere (in the sense of k-dimensional volume) in the surrounfing manifold. We set $\varkappa_k(v, x, \Pi) = \lim_{\varepsilon \to 0} \frac{\mathrm{vol}_k C_\varepsilon}{\mathrm{vol}_{k-1} S_\varepsilon}$. Let $\mathcal{X}_k(v, x) = \sup_{(\Pi)} \varkappa_k(v, x, \Pi)$. The function $\varkappa_k(v, x)$ is called a k-dimensional deformation coefficient of the field v. The deformation coefficient can easily be calculated in certain particular cases (see [3]). It turns out that this coefficient plays an important role in the theory of globally minimal surfaces.

Theorem 1. (See [1,2]). Let f be a Morse's function on a manifold M, $0 \leq f \leq 1$, $\partial M = M_1 \cap M_2$, $f|_{M_1} = 0$, $f|_{M_2} = 1$, and let v be an f-monotonic field on the manifold, all the critical points of the function being singular points of the field. We assume that all the singular points are non-degenerate and their indices are positive and do not exceed the number k-2, where $k < n$ is an integer. Let X^k be a globally minimal surface in the manifold that belongs to the class O(L), $\partial_1 X \neq \emptyset$, $L \neq 0$ (see the definitions given above). Then the following inequality is satisfied: $\Psi(X, f, r) \geq q(r) \cdot \lim_{\varepsilon \to 0} \Psi(X, f, a)/q(a)$, where the function $q(r) = \exp \int (\max_{x \in \{f=r\}} \varkappa_x(v,x) \cdot |\mathrm{grad}\, f|)dr$. Thus, the behaviour of the function $\Psi(X, f, r)$ is determined by its behaviour at the initial moment at r = a = 0, i.e. on the lower manifold boundary. Furthermore, the estimate depends on the geometry of the surrounding manifold. There exist, however, sufficiently rich series of quadruples (M, X, f, v) for which the above inequality turns to exact equality, thereby giving explicit formulas for the volume of such globally minimal surfaces, which in turn enables the minimal surface to be described explicitly.

Thus, the volume of any globally minimal surface of an arbitrary codimension that passes through the ball centre and whose boundary lies on the ball boundary is not less than the volume of the standard central plane section of the ball (of the same codimension), i.e. it is not less than the volume of the standard flat disk of the same dimension.

The author has formulated the following general hypothesis. Let B^n be a convex domain in $\mathbb{R}^n$ with a piecewise-smooth boundary, which is centrosymmetric relative to a point O. Let X^k be a globally minimal surface without boundary that tends to infinity in $\mathbb{R}^n$ and passes through the point O. Then the volume of the intersection of this surface with the domain is always not less than the volume of the smallest plane section of this domain by planes of dimension k through the centre of the domain. It is assumed that the surface X^k does not have "whiskers" of zero volume. This obvious assumption can be written as $\Psi_k(x) \geqslant 1$ at each point $x \in X^k$, where Ψ_k is the function of spherical density.

The author has proved this hypothesis for a ball domain (see above). Furthermore, certain particular cases of this statement, e.g. Lelon's theorem, have been known previously. If the domain is a standard 4-dimensional cube in $\mathbb{R}^4 = \mathbb{C}^2$, and X^2 is a complex-analytic surface of codimension 1 (i.e. a complex curve) through the cube centre, the validity of the hypothesis follows from the result of B.E. Katsnelson and L.I. Ronkin. Le Hong Van, a student at Moscow State University, has recently proved this hypothesis for a sufficiently large class of domains in $\mathbb{R}^n$. Her result is as follows. Let B^n be a straight parallelepiped or an ellipsoid with centre at a point O in $\mathbb{R}^n$. Then the area of the intersection of any two-dimensional simply connected minimal surface through the centre of B^n with this domain is always not less than the area of the minimal two-dimensional central plane section of the domain. Apparently, the condition that the surface is simply connected can easily be lifted. We should stress that the hypothesis suggested by the author is concerned with a globally minimal surface without boundary that goes to infinity. Not every minimal surface going to infinity is globally minimal. For example, a two-dimensional catenoid is locally minimal, though it is not a globally minimal surface when "extended to infinity".

§2. Harmonic mappings

The results presented in this section were obtained by post-graduates A.I. Pluzhnikov and A.V. Tyrin. It turns out that harmonic mapping of a manifold into a Lie group is closely related to forms taking values in the Lie algebra of this group. This opens an opportunity for a more detailed analysis of harmonic mappings.

Let $\mathfrak{G}$ be a Lie group, G its algebra, Ad left adjoint action of the group, $ad: G \to G$. Let $\omega \in \Lambda^1(T^*\mathfrak{G}) \otimes G$ be a structural form of the group. If $v \in T_g\mathfrak{G}$, then $\omega(v) = dL_{g-1}(v) \in G = T_e\mathfrak{G}$. Let $C(M, \mathfrak{G})$ be the space of smooth mappings of a smooth manifold M into the group $\mathfrak{G}$ that send a fixed point of the manifold into the unity of the group. Consider the following set of linear forms on M that take values in the Lie algebra of the group $\mathfrak{G}$: $Q = \{\Omega \in \Lambda^1(T^*M) \otimes G \mid d\Omega = -\frac{1}{2}[\Omega, \Omega]\}$. The structural form ω has the following properties: $L^*_g\omega = \omega$, $R^*_g\omega = Ad_{g-1}\omega$, $d\omega = -\frac{1}{2}[\omega, \omega]$. This implies that the relation $\alpha(f) = f^*\omega$ defines the mapping $\alpha: C(M, \mathfrak{G}) \to Q$. This mapping is bijective, and in the set Q one can choose a subset corresponding to harmonic mappings of the manifold into a Lie group. Let us consider, on the group $\mathfrak{G}$, a bi-invariant Riemannian metric and a Riemannian connection consistent with this metric. Then the following relations hold true:

$$S\nabla(f^*\omega) = \omega \circ Bf \quad \text{and} \quad tr\,\nabla(f^*\omega) = \omega \circ Kf,$$

where S is symmetrization functor, Bf is the second fundamental form of mapping f, and Kf is the mean curvature of mapping f.

Theorem 2 (A.I. Pluzhnikov). Let M be a smooth, orientable, simply connected Riemannian manifold. Assume that the Riemannian metric on the Lie group $\mathfrak{G}$ is bi-invariant and the Riemannian connection consistent with this metric is symmetric. Let $H(M, \mathfrak{G})$ and $T(M, \mathfrak{G})$ be spaces of harmonic mappings and completely geodesic mappings of the manifold M into the group $\mathfrak{G}$. Then the following relations are valid:

$$H(M,\mathfrak{G}) = \{F \in \Lambda^1(T^*M) \otimes G \mid dF = -\tfrac{1}{2}[F, F],\ tr\,\nabla F = 0\},$$

$$T(M,\mathfrak{G}) = \{F \in \Lambda^1(T^*M) \otimes G \mid dF = -\tfrac{1}{2}[F, F],\ S\nabla F = 0\}.$$

This result can also be formulated in a different form. The mapping

$f:M \longrightarrow \mathcal{G}$ is called irreducible if the image f(M) does not lie in any proper subgroup of the group $\mathcal{G}$. Introduce a set of pairs $\Omega(M) = \{(f, \mathcal{G}), \mathcal{G}$ is connected and simply connected Lie group, $f \in C(M, \mathcal{G})$ is irreducible$\}$ and introduce in the set $\Omega(M)$ an equivalence relation: $(f_1, \mathcal{G}_1) \sim (f_2, \mathcal{G}_2)$ if there exists the isomorphism $\varphi : \mathcal{G}_1 \longrightarrow \mathcal{G}_2$ such that $\varphi f_1 = f_2$. The set of equivalence classes in $\Omega(M)$ is denoted by $\Omega_0(M)$.

Let G^* be a finite-dimensional linear space of 1-forms on the manifold M and let $dG^* \subset \Lambda^2 G^*$. Then the space G^* is called differentiably decomposable. The set of such spaces is denoted by $\tilde{\Omega}(M)$. It turns out that $\Omega_0(M) = \tilde{\Omega}(M)$, i.e. with each differentially decomposable space $\in C^\infty(\Lambda^1 M)$ there is associated a uniquely defined (to within an isomorphism of the group $\mathcal{G}$) irreducible mapping $f:M \rightarrow \mathcal{G}$, where $\mathcal{G}$ is a certain connected and simply connected Lie group. The converse is also true: each mapping $f:M \longrightarrow \mathcal{G}$ uniquely defines a differentiably decomposable space $\in C^\infty(\Lambda^1 M)$. This means that there exists a correspondence between harmonic mappings of a manifold into a group and differentially decomposable spaces completely contained in the space of co-closed forms.

Here is a sketch of the proof. If G^* is a differentially decomposable space, a structure of a Lie algebra can be introduced in the adjoint space G. Then, as the group, which is used to construct the mapping in question, we can take a uniquely define connected and simply connected Lie group whose Lie algebra coincides with G. Let $L(\Lambda^1 \mathcal{G})$ be a left-invariant form on on the group $\mathcal{G}$. Then the space G^* and mapping f are related by $G^* = f^* L(\Lambda^1 \mathcal{G})$. Since the manifold M is simply connected, any mapping of M into a certain Lie group $\mathcal{G}$ is "passed" through the universal covering group $\tilde{\mathcal{G}}$. Since the study of irreducible mappings is equivalent to the study of mappings into Lie subgroups, the description of differentially decomposable spaces lying in the space of co-closed forms is equivalent to the description of all harmonic mappings of the manifold M into all connected Lie groups that have a bi-invariant metric. There is a correspondence between completely geodesic mappings and differentially decomposable spaces lying in $I(M)^*$, where I(M) is the algebra of vector fields corresponding to isometries.

A smooth Riemannian manifold M provided with the Levi-Civita connection is called an Einstein's space if the Ricci operator in TM is a scalar, i.e. $\mathrm{Ricci}_M = c \cdot \mathrm{id}$, c = const. Consider a harmonic mapping

$f: M \longrightarrow N$ of an orientable, compact, closed, simply connected Einstein's space M into an arbitrary Riemannian manifold N provided with the Levi-Civita connection. Let ind(f) denote the index of this mapping considered as a critical point in the functional space of mappings on which the functional of energy is defined. In other words, ind(f) is a maximal dimension of the space of vector fields "along f" on which the operator of the second variation of the functional of energy is negative definite. Let C(M) and I(M) be vector field algebras corresponding to groups of conformal diffeomorphisms and manifold isometries, respectively. Introduce the concept of the index of a manifold M: ind(M) = = dim C(M) - dim I(M). The following statement is valid.

Theorem 3 (A.I. Pluzhnikov). In dimensions greater than 3, the inequality $\operatorname{ind}(f) \geqslant \operatorname{ind}(M)$ is valid, i.e. if $\operatorname{ind}(M) > 0$, all critical points of the functional of energy in C(M, N) are of saddle type (i.e. they are not points of a local minimum).

These conditions are satisfied, for instance, for a standard sphere S^n. In this case, $\operatorname{ind}(S^n) = n+1$, which implies that the inequality $\operatorname{ind}(f) \geqslant n+1$ holds true for any harmonic mapping of S^n with dimension not less than 3 into the manifold N. In other words, the functional of energy for a sphere does not have local minima at all. The proof of Theorem 3 is based on the following statement (which is in itself of considerable interest): $C(M) = I(M) \oplus G_0(M)$, where $G_0(M) = C(M) \cap \operatorname{Grad}(M)$ is a gradient conformal vector field. This equality does not hold true only for $\dim M \geqslant 3$. The space $G_0(M)$ is the space of vector fields with scalar gradients $(\nabla v = \lambda \cdot \mathrm{id},\ \lambda \in C^\infty(M))$, the index of the manifold M being equal to $\dim G_0(M)$. If M is a Euclidean sphere, $G_0(S^n)$ is the gradient of the restriction, on the sphere S^n, of linear functions in the Euclidean space $\mathbb{R}^{n-1}$. In all earlier statements it was, of course, assumed that the mapping f is not trivial, i.e. the manifold is not mapped into a single point.

We now report the result of A.I. Pluzhnikov which extends the familiar results obtained previously by J. Eells, J.H. Sampson, L. Lemaire, J. Sacs, K. Uhlenbeck, and R.T. Smith. This result permits topological description of the class of manifolds that behave (in the above sense) as spheres in dimensions higher than two.

Theorem 4 (A.I. Pluzhnikov). Let M be a smooth, connected, closed, orientable manifold. Then the following statements are equivalent.

(1) The manifold M is doubly connected, i.e. its first two homotopic

groups are equal to zero.

(2) In at least one Riemannian metric on M, the minimum of the Dirichlet functional is zero on the homotopic class of the identity mapping of M into itself.

(3) For any smooth Riemannian manifold N and for an arbitrary Riemannian metric on the manifold M, the minimum of the functional of energy is zero on all connection components of the mapping space $C^\infty(M, N)$.

(4) For any smooth, compact, orientable Riemannian manifold N and for an arbitrary Riemannian metric on the manifold M, the minimum of the functional of energy vanishes on all connection components of the mapping space $C^\infty(N, M)$.

(5) For any smooth, compact, orientable Riemannian manifold N and for an arbitrary Riemannian metric on the manifold M, the functional of energy has a global minimum only in one homotopic class of mappings from N into M, namely, on locally constant mappings.

Theorem 4 suggests that in dimensions higher than 2 this class of mappings includes, besides spheres, all compact, connected and simply connected Lie groups because their second homotopic group is equal to zero, and also homogeneous Grassmann and Stiefel manifolds over the quaternion body.

The results just described are intimately linked with the result obtained by A.V. Tyrin. Besides the functional of energy $E(f) = \int_M g^{ij} f_i^\alpha f_j^\beta g_{\alpha\beta}\, d\sigma$, we shall consider the Dirichlet functional $D(f) = \int_M (g^{ij} f_i^\alpha f_j^\beta g_{\alpha\beta})^{p/2} d\sigma$, where $p = \dim M$. The latter functional will be considered only for $p \geqslant 2$.

<u>Theorem 5 (A.V. Tyrin)</u>. Let $N = S^q$, where $q \geqslant 3$. Then the smooth mapping $f: M \to S^q$ is a local minimum for the functional of energy (action) or for the Dirichlet functional if and only if the mapping f is trivial, i.e. it maps the manifold into a single point.

If the manifold N is a two-dimensional sphere ($q = 2$), the theorem is not valid.

In conclusion, we shall discuss several simple facts concerning local minimality of a subgroup in a Lie group. It is well known that any (connected) Lie subgroup in a Lie group provided with a bi-invariant Riemannian metric is a completely geodesic (and therefore locally minimal) submanifold. In view of applications, it is useful to consider local minimality of a subgroup in a Lie group provided with a left-invariant Riemannian metric (which is not, generally, bi-invariant).

So, let a left-invariant metric be defined on a Lie group $\mathcal{G}$. Then the criterion that the subgroup $\mathcal{H} \subset \mathcal{G}$ is locally minimal can be written in terms of the Lie algebra G of the group. Let H stand for the Lie algebra of the subgroup $\mathcal{H}$, let ρ be the orthogonal projection $G \to H$, and let $H^{\perp}$ be the orthogonal complement of H in G.

<u>Theorem 6 (A.V. Tyrin)</u>. (a) The subgroup $\mathcal{H}$ is locally minimal in the group $\mathcal{G}$ (relative to the left-invariant metric) if and only if $tr(\rho \circ ad_h) = 0$ for any $h \in H^{\perp}$ (here $\rho \circ ad_h$ is considered as a linear operator $H \to H$); (b) the subgroup $\mathcal{H}$ is a completely geodesic submanifold in the group $\mathcal{G}$ if and only if the operator $\rho \circ ad_h$ is skew-symmetric for any $h \in H$.

<u>Remark</u>. This theorem is a generalization of the earlier result obtained by A.I. Pluzhnikov.

<u>Corollary</u>. Let $\dim \mathcal{G} = \dim \mathcal{H} + 1$ and let $h \in H^{\perp}$, $h \neq 0$. Then the subgroup $\mathcal{H}$ is locally minimal if and only if $tr(ad_h) = 0$.

Note that the equality $tr(ad_h) = 0$ is equivalent to the condition that the form of volume on the group $\mathcal{G}$ (which is left-invariant, since the initial metric on the group is left-invariant) is invariant relative to a right-hand shift by an element of the one-parameter subgroup $\exp(th)$, i.e. the subgroup generated by the normal to H. Here is another useful corollary: if a bi-invariant metric is defined on the group $\mathcal{G}$, each subgroup of co-dimension 1 is (automatically) a locally minimal submanifold with respect to any left-invariant metric.

REFERENCES

1. Fomenko A.T. Multidimensional variational methods in the topology of extremals.- Usp. Matem. Nauk, 1981, <u>36</u>, No. 6, 105-135 (in Russian).
2. Fomenko A.T. On a minimal volume of topological globally minimal surfaces in cobordisms.- Izv. AN SSR, 1981, <u>45</u>, No. 1, 187-212 (in Russian).
3. Eells J. and Sampson H. Harmonic mappings of Riemannian manifolds.- Amer. J. of Math., 1964, <u>86</u>, No. 1, 109-160.

MINIMAL MORSE FUNCTIONS

V.V.Sharko
Institute of Mathematics
Academy of Sciences of the Uk.SSR
252601, Kiev, USSR

The paper deals with a range of questions associated with minimal Morse functions on a smooth manifolds of dimension greater than 5. In the first section we describe some stratification of smooth function space on the manifolds which is very convenient in investigating problems connected with pseudo-isotopy. The second section deals with minimal Morse functions on simply-connected manifolds. Criteria are given for equivalence of two minimal Morse function in terms of some marked elements of homology groups of the manifolds. In the third section the main attention is given to critical points of index I and 2. The relation between some unsolved problems of group theory and two-dimensional polyhedrons is described. The fourth section is devoted to minimal Morse functions on non-simply connected manifolds. The manifolds of dimensions 3 and 4 are not considered in the paper, because R.Mandelbaum's review [29] has appeared recently where related questions are considered. Notice that Morse functions are effectively applied in [39,45] to the study of dynamical systems.

I. "Natural" stratification space of smooth function on the manifolds.

Let M^n be a compact manifold without boundary of class C^∞. We shall denote by $F(M^n)$ the space of infinitely differentiable functions $f: M^n \longrightarrow I, I = 0,I$, $f^{-I}(0), f^{-I}(I) \neq \emptyset$. It is known, that $F(M^n)$ is a Frechet manifold [3]. On the manifolds $F(M^n)$ the group of diffeomorphisms $G = \text{Diff}\ (M^n) \times \text{Diff}\ (I)$ acts by the formula

$$Q : G \times F(M^n) \to F(M^n);\ (g_I, g_2), f = g_2 0 f\ g_I,$$

where $g_I \in \text{Diff}\ (M^n)$, $g_2 \in \text{Diff}\ (I)$. This action can be used for the stratification $F(M^n)$ [36]. Let $T_e(\text{Diff}\ (M^n) \times \text{Diff}\ (I))$ be a tangent space in a point $(id(M^n), id(I))$, $T_f(F(M^n))$ be a tangent space in a point f,

$$DQ : T_e(\mathrm{Diff}\ (M^n) \times \mathrm{Diff}\ (I)) \longrightarrow T_f(F(M^n)$$

be differential Q.

Definition I [35] . The codimension of the image

$$DQ(T_e(\mathrm{Diff}\ (M^n) \times \mathrm{Diff}\ (I)) \subset T_f(F(M^n)).$$

is called the codimension c(f) of the function $f \in F(M^n)$.

It is obvious that the orbit f under the action of G is the submanifold in $F(M^n)$ of codimension c(f). We shall show how it can be determined using its local ring.

Let $x \in M^n$, C_x be the ring of function germs from $F(M^n)$ in the point x. $J_x(f)$ is the ideal, generated by the germs $(\frac{df}{dx_i})$, i = I,2, ...,n. Denote

$$c(f,x) = \dim_R(C_x^{\infty} \quad ,J_x(f)),\ d(f,x,a) = \min(k/(f-a)^k \in J_x(f)),$$

$$d(f,a) = \sup_{f(x) = a} (f,x,a),\ d(f,a) = 0,\ \text{if } f^{-I}(a) = \emptyset.$$

Proposition I [35] . Let $f \in F(M^n)$ have finite codimension c(f), x_i be a finite number of critical points f. Then $c(f,x_i)$, $d(f,x_i,a_i)$, $d(f,a_i)$ are finite and

$$c(f) = \sum_i c(f,x_i) - \sum_i d(f,a_i),\ a_i = f(x_i).$$

Let X be a topological space.

Definition 2. A set of subsets $X^o, X^I, \ldots, X^i, \ldots, X^{\infty}$ of X such that;

I) $X^i \cap X^j = \emptyset$ if $i \neq j$;

2) $\bigcup_{o \leq i} X^i = X$;

3) $\bigcup_{o \leq i \leq k} X^i$ is open set in X for every k —

is called the stratification $\Sigma(X)$ of the space X.

Definition 3. Stratification Σ (X) of the space X is said to be locally trivial if for every point $x \in X$ there exist a stratified set E with a point strat $\{0\}$, a topological space Y with trivial stratification and a marked point y, and a morphism of stratified space $L : E \times Y \longrightarrow X$ is given such that

I) L(0,y) = x;

2) $L(E \times Y)$ is open in X;

3) L is homeomorphism into its image.

Denote by F^i the subspace from $F(M^n)$ consisting of functions of codimension i.

<u>Definition 4</u> [35] . The set of subspaces F^o, F^I,..., F^i,...,F^{∞} is called the "natural" stratification $(F(M^n))$.

It is known, that the "natural" stratification possesses a number of nice properties:

I) $\sum(F(M^n))$ is locally trivial;
2) For every i, F^i has codimension i in $F(M^n)$;
3) The stratification is invariant with respect to the action of the group G;
4) If $f \in F^i$ $(i < \infty)$, then Q_f determines the locally trivial fiber bundle $G \longrightarrow$ orbit f;
5) If $i \leq 5$, then F^i coincides in a neighbourhood of f with the orbit f;

Notice, that the example constructed by H.Hendrics shows that in the codimensions greater then 5 the last condition may not be fulfilled [42] . More detailed description $S(F(M^n))$ can be found in [35] . We shall describe the strat in codimension I and O. The strat in codimension O is Morse functions in general position. The Morse functions are the functions all critical points of which are non-degenerate. The critical point x of function f is called non-degenerate if in a neighbourhood of x there exists a system of coordinates in which f is a quadratic form. The number of minuses of the quadratic form is called the index of the critical point. The Morse function of general position is such a Morse function which admits different values in critical points. The strat in codimension I consists of two component $F^I = F^I_a \cup F^I_b$.

F^I_a is a "birth" function, F^I_b is a Morse function such that its values in two critical points coincide but in others they are different. The function f is called "birth" if all critical points are non-degenerate except for one which is "birth", and the values of the function in critical points are different. The critical point x of a function f is called "birth" if in a neighbourhood of x it is possible to represent the function f in some system of coordinates in the form $f = -x_I^2 - \dots -x_k^2 + x_{k+I}^2 + x_{n-I}^2 + x_n^3$.
The strat in codimension 2 consists of six components, a more detailed description see in [20,2I,27] . Notice, that the definitions abo-

ve are easily carried over the manifolds with boundary. For this it is necessary to require that all critical points of the function belong to the interior of the manifold and the values of the function on connected components of the boundary are constant.

D e f i n i t i o n 5 [20] . A differentiable mapping $\gamma : I \longrightarrow F(M^n)$ such that

I) $\gamma(0)$, $\gamma(I)$ is a Morse function of general position;

2) $\gamma(I) \cap F^2 = \emptyset$;

3) $\gamma(I) \cap F^I$ consists of finite number of points and $\gamma(t)$ intersects F^I in these points transversally, is called the curve of general position.

L e m m a I [20] . If $\gamma(0)$, $\gamma(I)$ are Morse functions of general position, then the curve of general position in $F(M^n)$ connecting $\gamma(0)$ and $\gamma(I)$ exists.

Notice that if $\gamma(I) \cap (\bigcup_{i=I}^{\infty} F^i) = \emptyset$ then $\gamma(0)$ and $\gamma(I)$ coincide within a diffeomorphism of M^n.

We shall describe what happens with the functions under the transformation of strat in codimension I. Let $\gamma(t)$ - be curve of general position intersecting F^I along F^I_b for the value t_o. Futher, let x_o be a critical point of the function $\gamma(0)$, then in M^n we can find a curve $l_{x_o} : I \longrightarrow M^n$, such that $l_{x_o}(0) = x_o$, $l_{x_o}(t) = x_t$, where x_t is a critical point of function $\gamma(t)$. Thus, for every critical point of the function $\gamma(0)$ we can put unique critical point x_t of the function $\gamma(t)$. Let $x_o^I, x_o^2, \dots, x_o^p$ be critical points of the function $\gamma(0)$ which are ordered in such a way that $\gamma(0)(x_o^i) < \gamma(0)(x_o^r)$ for $i < r$. Then we can find exactly two critical points x_o^u, x_o^v of the function $\gamma(0)$ such that $\gamma(0)(x_o^u) < \gamma(0)(x_o^v)$, $\gamma(t)(x_t^u) < \gamma(t)(x_t^v)$, $t \in [0, t_o]$;

$$\gamma(t_o)(x^u_{t_o}) = \gamma(t_o)(x^v_{t_o}),$$

$$\gamma(t)(x_t^u) > \gamma(t)(x_t^v), \ t \in (t_o, I] \ ;$$

for all other critical points if $\gamma(0)(x^{\hat{u}}) < \gamma(0)(x^{\hat{v}})$, then $\gamma(t)(x_t^{\hat{u}}) < \gamma(t)(x_t^{\hat{v}})$ for all $t = I$, $\hat{u} \neq u$, $\hat{v} \neq v$.

Let $\gamma(t)$ be curve of general position crossing F^I along F^I_a for t_o. Then $\gamma(t)$, $(t \neq t_o)$ are Morse functions of general position; $\gamma(t)$ for $t > t_o$ has two critical points more than $\gamma(t)$, $(t < t_o)$, and $\gamma(t_o)$ is "birth".

In transition to strats of more high codimensions, singularities with more high degenerations arise. Thus a number of questions occurs:

I) Is it always possible to connect two Morse functions of general position by a curve in $F(M^n)\setminus \bigcup_{i=I}^{\infty}(F^i)$, i.e. to calculate $\pi_0[F(M^n)\setminus \bigcup_{i=I}^{\infty}(F^i)]$? If no, which type of singularities can arise in a curve of general position?

2) Let $\Omega(M^n)$ be a component of connectedness of a Fréchet manifold $F(M^n)\setminus \bigcup_{i=I}^{\infty}(F^i)$, $\gamma_I(t)$ and $\gamma_2(t)$ are two curves in $\Omega(M^n)$ such that $\gamma_I(0) = \gamma_2(0)$, $\gamma_I(I) = \gamma_2(I)$. Is it possible inside $\Omega(M^n)$ to deform $\gamma_I(t)$ to $\gamma_2(t)$, i.e. to calculate $\pi_I(\Omega(M^n))$ or in general

$$\pi_i\ (\Omega(M^n))\ (i > I)?$$

Something about this is known. On a simply-connected manifold (hence also on not simply-connected one) of dimension greater than 5 it is not always possible to connect the pair of Morse functions of general position by a curve of general position without crossing F^I [II,I3,I6,2I,27,3I].

However, the following statement is valid.

Theorem I 20. Let M^n, $n > 5$, be a smooth, simply-connected closed manifold. Denote by $\mathcal{P}(M^n xI)$ the space of smooth functions on M^n, $f : M^n xI \longrightarrow I$ such that $f^{-I}(0) = M^n \times 0$, $f^{-I}(I) = M^n \times I$, and f has no critical points. Then $\mathcal{P}(M^n \times I)$ is connected.

From this theorem it follows that it is possible to deform arbitrary two pseudoisotopic diffeomorphisms to isotopic diffeomorphisms.

The proof of the Theorem I is rather complicated. Two functions without critical points are connected by a curve of general position and then this path is gradually removed from the strat of codimension I. There is a generalization of this theorem to not simply-connected case [2I,27,33,43] , in that case $\pi_0(\mathcal{P}(M^n \times I)) \neq 0$ depends on higher groups of Whitehead.

2. Ordered minimal Morse functions (simply connected case)

Let f be Morse function of general position. Say, that f is ordered, if $f(x_i^{\lambda}) < f(y_j^{\lambda+I})$, where x_i^{λ}, $y_j^{\lambda+I}$ are critical points of the index λ, $\lambda+I$.

Denote by $F_{\lambda} \subset F(M^n)$ the Morse functions which have minimal number of critical points of index λ .

Definition 6. $f \in \bigcap_{\lambda=0}^{n} F_{\lambda}$ is called minimal (exact) Morse function on M^n.

In general, minimal Morse functions do not exist on every manifold.

The property of ring $\mathbb{Z}\left[\pi_I(M^n)\right]$ to have stable-free modules can form an obstacle, as well as nontriviality of the group $Wh(\pi_I(M^n))$. The non-simply connected case we shall return to later.

The next theorem belongs to S.Smale. It generalizes the well-known Morse inequalities, which connect a number of critical points of smooth Morse function on the manifold with the minimal number of homology group generators of the manifold [5,30].

Theorem 2 [I0,38]. Let W^n be a smooth compact manifold, $\partial W^n = V_o \cup V_I$, $n \geqslant 6$, $\pi_I(W^n) = \pi_I(V_o) = \pi_I(V_I) = 0$. Then on W^n there exists a minimal Morse function whose number of critical points of the index λ is equal to

$$N_\lambda = \mu(H_\lambda(W^n, V_o, \mathbb{Z}) + \mu(\mathrm{Tors}H_{\lambda-I}(W^n, V_o, \mathbb{Z})),$$

where $\mu(H)$ is the minimal number of generators of the group H.

Here V_o and V_I may be empty. There is a generalization of this theorem (belonging to D.Barden) to simply-connected 5-dimensional closed manifolds [I9]. Among the large number of corollaries from Smale 's Theorem we mention the generalized hypothesis of Poincaré in dimensions greater than or equal to 5 (n - dimensional smooth simply-connected manifold, which is homotopically equivalent to n-dimensional sphere, is homeomorphic to n-sphere) and the Theorem on h-cobordism (simply connected h-cobordant manifolds of dimension greater then 4 are diffeomorphic; the manifolds V_o and V_I are called h-cobordant if there exists a manifold W^n such that $\partial W^n = V_o \cup V_I$, and V_o and V_I are homotopically equivalent to the manifold W^n).

There is a generalization of Smale's theorem to the simply connected manifolds with non-simply connected boundary [I5].

Theorem 3. Let W^n be a smooth compact manifold with the boundary $\partial W^n = V_o \cup V_I$, $n \geqslant 6$, $\pi_I(W^n) = 0$. On W^n there exists a minimal Morse function without critical points of indexes 0, I, n - I, n, and with $\mu(\pi_2(W^n, V_o))$ critical points of the index 2, $\mu(\pi_2(W^n, V_o)) + \mu(H_3(W_I, V_o, \mathbb{Z})) - \mu(H_2(W^n, V_o, Q)$ of the index 3, $\mu(\pi_2(W^n, V_I)) + \mu(H_{n-3}(W^n, V_o, Q) - \mu(H_{n-2}(W^n, V_o, \mathbb{Z}) + \mu(\mathrm{Tors}\, H_{n-4}(W^n, V_o, \mathbb{Z})$ - of the index n-3, $\mu(\pi_2(W^n, V_I))$ of the index n-2, $\mu(H_\lambda(W^n, V_o, \mathbb{Z}) + \mu(\mathrm{Tors}\, H_{\lambda-I}(W^n, V_o, \mathbb{Z})$ of the index λ $(4 \leqslant \lambda \leqslant n-4)$.

Here V_o and V_I may be empty.

Let $f: M^n \to I$ be proper Morse function on M^n. The proof of existence of such Morse function can be found in [6].

Assume that $f(x_i^\lambda) < C_\lambda < f(y_j^{\lambda+I})$, where $x_i^\lambda (y_j^{\lambda+I})$ are critical points of the index λ (λ+I). Let $M_\lambda^f = f^{-I}[0, C_\lambda]$, and $\xi(f)$ be gradient-like vector field on M^n for a function f [6]. Given f and $\xi(f)$ determine the expansion of the manifold M^n in the handles. Because of technical reason it is more convenient to use the expansion of the manifold in the handles instead of Morse functions [9] . We have the filtration $M_0 \subset M_1 \subset \ldots \subset M_h = M^n$ $\overline{M_k \setminus M_{k-1}} = \bigcup_i D_i^{\lambda} \times D_i^{n-\lambda}$ constructed with respect to proper Morse function f. It is well-known that

$$H_i(M_{\lambda}, M_{\lambda-1}, \mathbb{Z}) = \begin{cases} 0 & i \neq \lambda \\ \underbrace{\mathbb{Z} \oplus \ldots \oplus \mathbb{Z}}_{k} & \end{cases}$$

where k is the number of critical points (handles) of the index λ . The basis $H_{\lambda}(M_{\lambda}, M_{\lambda-1})$ is defined by middle disks of the handles of the index λ with fixed orientations. Consider a chain complex

$$\{C, \partial\} ; \longleftarrow C_{\lambda-1} \xleftarrow{\partial_{\lambda}} C_{\lambda} \xleftarrow{\partial_{\lambda+1}} C_{\lambda+1} \longleftarrow$$

where $C_{\lambda} = H_{\lambda}(M_{\lambda}, M_{\lambda-1}, \mathbb{Z})$. The homology groups of $\{C, \partial\}$ coincide with the homologies of M^n. The known principle of addition of handles enables us to perform elementary transformations over the matrix of homomorphism ∂_{λ} . Further we shall need some statements on epimorphism of a free group into a finite-generated abelian group.

Let G be a finite-generated abelian group, $\mu(G) = k + l$ be minimal number of its generators.

D e f i n i t i o n 7. Let $\varphi_i : F_i \longrightarrow G$ be epimorphism, where F_i are free abelian groups of the range $k + l$ ($i = 1,2$). Say that φ_1 is equivalent to φ_2, if there exists an isomorphism $f : F_1 \longrightarrow F_2$, such that $\varphi_1 = \varphi_2 \circ f$.

Assume that $g_1, \ldots g_k, \bar{g}_1, \ldots \bar{g}_l$ is a fixed minimal system of generators of the group G, where g_i are free generators, $\bar{g}_i$ are generators of the torsion such that the order of g_i divides the order of g_{i+1}. Suppose that epimorphism $\varphi : F \longrightarrow G$ is given, where F is free group of the range $k + l$. The following Lemma is proved in [11,13,40] .

L e m m a 2. In F there exists a basis $f_1, \ldots, f_k, f_{k+1}, \ldots, f_{k+l}$ such that φ is represented in the form $\varphi(f_i) = g_i$, $(1 \leq i \leq k)$, $\varphi(f_{k+1}) = \alpha \bar{g}_1$, $\varphi(g_{k+j}) = \bar{g}_j$ $(2 \leq j \leq l)$, where α is a generating element in $\mathbb{Z}_{m_1}$ (m_1 is the order of the group, generated by the element $\bar{g}_1$). We shall call α the determinant of epimorphism φ , $|\alpha| < \frac{m_1}{2}$.

Theorem 4 [11,13,40] . Let $\varphi_i : F_i \longrightarrow G$ be epimorphisms into G of free abelian groups of the range $k + l$ ($i = 1,2$). φ_1 is equivalent to φ_2 if and only if their determinants coincide.

Thus, the determinant is naturally correspondant to each epimorphism. We shall use this fact later in classification of minimal Morse functions.

Notice that if $\mathbb{Z}_{m_I}$ is cyclic group of an order m_I and G can be represented in the form $G = \mathbb{Z} \oplus \ldots \oplus \mathbb{Z} \oplus \mathbb{Z}_{m_I} \oplus \ldots \oplus \mathbb{Z}_{m_l}$ where m_i divides m_{i+I}, then there exist $\varphi\left(\left[\frac{m_I}{2}\right]\right)$ of non-equivalent epimorphisms of free abelian groups of the range k + l.

We shall fix certain minimal system of generators $H_\lambda(M^n,\mathbb{Z})$, $g_I^\lambda, \ldots g_k^\lambda, \bar{g}_I^\lambda, \ldots \bar{g}_l^\lambda$, where g_i^λ are free generators, $\bar{g}_j^\lambda$ are generators of torsion, the order $\bar{g}_j^\lambda$ is equal to m_j^λ and divides m_{j+I}^λ. Consider the exact homology sequence of the pair $(M_\lambda^n, M_\lambda^f)$:

$$\longrightarrow H_{\lambda+I}(M_\lambda^f\ \mathbb{Z}) \xrightarrow{i_*} H_\lambda(M^n,\mathbb{Z}) \longrightarrow H_\lambda(M^n, M_\lambda^f, \mathbb{Z}) \longrightarrow 0$$

It is obvious that $H_\lambda(M^n, M_\lambda^f) = 0$ since there are no critical points of the index λ on $M^n \setminus M_f^\lambda$. Hence i_* is epimorphism. It is easy to see that $H_\lambda(M_\lambda^f,\mathbb{Z})$ is free group of the range $\mu(H_\lambda(M^n,\mathbb{Z})$. From the Lemma 2 in $H_\lambda(M_\lambda^f,\mathbb{Z})$ there exists a basis in which i_* is represented in the form

$$\begin{pmatrix} 1_{1_1} & & & 0 \\ & \ddots \alpha_f^\lambda & & \\ & & \ddots & \\ 0 & & & 1_{1_1} \end{pmatrix}$$

α_f^λ is called the λ-determinant of ordered minimal Morse function f.

D e f i n i t i o n 8. Let f_I, f_2 be ordered minimal Morse functions on the manifold M^n. Say, that f_I is equivalent to f_2 if there exists a diffeomorphism $h\colon M^n \longrightarrow M^n$, which is isotopic to the identical, such that $M_\lambda^{f_I} = h(M_\lambda^{f_2})$ for all λ.

Theorem 5 [II,I3]. Ordered minimal Morse functions on M^n are equivalent if and only if all of their λ - determinants ($\pi_I(M^n) = 0$, $n \geqslant 6$) coincide.

For manifolds without torsion in the homology groups all ordered minimal Morse functions are equivalent. This result is due to M.Agoston [I6], also see [3I].

Iet $H(\bar{g}_I^\lambda)$ be a group, generated by $\bar{g}_I^\lambda \in H_\lambda(M^n,\mathbb{Z})$. The generators $H(\bar{g}_I^\lambda)$ we call the marked elements in $H_\lambda(M^n,\mathbb{Z})$.

Theorem 6 [II,I3]. Let M^n be a closed manifold, $\pi_I(M^n) = 0$, $n \geqslant 6$. For every group $\{\alpha_\lambda\}$ of marked elements in $H_\lambda(M^n,Z)$ the ordered minimal Morse function can be constructed on M^n, λ-determinants of which coincide with α_λ.

Using Poincaré's duality it is easy to calculate, that on M^n there exist $G(M^n)$ non-equivalent ordered Morse functions:

$$G(M^n) = \prod_{\lambda=2}^{k-I} \left(\left[\frac{\varphi(m_\lambda) + I}{2} \right] \right)^2 \qquad (n = 2k)$$

$$G(M^n) = \varphi(m_k) \prod_{\lambda=2}^{k-I} \left(\left[\frac{\varphi(m_\lambda) + I}{2} \right] \right)^2 \qquad (n = 2k+I)$$

where m_λ is the order of $H(\bar{g}_I^\lambda) \in H_\lambda(M^n, Z)$; $\varphi(m_\lambda)$ is Euler's function; $[n]$ is the integer part of n:

Notice, that if f_I and f_2 are non-equivalent ordered Morse functions then they can not be connected by an orbit in $F(M^n)$ without crossing the strat of codimension I. See details in [II,3I] .

3. The handles of indices I and 2.

Here we dwell on studying the handles of indices I and 2. We need to fix a marked point in the manifold and to consider marked handles, i.e. a handle with certain fixed path which connects the marked point of the middle sphere of this handle with the marked point of the manifold. The handles of indices I and 2 are of special interest because they define the fundamental group of the manifold. For example, if the homotopic type of the manifold is "good", the handles of the index I are excluded with the help of the handles of indices 2 and 3. The following lemma express this fact more clear.

Lemma 3 [9] . If the manifold W^n is connected, $\pi_I(W^n, V_o) = 0$, $n \geq 6$, then for every expansion of W^n from V_o, which has no handles of the index 0, one can construct new expansion without handles of the index $\leq I$ such that the number of the handles of the index $t = I,2$ is equal to the number of handles of the index i_t in the initial expansion, and the number of the index 3 is equal to $i_I + i_3$.

A question arises, why it is necessary to use the handles of index 3 for exclusion of the handles of the index I? This question is connected with the unsolved hypothesis of Andrews-Curtis on the representation of a trivial group.

Let $G = \{x_I, \ldots, x_n, r_I \ldots, r_m\}$ be the representation of group G. Consider the following transformations over r_i:

I) replacement of r_i by $g \cdot r_i \cdot g^{-I}$, where g is a word in the generators x_i;

2) replacement of r_i by $r_i \cdot r_j^\varepsilon$ or $r_j \cdot r_i^\varepsilon$ $(i \neq j)$, $\varepsilon = \pm I$;

3) introduction of new generator x_{n+I} and new relation $r_{m+I} = x_{m+I}$.

Hypothesis of Andrews-Curtis [I8] . Let $x_I, \ldots, x_n, r_I \ldots, r_m$ be the representation of a trivial group. Then with the help of transformations I) - 3) the representation of the form $\{x_I, \ldots, x_n; x_I = I, \ldots$

$x^n = I; \underbrace{I = I, \ldots I = I}_{m \cdot n} \}$ can be found.

The transformations of the form I)-3) have the following geometric interpretation in terms of handles:

I) chose of new orbit, connecting the marked point with the marked point of the handle of index 2;

2) combination of one handle of the index 2 with another;

3) introduction of new handle of the index I and the additional handle of index 2;

R e m a r k I. If to the transformations of the form I)-3) we add one more transformation: 4) introduction of a new relation $r_{m+I} = I$, I)-4) are equivalent to Tits transformations [4] . It is well known that two transformations of the same group can be transformed one into another with the help of transformation of the form I)-4) [4] . Lemma 3 is a geometric interpretation of this assertion.

R e m a r k 2. There exist representations of isomorphic groups, which are not equivalent with respect to the transformations of the form I)-3) [24 , 32] . As an example of such groups one can take the group of three-leaved figure or a finite abelian group.

These questions border on the following problem due to Likorish [33,4I] .

Let K be contractible finite two-dimensional complex:

A) hypothesis (Zieman): complex K is 3-collapsed into the point [44] ;

B) hypothesis: K is 3-deformed into the point, i.e. there exists a three-dimensional complex L, such, that L is deformed into K and L is deformed into the point;

C) the unique regular neighbourhood of K in $\mathbb{R}^5$ is 5-dimensional disk D^5;

D) the hypothesis of Andrews-Curtis.

It is known that from A) follows three-dimensional hypothesis of Poincaré, A) $\Rightarrow$ B) $\Rightarrow$ C) and B) $\Leftrightarrow$ D) [4I] .

Let (W^n, V_o, V_I) be h-cobordism, $n \geqslant 5$. It is well known [9] , that W^n has the expansion in the handles from the manifold V_o so that there exist only the handles of indices λ and $\lambda + I$ in equal quantities ($2 \leqslant \lambda \leqslant n-4$). It is known also [9] , that class of manifolds, diffeomorfic to W^n is determined by Whitehead's torsion.

Recall the construction of Whitehead group Wh(G).

Let G be the group, $\mathbb{Z}[G]$ be its integer group ring. Then the group $K_I(\mathbb{Z}[G])$, [7], by the definition is the abelinisation of infinite linear group $GL(\mathbb{Z}[G]) = \varinjlim GL(\mathbb{Z}(G), n)$. Whitehead's group $Wh(G) = K_I$

$(\mathbb{Z}[G])/\{\pm G\}$. The intersection of marked handles of the index λ and $\lambda+I$ in h-cobordism (W^n, V_o, V_I), which is considered in the terms of group ring $\mathbb{Z}[\pi_I(V_o)]$, determines some nondegenerate matrix A of the demension nxn(n is number of handles of index λ (λ +I)) and thus defines the torsion $\tau(W^n, V_o)$. On the another hand, for every element τ_o of the group Wh(G) (G is fundamental group of the manifold V_o, $n \geqslant 5$) one can construct h-coborsism (W^n, V_o, V_I) such that $\tau_o = \tau(W^n, V_o)$. A natural question arises how many handles of the ind es λ and λ + I are necessary for the construction of h-cobordism. Before answering this question we need the following definition.

Definition 9. Say that the representable dimension G is $\leqslant$ i (r-dimG $\leqslant$ i), if every element Wh(G) can be represented by a matrix from GL ($\mathbb{Z}[G]$,i).

Let $U(\Lambda)$ be the group of units of the ring Λ . Assume $SK_I(\mathbb{Z}(G))=$ = Ker $K_I(\mathbb{Z}[G]) \longrightarrow K_I(Q[G])$ [38] . If G is the abelian group, then $Wh(G) \approx U(\mathbb{Z}[G]/\{\pm G\} \oplus SK_I(\mathbb{Z}[G]))$ [22] . Thus, if G is abelian then r-dimG = I if and only if $SK_I(\mathbb{Z}[G]) = 0$. We shall sum the well-known results about r-dimG in the following theorem.

Theorem 7. a). Let G be a finite group, then r-dimG $\leqslant$ 2 [I,28] ;

b). If $G \approx \mathbb{Z}_n$ or $\mathbb{Z}_p \oplus \mathbb{Z}_{p^i}$ for certain prime p, then r-dimG = I [29] ;

c) Let G be finite abelian group, which has subgroup, isomorphic to $\mathbb{Z}_p \oplus \mathbb{Z}_p \oplus \mathbb{Z}_p$ (p is prime number) or isomorphic to $\mathbb{Z}_2 \oplus \mathbb{Z}_2 \oplus \mathbb{Z}_2$, or to $\mathbb{Z}_4 \oplus \mathbb{Z}_4$, then r-dimG = 2 [I7] .

At present, the examples of finitely representable groups for which r-dim u > 2 are unknown. On the other hand, it is not known whether there exists an integer n=n (G) such that r-dimG $\leqslant$n.

Thus, the theorem 7 provides the estimate of the handle number, which is necessary for the construction of h-cobordism.

On the other hand, it is naturally to put the question whether every element $\tau_o = \tau(W^n, V_o) \in Wh(\pi_I(V_o))$ can be realized by the expansion of W^n in handles of the index 2 and I from the manifold V_o. A partial answer is given by the following theorem.

Theorem 8 [23] . There exists a group G and a torsion element $\tau_o \in Wh(G)$ such that if (W^n, V_o, V_I) is h-cobordism with $\tau(W^n, V_o) = \tau_o$ and $\pi_I(V_o) \approx G$, then every expansion of W^n in handles from the manifold V_o contains handles of the index 3 and more.

Since for every element $\tau_o \in Wh(G)$ there exists h-cobordism (W^n, V_o, V_I) with $\tau(W^n, V_o) = \tau_o$, we denote by $Wh_2(G)$ those elements from Wh(G), for which the corresponding to them h-cobordisms have in their expansion only handles of the indexes I and 2. It is not known whether $Wh_2(G)$ is the subgroup in Wh(G). For details see in [23,32,34] .

4. Minimal Morse functions (not-simply connected case)

Simple examples show that minimal Morse functions (in sense of Definition 6) do notnecessarily exist on not simply-connected manifolds. At least three objects can make obstruction for the existence of minimal Morse function on not-simply connected manifold M^n: fundamental group of M^n, homologies of universal covering manifold M^n and Whitehead torsion. Some related results can be found in [2,IO,I2,I4,25, 26] .

We dwell on the difficulties which here are encountered. We start from Whitehead's torsion. The examples with h-cobordism show, that in the not-simply connected case certain ambiguity occurs in the expansion of manifold in the handles. For example, there exists a h-cobordism (W^n,V_o,V_I), which admits the expansion in handles of the indexes 2 and 3 in one case and in the handles of the indexes 4 and 5 in another one. Analogous examples can be constructed for closed manifolds.

Consider now the influence of the manifold fundamental group. Let $G = \{x_I,\dots,x_n,r_I\dots,r_m\}$ be representation. It is minimal, if for arbitrary another representation $G = \{\bar{x}_I,\dots,\bar{x}_{\bar{n}},\bar{r}_I\dots,\bar{r}_{\bar{m}}\}$ the inequalities $n \leqslant \bar{n}$, $m \leqslant \bar{m}$ hold. It is unk wn for what class of groups there exists minimal representation.

<u>D e f i n i t i o n IO</u>. Let Λ be iBN-ring (the ring over which a finite generated module has uniquely defined range). Say, that Λ is S-ring, if arbitrary finitely generated stable free module is free.

Recall, that the module M is called stable free if $\Lambda \oplus \dots \oplus \Lambda$ is free.

<u>Hypothesis</u>. Let G be finitely defined group, $\mathbb{Z}[G]$ be its integer number group ring. Then for G there exists minimal representation, if $\mathbb{Z}[G]$ is S-ring.

Since on the closed manifold of a handle of the indices I and 2 the fundamental groups give representation, then for the construction of minimal Morse functions first it is necessary to know whether there exists the minimal repsentation for the fundamental group or not.

Notice, that all group rings for finitely representable groups are iBN-rings.

Let M^n be the closed manifold, $\pi_I(M^n) \approx G$. For each expansion M^n in the handles one can construct the chain complex of free modules over $\mathbb{Z}[G]$ of the form

$$\{\mathbb{C},\partial\}\ ,\ 0 \leftarrow \mathbb{Z} \xleftarrow{\varepsilon} \mathbb{Z}[G] \xleftarrow{\partial_1} C_I \leftarrow \dots \xleftarrow{\partial_n} C_n \leftarrow \mathbb{Z} \leftarrow 0$$

the homology modules of which coinside with the homology modules of universal covering manifold M^n.

D e f i n i t i o n II. Say that the chain complex $\{\mathbb{C},\partial\}$ is minimal if for any chain complex

$\{\bar{\mathbb{C}},\bar{\partial}\}$, $0 \leftarrow \mathbb{Z} \xleftarrow{\varepsilon} \mathbb{Z}[G] \xleftarrow{\bar{\partial}_1} \bar{C}_I \leftarrow \dots \xleftarrow{\partial_n} \bar{C}_n \leftarrow \mathbb{Z} \leftarrow 0$, which is chain equivalent to $\{\mathbb{C},\partial\}$, the inequality $\mu(C_i) \leq \mu(\bar{C}_i)$ holds, where $\mu(C)$ is minimal number of the generators of module C.

In general, if $\mathbb{Z}[G]$ is not S-ring, the minimal chain complex may not exist in the class of chain-equivalent chain complexes. It is not difficult to show this by use of stable free modules. On the basis of this fact a closed manifold can be constructed such that its fundamental group is isomorphic to G, but it is impossible to define minimal Morse functions on it.

Theorem 9. Let (W^n, V_o, V_I) be the smooth compact manifold with boundary $\partial W^n = V_o \cup V_I$, $n \geqslant 6$. Suppose, that $\pi_I(W^n) \approx \pi_I(V_o) \approx \pi_I(V_I)$, $\mathbb{Z}[G]$ is Noether S-ring, Wh(G) = 0. Then on W^n there exists minimal Morse function.

The rest of this section is devoted to the discussion of the proof of this Theorem.

D e f i n i t i o n I2. Let $f : F \longrightarrow G$ be homomorphism of free modules, $n\Lambda$ be free module of the range n. Homomorphism $f(n) : F \oplus n\Lambda \xrightarrow{f \oplus jd} G \oplus n\Lambda$ is called the stabilisation of homomorphism f.

D e f i n i t i o n I3. Let $f : F \longrightarrow G$ be homomorphism of free modules with fixed basises. If with the help of elementary transformations of the basises it is possible to represent f as the stabilisation of some homomorphism $\bar{f}$ with the help of module $n\Lambda$, we say that $\bar{f}$ is the contraction of the homomorphism f.

By the elementary transformation of the basis we mean the transformation which is realized by elementary matrix.

The following theorem is valid.

Theorem IO. Let $\mathbb{Z}[G]$ be Noether S-ring, $K_I(\mathbb{Z}[G]) = 0$,

$$\{\mathbb{C},\partial\}\ ,\ 0 \leftarrow \mathbb{Z} \xleftarrow{\varepsilon} \mathbb{Z}[G] \xleftarrow{\partial_1} C_I \leftarrow \dots \xleftarrow{\partial_n} C_n \leftarrow \mathbb{Z} \leftarrow 0$$

be chain complex of free modules over $\mathbb{Z}[G]$. Then in the class of chain complexes chainly equivalent to $\{\mathbb{C},\partial\}$, it is possible to find minimal chain complex with the help of stabilisation and reduction of boundary homomorphisms (we assume, that some bas s are fixed in C_i).

The Proof of Theorem 9. Since $\pi_I(W^n,V_o)\approx\pi_I(W^n,V_I) = 0$ on W^n it is possible to give the expansion in the handles without handles of indexes 0,I [9, Lemma 6.I5] . We choose marked points on V_o and on the middle spheres of each hande and then connect them with non-crossing curves. Using universal covering manifold we shall construct a chain complex of free modules over $\mathbb{Z}[G]$. Notice that homologies of the chain complex in this case coinside with those of the universal covering. The introduction and reduction of additional handles of the indexes λ and $\lambda+I$ $(2\leq\lambda\leq n-4)$ correspond to stabilisation and reduction over boundary homomorphisms in $\{\mathbb{C},\partial\}$. Using theorem IO, we find minimal expansion in the handles on (W^n,V_o,V_I).

The proof of Theorem IO is rather cumbersome; we give only key moments of it.

D e f i n i t i o n I4. Let $f : F \longrightarrow M$ be homomorphism, where F is free module. Say that f is minimal, if $\mu(F) = \mu(M)$.

D e f i n i t i o n I5 [I] . Let N be submodule of the module M. Define f-rank (N,M) as maximal value of those integers $r \geqslant 0$, for which N contains direct summand of the module M isomorphic to $r\Lambda$.

D e f i n i t i o n I6. Say that f-rank (N,M) is additive, if f-rank $(N\oplus k\Lambda\ , M\oplus k\Lambda\ )$ = f-rank (N,M) + k.

Lemma 4. Let Λ be iBN-ring such that for every module **M** the equility $\mu(M\oplus n\Lambda\ ) = \mu(M) + n$ is fulfilled. Suppose that N is submodule of M. Then there exists such n_o that for $n \geqslant n_o$ f-rank $(N\oplus n\Lambda\ , M\oplus n\Lambda)$ is additive.

Notice, that for S-ring the equality $\mu(M\oplus n\Lambda\ ) =\mu(M)+ n$ is always fulfilled [I4] .

Lemma 5. Let Λ be Noether S-ring, $g : G \longrightarrow M$ is epimorphism of free Λ-module G onto Λ-module M. g is minimal epimorphism if and only if f-rank (Ker G,G) = 0 and is additive.

Lemmas 4 and 5 play an important role in proving the Theorem IO. Using the operations of the stabilisation and reduction of boundary homomorphisms of the chain complex one can find the chain complex

$$0 \longleftarrow \bar{C}_2 \xleftarrow{\ \partial\ } \bar{C}_3 \longleftarrow \cdots \xleftarrow{\partial_{n-2}} \bar{C}_{n-2} \longleftarrow 0,$$

for which f-rank $(\partial_{i+I}(\bar{C}_{i+I}),\bar{C}_i) = 0$ and is additive. Then it is possible to show that such chain complex is automatically minimal.

In conclusion we note the paper of S.P.Novikov [8] , where certain many-valued Morse theory is investigated. It leads to construction problem of minimal Morse function for a mapping of a manifold in to the circle.

References

I. H.Bass. Algebraic K-theory, Benjamin, New-York, I968.

2. O.I.Bogoyavlenskiĭ. On exact function on manifold. - Mat.zametki, 8(I970), 77-83 (in Russian).

3. M.Golubitsky, V.Guillemin. Stable mappings and their singularities Graduate text in Mathematics I4, Springer, I973.

4. R.Lyndon, P.Schup. Combinatorial group theory. Ergebnisse der Mathematik und ihrer Grenzgebiete 69, Springer-Verlag, I972.

5. J.Milnor. Morse theory. Princeton University Press, I965.

6. J.Milnor. Lectures of the h-cobordism theorem. Princeton University Press, I963.

7. J.Milnor. Introduction to algebraic K-theory. Princeton University Press, I97I.

8. S.P.Novikov. Many-valued functionals and functions. Analogy of Morse theory.-DAN SSSR, 260(I98I), 3I-34 (in Russian).

9. C.Rourrke, B.Sanderson. Introduction to picewise-linear topology. Ergebnise der Mathematik und ihre Grenzgebiete 69, Springer-Verlag I972.

IO. V.V.Sharko. Exact Morse functions. Preprint IM-75-9, Kiev, Institut matematiki AN Ukr.SSR, I975 (in Russian).

II. V.V.Sharko. On smooth functions on manifolds. Preprint IM-79-22 Kiev, Institut matematiki AN Ukr.SSR, I979 (in Russian).

I2. V.V.Sharko. Minimal resolvents and Morse functions. - Doklady AN Ukr.SSR, ser.A, No II (I980), 3I-33 (in Russian).

I3. V.V.Sharko. On the equivalence of exact Morse functions. - Doklady AN Ukr.SSR, ser.A, No I2 (I980), I8-20 (in Russian).

I4. V.V.Sharko. Stable algebra and Morse theory. - Ukr.mat.zhurnal, 32 (I980), 7II-7I3 (in Russian).

I5. V.V.Sharko. Exact Morse functions on simply connected manifolds with not simply connected boundary. - Uspehi mat.nauk, 36 (I98I), 205-206 (in Russian).

16. M.Agoston. On handle decomposition and diffeomorphism. - Trans.AMS 198(1969), 21-26.
17. R.Alperin, K.Dennis, M.Stein. The non-triviality of $SK_1(Zn)$. Lecture Notes in Mathematics, Springer-Verlag, New-York, 1973.
18. J.Andrews, Curtis M. Free group and handelbodies. - Proc.Amer.Math Soc., 16 (1965), 263-268.
19. D.Barden. Structure of manifolds. Thesis, Cambridge University, 1963.
20. J.Cerf. Stratification naturelle des espaces de fonctions differentiables reeles et le theoreme de la pseudo-isotopie. - Publ.Math. I.E.H.S., 39 (1970), 3-280.
21. A.Chenciner, F.Laudenbach. Contribution a une theorie de Smale a un parametre dans la cas non simplement connexe. - Ann.Scient.Ec. Norm.Sup., 3 (1970), 409-478.
22. M.Cohen. A course in simple-homotopy theory. Graduate Text in Mathematics, Springer-Verlag, New-York, 1973.
23. H.Cohen. Whitehead torsion, group extension and Zeeman's conjecture in high dimensions. - Topology, 15 (1976), 79-88.
24. M.Dunwoody. The homotopy type of a two-dimensional complex. - Bul London, Math.Soc., 8 (1976), 218-229.
25. B.Hajduk. Comparing handle decompositions of homotopy equivalent manifolds. - Fund.Math., 95 (1977), No 1.
26. B.Hajduk. Presentations of the fundamental group of manifolds. - Collq.Math., 1977.
27. H.Hatcher, J.Wagoner. Pseudo-isotopies of compact manifolds. - Asterisque, 6(1973), 3-274.
28. T.Lam. Introduction theorem for Grotendieck groups and Whitehead groups of finite groups. - Ann.Sci.Ecole Norm.Sup., 1 (1968),No 1.
29. R.Mandelbaum. Four-dimensional topology: an introduction. - Bul. Amer.Math.Soc., 2 (1980), 1-159.
30. M.Morse. The calculus of variations in the large. New York, 1934.
31. T.Matumoto. On the minimal ordered Morse functions on compact manifold. - Publ.RIMS, Kyoto Univ., 14 (1978), 673-684.
32. W.Metzler. Uber den homotopie Typ zweidiemensionaler CW-komplexe und elementar Transformationen bei Darstellung von Gruppen durch erceugende und defienierende Relationen. - J.rei.angew.Math.,285 (1976), 7-23.
33. Algebraic and Geometric Topology. - Proc. of Sym. in Pure Math., 32 (1978), 311.
34. O.Rothaus. On the non-triviality of some group extensions given by generators and relations. - Ann. of Math. 106 (1977), No 3.

35. F.Sergereart. Un Theoreme de functions impsites sur cartains espaces de Frechet et quelqes applications. - Ann.Sci.Ecole Norm.Sup. 5 (I972), 599-660.
36. S.Smale. Generalised Poincare's conjecture in dimensions greater then four. - Ann. of Math., 74 (I96I), 39I-406.
37. S.Smale. On the structure of manifolds. - Amer.J.Math., 84 (I962) 387-399.
38. M.Stein. Whitehead groups of finite groups. - Bul.Amer.Math.Soc., 84 (I978), No 2.
39. M.Shub, D.Sullivan. Homology theory and Dynamical Systems. - Topology, I4 (I975), I09-I32.
40. J.Webb. The minimal relation modules of a finite abelian group. - J. of Pure and Appl.Algebra, 2I (I98I), No 2.
4I. P.Wright, Group presentations and formal deformations. - Trans. Amer.Math.Soc., I975, 208, I6I-I69.
42. H.Hendrics. La stratification naturelle de espace des functions differentiables reeles ne est pas la bonne. - C.R.Acad.Sci., Paric, 244 (I972), 6I8-620.
43. I.A.Volodin. Generalized Whitehead groups and pseudoisotopics. - Uspehi mat.nauk, 27 (I972), 229-230 (in Russian).
44. E.Zeeman. On the dunce hat. - Topology, 2 (I964), 34I-358.
45. M.Maller. Thesis. University of Warwick, I978.

TOPOLOGICAL METHODS OF INVESTIGATION OF GENERAL NONLINEAR ELLIPTICAL BOUNDARY VALUE PROBLEMS

I.V.Skrypnik
Institute of Applied Mathematics and Mechanics
Universitetskaya,77,Doneck,USSR

INTRODUCTION

One of the fundamental methods of qualitative study of nonlinear elliptic and parabolic boundary value problems, which facilitates the study of solvability, branching, bifurcation of solutions problems of eigenfunctions, is the topological method, based on the theory of degree of nonlinear mappings in Banach spaces. The origin of this method goes back to the remarkable paper by Leray-Schauder [I], where the authors, using Brouwer's theory of degree of finite-dimensional mappings, introduced the degree of a map $I-F: X \to X$, where I is the identity, F a totally continuous mapping of a Banach space X (or its part) into itself. In the same paper the authors gave a method of reducing the quasilinear Dirichlet problem to the operator equation $u-Fu=0$ with a totally continuous operator F. Applying the theory of degree of a map the authors found that in order to prove an existence theorem it is sufficient to establish an apriori estimate of solutions of s certain parametric family of boundary value problems.

On the other hand, the limited possibilities of application of the Leray-Schauder degree to more general boundary value problems were soon recognized. The application of these methods to the Dirichlet problem for the general nonlinear equation is cumbersome and requires additional restrictions [2]. When studying the second fundamental problem for quasilinear elliptic equations - the Neumann problem - it was found [3] that by applying the Leray-Schauder scheme the differential problem reduces to the equation $u-\Phi u=0$ with a continuous but not compact operator Φ. Consequently, applying the topological methods in this case we have either to change the way in which we reduce the problem, or to reduce the problem to other classes of operators.

In this way there appeared a demand for topological methods of investigation of more general classes of operator equations in Banach spaces, namely, of such classes to which it is possible to reduce differential boundary value problems for general nonlinear differential elliptic and parabolic equations with general nonlinear boundary value conditions.

The necessity of developing a theory of degree of more extensive classes of mappings also arose from the problem of existence of generalized solutions of boundary value problems for divergent equations. An essential feature of the study of generalizeg solvability is the sufficiency of considerably weaker apriori estimates of solutions - estimates involving the energy norm. At the beginning of the sixties, Browder, Minty and others (see survey papers [4] - [7]) discovered new classes of operators - the monotone operators - whose essential property is that they preserve the weak convergence under Galerkin's approximations. Problems of finding generalized solutions for divergent elliptic and parabolic equations naturally lead to equations with similar operators. The application of the methods of monotone operators led to a considerable progress in the theory of nonlinear boundary value problems.

The results of various authors concerning solvability of equations with monotone operators (in the coercive case) (see [4] - [7]) were entailed by the establishment of the theory of degree of various classes of monotone operators and their generalizations, given simultaneously and independently by the author [8] and Browder, Petryshyn [9]. A survey of these and other related results is founf in [10]. Browder and Petryshyn introduced a multivalued degree of A - proper mappings. The non-uniqueness of the degree (the degree of a mapping is a subset of the set $Z \cup \{-\infty\} \cup \{+\infty\}$) as well as the fact that this degree fails to possess all the properties of the degree of finite-dimensional mappings, make its application to differential problems difficult.

The author introduced a single-valued degree of mapping satisfying a certain condition (α_0). The degree introduced exhibits all the properties of the degree of finite-dimensional mappings. Even the analogue of Hopf's theorem asserting that the degree is the unique homotopic invariant of the class of mappings considered, is valid.

The above mentioned degree of a mapping satisfying the condition (α_0) was in the beginning emploed in the study of boundary value problems for divergent quasilinear equations [11]. Further, the author demonstrated [10, 12, 13] that this degree can serve as the basis for developing topological methods of investigation of boundary value problems for general essentially nonlinear elliptic and parabolic equations with general nonlinear boundary value conditions. Explicit constructive methods of reducing boundary value problems to the corresponding classes of operator equations in Sobolev spaces were given. When veryfying the condition (α_0) for the resulting operators, apriori L_p -estimates of elliptic and parabolic linear problems are essentially involved.

In the case of the general nonlinear Dirichlet problem it is possible to give simpler methods of reducing the original problems to operator equations with operators satisfying the condition (α_0). These simplifications are based on coercive estimates for pairs of linear elliptic operators, established in [I4] under weak assumptions.

In the present paper we shall expound the ways of reducing the general nonlinear elliptic boundary value problems to the operator equations with operators satisfying the condition (α_0) and construct the theory of degree of mapping satisfying the condition (α_0), and establish the different properties of degree. There are given the examples of utilizing of the developed topological methods to the proof of solvability of the Dirichlet problem for Mongue - Ampere equation and the general nonlinear equation of the thin layer. This paper gives more detailed and systematic exposition of the results printed in [8, I2, I3, I9].

Note, that these developed methods allows to study the behaviour of solutions for a group of problems, in particular, brunching, bifurcation of solution, eigenvalues problems (see [IO, II]). We did not touch upon these problems in this paper.

Note also, that for the studying of the general nonlinear elliptic problems it is possible to apply the theory of degree of Fredholm's operators (the review of these results see in [I5]). However the definition of the last degree is of nonapproximative character and it is obligatory supposed the differentiability of abstract mapping. The explicit method of reducing boundary value problems to operator equations with generalized monotone operators enables to point out the concrete results on solvability (the analog of coercive condition) and to consider quasilinear boundary problems without condition of differentiability of coefficients.

I. THE DEGREE OF GENERALIZED MONOTONE MAPPINGS

I. In this section the abstract general monotone operators are considered and the degree theory for corresponding mappings is con -

structed. The numerous examples of concrete mappings of that kind connecting with elliptical boundary problems one can find in [IO, II] and further throughout the paper. Here we give only a scheme of definition of the degree and formulate the main assertions without prooving them. The proof of all these assertions one can find in [IO, II] .

Further we shall denote by X an infinite-dimentional real separable reflexive Banach space and by X^* its adjoint. We denote the strong and the weak convergence by $\rightarrow$ and $\rightharpoonup$, respectively; it will be clear from the context in which space the convergence is considered. Given elements $u \in X$ and $h \in X^*$ we denote by $\langle h, u \rangle$ the value of functional h at the element u .

We will consider operators A , defined on a set $D_A \subset X$,with values in X^*. An operator A will be called demicontinuous if it maps strongly convergent sequences into weakly convergent ones.

DEFINITION I. We shall say an operator A satisfies the condition (α_0) if for any sequence $u_n \in D_A$,the relations $u_n \rightharpoonup u_0$, $Au_n \rightharpoonup 0$ and

$$\lim_{n\to\infty} \langle Au_n, u_n - u_0 \rangle \leq 0 \tag{I}$$

imply the strong convergence of u_n to u_0 .

The set of operators satisfying the condition (α_0) is a certain extension of the set of operators satisfying the condition $(S)_+$ [16]. We say that an operator A satisfies the condition $(S)_+$ if for every sequence $u_n \in D_A$ that satisfies (I), the weak convergence $u_n \rightharpoonup u_0$ implies $u_n \rightarrow u_0$.

In this chapter we shall define the degree of a mapping A satisfying the condition (α_0) .

Let $\{v_i\}$ $i = 1, 2, \ldots$ be an arbitrary complete system of the space X and assume that the elements $v_1, \ldots, v_N$ are linearly independent for every N . Denote by F_N the linear hull of the elements $v_1, \ldots, v_N$.

Let us assume that the interior of the set D_A is nonempty and let D be an arbitrary bounded open subset of the space X , such that $\bar{D} \subset D_A$. We will define the degree $Deg(A, D, 0)$ of the mapping A of the set $\bar{D}$ with respect to the point 0 of the space X^* .

Let us notice that in the author's papers [IO, II] the term of degree of a mapping is replaced by the term "rotation of a vector field on the boundary of a set". These two notions are equivalent .

For every $n = 1, 2, \ldots$ we shall introduce finite-dimensional approximations A_n of the map A, $A_n : D_n = \bar{D} \cap F_n \rightarrow F_n$, by

(2) $$\mathcal{A}_n(u) = \sum_{i=1}^{n} \langle \mathcal{A}u, v_i \rangle v_i \,.$$

LEMMA I. Let $\mathcal{A}: \mathcal{D}_{\mathcal{A}} \to X^*$ be a demicontinuous bounded operator satisfying the condition (α_0), $\bar{\mathcal{D}} \subset \mathcal{D}_{\mathcal{A}}$ and $\mathcal{A}u \neq 0$ for $u \in \partial\mathcal{D}$. Then there is a number N_1 such that Brouwer's degree $deg(\mathcal{A}_n, \mathcal{D}_n, 0)$ is defined provided $n \geq N_1$.

Further, we shall establish the stabilization of Brouwer's degree of the mappings $\mathcal{A}_n$.

LEMMA 2. The limit $\lim_{n\to\infty} deg(\mathcal{A}_n, \mathcal{D}_n, 0)$ exists.

Denote

(3) $$D\{v_i\} = \lim_{n\to\infty} deg(\mathcal{A}_n, \mathcal{D}_n, 0).$$

We shall further prove that the limit $D\{v_i\}$ is independent of the choice of the system of elements $\{v_i\}$. Let $\{w_i\}$ be another system possessing the same properties as the system of elements $\{v_i\}$. Denote by E_n the linear hull of the elements $w_1, \dots, w_n$ and define the mapping

$$\mathcal{A}_n' : \mathcal{D}_n' = \bar{\mathcal{D}} \cap E_n \to E_n \,, \qquad \mathcal{A}_n'(u) = \sum_{i=1}^{n} \langle \mathcal{A}u, w_i \rangle w_i \,.$$

LEMMA 3. Let the assumptions of lemma I be fulfilled. Then $D\{v_i\} = D\{w_i\} = D$, where $D\{v_i\}$ is defined by (3) and $D\{w_i\}$ is defined analogously.

Lemmas I – 3 enable us to introduce the following definition.

DEFINITION 2. The number D from lemma 3 is called the degree of the mapping $\mathcal{A}$ of the set $\bar{\mathcal{D}}$ with respect to the point $0 \in X^*$ and is denoted by $Deg(\mathcal{A}, \bar{\mathcal{D}}, 0)$.

Hence we have by definition

$$Deg(\mathcal{A}, \bar{\mathcal{D}}, 0) = D = D\{v_i\} = \lim_{n\to\infty} deg(\mathcal{A}_n, \mathcal{D}_n, 0).$$

The introduced degree of a mapping $Deg(\mathcal{A}, \bar{\mathcal{D}}, 0)$ possesses all the properties of the degree of finite-dimensional mappings (see [IO, II]). We mention only the most important of them, omitting the proofs.

DEFINITION 3. Let $\mathcal{D}_0$ be an arbitrary set in the space X and $\mathcal{A}_t : \mathcal{D}_0 \to X$, $t \in [0,1]$, a parametric family of mappings. The family $\mathcal{A}_t$ is said to satisfy the condition $(\alpha_0^{(t)})$, if for any sequences $u_n \in \mathcal{D}_0$, $t_n \in [0,1]$, the relations $u_n \rightharpoonup u_0$, $t_n \to t_0$, $\mathcal{A}_{t_n}(u_n) \rightharpoonup 0$ and

$$\lim_{n\to\infty} \langle \mathcal{A}_{t_n}(u_n), u_n - u_0 \rangle \leq 0$$

imply the strong convergence of u_n to u_0.

DEFINITION 4. Let $A', A'' : \mathcal{D}_0 \to X$ be bounded demicontinuous operators satisfying the condition (α_0), $\mathcal{D}$ a bounded open set such that $\bar{\mathcal{D}} \subset \mathcal{D}_0$, and let $A'u \neq 0$, $A''u \neq 0$ for $u \in \partial\mathcal{D}$. The mappings A', A'' are called homotopic on $\bar{\mathcal{D}}$, if there is a parametric family of mappings $A_t : \bar{\mathcal{D}} \to X$ satisfying the condition $(\alpha_0^{(t)})$ and the following conditions:

(a) $A_t(u) \neq 0$ for $u \in \partial\mathcal{D}$, $t \in [0,1]$; $A_0 = A'$, $A_1 = A''$;

(b) for each $t \in [0,1]$ the operator A_t is continuous;

(c) there is function $\omega : [0,1] \to R^1$, $\omega(\rho) \to 0$ with $\rho \to 0$, such that $\sup_{u \in \bar{\mathcal{D}}} \| A_t u - A_s u \| \leq \omega(|t-s|)$.

The following theorems provide a classification of mappings that satisfy the condition (α_0), in terms of the degree of a mapping.

THEOREM 1. Let $A' : \bar{\mathcal{D}} \to X^*$, $A'' : \bar{\mathcal{D}} \to X^*$ be two mappings that are homotopic to each other in the sense of Definition 4. Then $Deg(A', \bar{\mathcal{D}}, 0) = Deg(A'', \bar{\mathcal{D}}, 0)$.

THEOREM 2. Let $\mathcal{D}$ be a convex bounded open set in the space X and $A' : \bar{\mathcal{D}} \to X$, $A'' : \bar{\mathcal{D}} \to X$ bounded demicontinuous mappings satisfying the condition (α_0), such that $A'u \neq 0$, $A''u \neq 0$ for $u \in \partial\mathcal{D}$ and $Deg(A', \bar{\mathcal{D}}, 0) = Deg(A'', \bar{\mathcal{D}}, 0)$. Then the mappings A', A'' are homotopic on $\bar{\mathcal{D}}$.

The application of the theory of degree of a mapping to the solvability of the operator equation

(4)
$$Au = 0$$

is based on the following principle.

Principle of Nonzero Rotation.

Let $A : \bar{\mathcal{D}} \to X^*$ be a bounded demicontinuous operator satisfying the condition (α_0). A sufficient condition for the equation (4) to be solvable in $\mathcal{D}$ is that $Deg(A, \bar{\mathcal{D}}, 0) \neq 0$.

Let us present two criteria of non-vanishing of the degree of a mapping.

THEOREM 3. Let $B(0,r) = \{ u \in X : \|u\| \leq r \}$, $A : B(0,r) \to X^*$ be a bounded demicontinuous operator satisfying the condition (α_0). Assume that for every $u \in \partial B(0,r)$ the inequalities

(5)
$$Au \neq 0, \quad \frac{Au}{\|Au\|_*} \neq \frac{A(-u)}{\|A(-u)\|_*}$$

hold. Then $Deg\{A, B(0,r), 0\}$ is an odd number.

THEOREM 4. Let $\mathcal{D}$ be an arbitrary bounded domain in the space X

$0 \notin \partial\mathcal{D}$ and $\mathcal{A}: \bar{\mathcal{D}} \to X^*$ a bounded demicontinuous operator satisfying the condition (α_0). Assume that for $u \in \partial\mathcal{D}$ the inequalities

(6) $$\mathcal{A}u \neq 0, \quad \langle \mathcal{A}u, u \rangle \geq 0$$

hold. Then $Deg(\mathcal{A}, \bar{\mathcal{D}}, 0) = 1$ for $0 \in \mathcal{D}$, $Deg(\mathcal{A}, \bar{\mathcal{D}}, 0) = 0$ for $0 \notin \bar{\mathcal{D}}$.

Let us formulate an easy consequence of Theorem I and the principle of nonzero rotation.

THEOREM 5. Let $\mathcal{D}$ be a bounded domain in X and $\mathcal{A}_t: \bar{\mathcal{D}} \to X^*$ a family of operators that realize the homotopy between $\mathcal{A}_0$ and $\mathcal{A}_1$ in the sense of Definition 4. Assume that $Deg(\mathcal{A}_0, \bar{\mathcal{D}}, 0) \neq 0$. Then the equation $\mathcal{A}_1 u = 0$ has in $\mathcal{D}$ at least one solution.

2. TOPOLOGICAL CHARACTERISTICS OF GENERAL NONLINEAR ELLIPTIC PROBLEMS

I. In this chapter it is shown how general elliptic problems can be reduced to operator equations of the form (4) with the operator $\mathcal{A}$ satisfying the condition (α_0).

In what follows, Ω is a bounded domain in R^n with an infinitely differentiable boundary $\partial\Omega$, $n_0 = [\frac{n}{2}] + 1$, $m, m_1, \dots, m_m$ are nonnegative integers, $m \geq 1$. We denote by $M(q)$ the number of mutually distinct multiindices $\alpha = (\alpha_1, \dots, \alpha_n)$ with nonnegative integer coordinates α_i and the length $|\alpha| = \alpha_1 + \dots + \alpha_n$ not greater than q.

Let $\ell_0 = max(2m, m_1 + \frac{1}{2}, \dots, m_m + \frac{1}{2})$ and assume that functions

$$F: \bar{\Omega} \times R^{M(2m)} \to R^1, \quad G_j: \bar{\Omega} \times R^{M(m_j)} \to R^1, \quad j = 1, \dots, m$$

are given, possessing continuous derivatives with respect to all their arguments up to the orders $\ell - 2m + 1$, $\ell - 2m_j + 1$, respectively, where the number ℓ satisfies the condition $\ell \geq \ell_0 + n_0$.

The functions $F(x, \xi)$, $G_j(x, \eta)$, $\xi = \{\xi_\alpha : |\alpha| \leq 2m\}$, $\eta = \{\eta_\beta : |\beta| \leq m_j\}$ will be also written in the form

$$F(x, \xi) = F(x, \xi_0, \xi_1, \dots, \xi_{2m}), \quad G_j(x, \eta) = G_j(x, \eta_0, \eta_1, \dots, \eta_{m_j})$$

with $\xi_k = \{\xi_\alpha : |\alpha| = k\}$, $\eta_k = \{\eta_\alpha : |\alpha| = k\}$.

We shall also use the notation $\mathcal{D}^\alpha u = (\frac{\partial}{\partial x_1})^{\alpha_1} \cdots (\frac{\partial}{\partial x_n})^{\alpha_n} u$ for any multiindex α, $\mathcal{D}^k u = \{\mathcal{D}^\alpha u : |\alpha| = k\}$.

Finally, let

(7) $$F_\alpha(x,\xi)=\frac{\partial F(x,\xi)}{\partial \xi_\alpha}, \qquad G_{j,\beta}(x,\eta)=\frac{\partial G_j(x,\eta)}{\partial \eta_\beta}.$$

In the present chapter we shall consider the boundary value problem

(8) $$F(x,u,\dots,\mathcal{D}^{2m}u)=f(x), \qquad x\in\Omega$$

(9) $$G_j(x,u,\dots,\mathcal{D}^{m_j}u)=g_j(x), \quad j=1,\dots,m, \quad x\in\partial\Omega$$

under the following conditions:

(i) for an arbitrary function $v(x)\in C^{l_0,\delta}(\bar\Omega)$, $0<\delta<1$ the operator

$$\mathfrak{A}(v): H^{l_0}(\Omega)\to H^{l_0}(\Omega,\partial\Omega)=H^{l_0-2m}(\Omega)\times H^{l_0-m_1-\frac12}(\partial\Omega)\times\dots\times H^{l_0-m_m-\frac12}(\partial\Omega)$$

defined by the identities

$$\mathfrak{A}(v)u=\{L(v)u, B_1(v)u,\dots,B_m(v)u\},$$

(10) $$L(v)u=\sum_{|\alpha|\le 2m}F_\alpha(x,v,\dots,\mathcal{D}^{2m}v)\mathcal{D}^\alpha u, \qquad x\in\Omega,$$

$$B_j(v)u=\sum_{|\beta|\le m_j}G_{j,\beta}(x,v,\dots,\mathcal{D}^{m_j}v)\mathcal{D}^\beta u. \qquad x\in\partial\Omega$$

is elliptic and Fredholm; here $H^l(\Omega)=W_2^l(\Omega)$;

(ii) there is a function $H:\bar\Omega\times R^{M(2m-1)}\to R^1$ of the class C^{l-2m}, such that the problem

(11) $$\begin{aligned} L(v)u+M(v)u&=0 \qquad && x\in\Omega\\ B_j(v)u&=0 \qquad j=1,\dots,m && x\in\partial\Omega\end{aligned}$$

has in $C^{l_0,\delta}(\bar\Omega)$ only the zero solution for an arbitrary function $v\in C^{l_0,\delta}(\bar\Omega)$. Here

$$M(v)u=\sum_{|\gamma|\le 2m-1}H_\gamma(x,v,\dots,\mathcal{D}^{2m-1}v)\mathcal{D}^\gamma u, \qquad H_\gamma(x,\xi)=\frac{\partial H(x,\xi)}{\partial\xi_\gamma}$$

The condition (i) expresses the ellipticity of the operator $L(v)$ and the fact that $L(v)$ and $B_j(v)$ at each point $x\in\partial\Omega$ satisfy the condition of Ya. B. Lopatinskii, while the fact that is a Fredholms operator indicates that its index vanishes.

Let $\mathcal{D}$ be an arbitrary bounded domain in the space $H^l(\Omega)$ with a boundary $\partial\mathcal{D}$, let $f(x)$, $g_j(x)$ be fixed elements of the spaces $H^{l-2m}(\Omega)$, $H^{l-m_j-\frac12}(\partial\Omega)$, respectively.

We define a nonlinear operator $A_1 : H^\ell(\Omega) \to [H^\ell(\Omega)]^*$ by

(I2)
$$\langle A_1 u, \varphi\rangle = \left(F(x,u,\dots,D^{2m}u) - f(x), L(u)\varphi + M(u)\varphi\right)_{\ell-2m,\Omega} + \sum_{j=1}^{m} \left(G_j(x,u,\dots,D^{m_j}u) - g_j(x), B_j(u)\varphi\right)_{\ell-m_j-\frac{1}{2},\partial\Omega},$$

where $(\cdot,\cdot)_{\ell,\Omega}$ and $(\cdot,\cdot)_{\ell,\partial\Omega}$ are the inner products in the spaces $H^\ell(\Omega)$, $H^\ell(\partial\Omega)$, respectively.

THEOREM 6. <u>If the conditions (i), (ii) are fulfilled and $\ell \geq \ell_0 + n_0$, then the operator A_1 defined by (I2) is continuous bounded and satisfies the condition $(S)_+$. If the problem (8), (9) has no solutions belonging to ∂D, then the degree $Deg(A_1, \bar{D}, 0)$ of the mapping A_1 of the domain $\bar{D}$ with respect to zero of the space $[H^\ell(\Omega)]^*$ is defined.</u>

Proof. The continuity and boundedness of the operator A_1 are easily verified. Let us prove the validity of the condition $(S)_+$.

Let u_n be an arbitrary sequence weakly converging to u_0, let $u_n \in D$ and

(I3)
$$\lim_{n\to\infty} \langle A_1 u_n, u_n - u_0\rangle \leq 0.$$

The weak convergence of u_n in $H^\ell(\Omega)$ implies the strong convergence of u_n to u_0 in $C^{\ell_0}(\bar{\Omega})$, since $\ell - \ell_0 \geq n_0 = [\frac{n}{2}] + 1$ and the corresponding imbedding operator $H^\ell(\Omega) \to C^{\ell_0}(\bar{\Omega})$ is compact. For $\ell_0 < j < \ell$ the Nirenberg-Gagliardo inequality yields

(I4)
$$\sum_{|\alpha|=j} \|D^\alpha(u_n - u_0)\|_{L_{\frac{2(\ell-\ell_0)}{(j-\ell_0)}}} \leq$$
$$\leq C \sum_{|\alpha|=\ell} \|D^\alpha(u_n - u_0)\|_{L_2}^{\frac{j-\ell_0}{\ell-\ell_0}} \cdot \|u_n - u_0\|_{C^{\ell_0}(\bar{\Omega})}^{1-\frac{j-\ell_0}{\ell-\ell_0}} + C\|u_n - u_0\|_{C^{\ell_0}(\bar{\Omega})}$$

and this implies that $u_n \to u_0$ in $W^j_{\frac{2(\ell-\ell_0)}{j-\ell_0}}(\Omega)$.

Consider $D^\alpha F(x,u,\dots,D^{2m}u)$ with $|\alpha| \leq \ell - 2m$. An easy computation yields

$$D^\alpha F(x,u,\dots,D^{2m}u) = \sum_{|\beta|\leq 2m} F_\beta(x,u,\dots,D^{2m}u) D^{\alpha+\beta}u + R_\alpha u,$$

Where $R_\alpha(u)$ satisfies the estimate

(I5) $$|R_\alpha(u)| \le C_0(M)\Big\{\sum_{j=l_0+1}^{l-1} |\mathcal{D}^j u|^{\frac{l-l_0}{j-l_0}} + 1\Big\}$$

provided the function $u(x)$ satisfies the inequality

(I6) $$\|u\|_{C^{l_0}(\bar\Omega)} \le M$$

with a constant M.

Analogously we verify that $\mathcal{D}^\alpha(M(u)v) = R_{1,\alpha}(u,v)$,

(I7) $$\mathcal{D}^\alpha(L(u)v) = L(u)\mathcal{D}^\alpha v + R_{2,\alpha}(u,v)$$

where $R_{i,\alpha}(u,v)$ satisfies the estimate

(I8) $$|R_{i,\alpha}(u,v)| \le C_0(M)\cdot\|v\|_{C^{l_0}(\bar\Omega)}\cdot\Big\{\sum_{j=l_0+1}^{l-1} |\mathcal{D}^j u|^{\frac{l-l_0}{j-l_0}} + 1\Big\} +$$
$$+ C_0(M)\sum_{j=l_0+1}^{l-1} |\mathcal{D}^j v|\Big\{\sum_{i=l_0+1}^{l+2m-1} |\mathcal{D}^i u|^{\frac{l-j}{i-l_0}} + 1\Big\}, \quad i=1,2,$$

provided the function $u(x)$ satisfies the condition (I6).

Using the above identities we can write

$$\big(F(x,u_n,\dots,\mathcal{D}^{2m}u_n) - f(x),\ L(u_n)(u_n-u_0) + M(u_n)(u_n-u_0)\big)_{l-2m,\Omega} =$$
$$= \sum_{|\alpha|\le l-2m}\int_\Omega \mathcal{D}^\alpha[F(x,u_n,\dots,\mathcal{D}^{2m}u_n) - f(x)]\cdot\mathcal{D}^\alpha[(L(u_n)+M(u_n))(u_n-u_0)]\,dx =$$

(I9)
$$= \sum_{|\alpha|\le l-2m}\int_\Omega [L(u_n)\mathcal{D}^\alpha u_n + R_\alpha(u_n) - \mathcal{D}^\alpha f(x)]\cdot[L(u_n)\mathcal{D}^\alpha(u_n-u_0) +$$
$$+ \sum_{j=1}^{2} R_{j,\alpha}(u_n, u_n-u_0)]\,dx = (L(u_0)u_n, L(u_0)(u_n-u_0))_{l-2m,\Omega} + R_1^{(n)},$$

where

$$R_1^{(n)} = \sum_{|\alpha|\le l-2m}\int_\Omega\Big\{[L(u_n)\mathcal{D}^\alpha u_n - L(u_0)\mathcal{D}^\alpha u_n - R_{2,\alpha}(u_0,u_n) + R_\alpha(u_n) - \mathcal{D}^\alpha f]\cdot$$
$$\cdot[L(u_n)\mathcal{D}^\alpha(u_n-u_0) + \sum_{j=1}^{2} R_{j,\alpha}(u_n,u_n-u_0)] + \mathcal{D}^\alpha(L(u_0)u_n)\cdot$$
$$\cdot[L(u_n)\mathcal{D}^\alpha(u_n-u_0) - L(u_0)\mathcal{D}^\alpha(u_n-u_0) + \sum_{j=1}^{2} R_{j,\alpha}(u_n,u_n-u_0) - R_{2,\alpha}(u_0,u_n-u_0)]\Big\}\,dx.$$

In virtue of the inequalities (I5), (I8) and the above mentioned strong convergence of u_n to u_0 in $C^{l_0}(\bar{\Omega})$ and $W^{j}_{\frac{2(l-l_0)}{j-l_0}}(\Omega)$ we easily obtain

(20)
$$\lim_{n\to\infty} R_1^{(n)} = 0$$

Now we pass to integrals over $\partial\Omega$ in (I2), denoting

(2I)
$$G_j(x, u_n, \ldots, \mathcal{D}^{m_j} u_n) = G_j(x, u_0, \ldots, \mathcal{D}^{m_j} u_0) + \int_0^1 B_j(u_0 + t(u_n - u_0))(u_n - u_0)\,dt$$

Then

(22)
$$\left(G_j(x, u_n, \ldots, \mathcal{D}^{m_j} u_n) - g_j(x), B_j(u_n)(u_n - u_0)\right)_{l-m_j-\frac{1}{2}, \partial\Omega} = \left(B_j(u_0)(u_n - u_0), B_j(u_0)(u_n - u_0)\right)_{l-m_j-\frac{1}{2}, \partial\Omega} + R_{2,j}^{(n)},$$

where
$$R_{2,j}^{(n)} = \left(B_j(u_0)(u_n - u_0), B_j(u_n)(u_n - u_0) - B_j(u_0)(u_n - u_0)\right)_{l-m_j-\frac{1}{2}, \partial\Omega} + \Big(G_j(x, u_0, \ldots, \mathcal{D}^{m_j} u_0) - g_j(x) + \int_0^1 B_j(u_0 + t(u_n - u_0))(u_n - u_0)\,dt - B_j(u_0)(u_n - u_0), B_j(u_n)(u_n - u_0)\Big)_{l-m_j-\frac{1}{2}, \partial\Omega}.$$

We shall prove that

(23)
$$\lim_{n\to\infty} R_{2,j}^{(n)} = 0.$$

Taking into account the boundedness of the imbedding operator $W_2^{l-m_j}(\Omega) \to W_2^{l-m_j-\frac{1}{2}}(\partial\Omega)$, we conclude, for example for the first summand:

(24)
$$\left|\left(B_j(u_0)(u_n - u_0), B_j(u_n)(u_n - u_0) - B_j(u_0)(u_n - u_0)\right)_{l-m_j-\frac{1}{2}, \partial\Omega}\right| \le C \|B_j(u_0)(u_n - u_0)\|_{l-m_j, \Omega} \cdot \|B_j(u_n)(u_n - u_0) - B_j(u_0)(u_n - u_0)\|_{l-m_j, \Omega}.$$

Further, writing the operator B_j in the form of (I7) we find that the first summand on the right hand side of (24) is uniformly bounded while the second tends to zero. Analoqously we show that the limit of the second summand for $R^{(n)}_{2,j}$ vanishes, which completes the proof of (23).

The above considerations imply

(25)
$$\langle \mathcal{A}_1(u_n), u_n - u_0\rangle = \|L(u_0)(u_n-u_0)\|^2_{\ell-2m,\Omega} + \sum_{j=1}^{m} \|B_j(u_0)(u_n-u_0)\|^2_{\ell-m_j-\frac{1}{2},\partial\Omega} + R^{(n)},$$

where

$$R^{(n)} = (L(u_0)u_0, L(u_0)(u_n-u_0))_{\ell-2m,\Omega} + R^{(n)}_1 + \sum_{j=1}^{m} R^{(n)}_{2,j}$$

and $R^{(n)} \to 0$ for $n \to \infty$.

Making use of the apriori estimates for linear elliptic operators [I7], we obtain

(26)
$$\|u_n - u_0\|^2_{\ell,\Omega} \le C\{\|L(u_0)(u_n-u_0)\|^2_{\ell-2m,\Omega} + \sum_{j=1}^{m}\|B_j(u_0)(u_n-u_0)\|^2_{\ell-m_j-\frac{1}{2},\partial\Omega} + \|u_n-u_0\|^2_{0,\Omega}\}.$$

Now (I3), (25) and (26) imply the strong convergence of u_n to u_0 in $H^\ell(\Omega)$, which completes the proof of the theorem.

REMARK I. The coefficients of the operators $L(u_0)$, $B_j(u_0)$ are not sufficiently smooth to allow for immediate application of the results of [I7] when establishing the inequality (26). In this case it is necessary to obtain more precise estimates, based on the form of the operators $L(u_0)$, $B_j(u_0)$, on the inclusion of u_0 in the space $H^\ell(\Omega)$ and on the Nirenberg-Gagliardo inequality. We are not going into details since the procedure is straightforward.

2. The degree of mapping $\mathcal{A}_1$ introduced in Theorem 7 enables us to apply topological methods when studying the problrm (8),(9). In particular, these methods are based on the investigation of a family of parametric problems and, in presence of apriori estimates, on the possibility of a homotopy between the problem (8), (9) and another simpler and more special problem of the same type.

We restrict ourselves to the formulation of a single one of the possible consequences.

THEOREM 7. Let $\widetilde{F}: [0,1]\times\bar{\Omega}\times R^{M(2m)} \to R^1$, $\widetilde{G}_j: [0,1]\times\bar{\Omega}\times R^{M(m_j)} \to R^1$, $j = 1,\dots,m$ be continuous mappings. Assume that for each $t\in[0,1]$, the functions $F_t(x,\xi) = \widetilde{F}(t,x,\xi)$, $G_{j,t}(x,\xi) = \widetilde{G}_j(t,x,\xi)$ satisfy the conditions (i), (ii) of Sec. I and, moreover,

(a) there exists a positive function $K: R^1_+ \to R^1$ such that for $t\in[0,1]$, $u\in H^\ell(\Omega)$, the relations

$$F_t(x, u, \dots, \mathcal{D}^{2m} u) = t f(x), \quad x \in \Omega,$$
(27)
$$G_{j,t}(x, u, \dots, \mathcal{D}^{m_j} u) = t g_j(x), \quad j = 1, \dots, m, \quad x \in \partial\Omega$$

<u>imply the estimate</u>

$$\|u\|_{\ell,\Omega} \leq K\Big(\|f\|_{\ell-2m,\Omega} + \sum_{j=1}^{m} \|g_j\|_{\ell-m_j-\frac{1}{2},\partial\Omega}\Big);$$

(b) $F_0(x, -\xi) = F_0(x, \xi)$; $G_j(x, -\eta) = -G_j(x, \eta)$.

<u>Then the problem (27) has at least one solution in</u> $H^{\ell}(\Omega)$ <u>for arbitrary</u> $f \in H^{\ell-2m}(\Omega)$, $g_j \in H^{\ell-m_j-\frac{1}{2}}(\partial\Omega)$, $t \in [0, 1]$.

The proof of the theorem follows from Theorems 5 and 3, provided that for fixed f, g_j we replace $\mathcal{D}$ by the ball $B(0,R)$ in $H^{\ell}(\Omega)$ of the radius

$$R = K\Big(\|f\|_{\ell-2m,\Omega} + \sum_{j=1}^{m} \|g_j\|_{\ell-m_j-\frac{1}{2},\partial\Omega}\Big) + 1.$$

The operators $\mathcal{A}_t$ corresponding to the problem (27) are introduced according to the identity (I2).

REMARK 2. The condition (a) of Theorem 8 may be weakened by requiring an apriori estimate of the problem (27) in $C^{\ell_0+\alpha}(\bar{\Omega})$.

3. The definition of the operator $\mathcal{A}_1$ introduced in Sec. I requires sufficient smoothness of the functions $F(x,\xi)$, $G_j(x,\eta)$. It is possible to weaken the conditions concerning smoothness of these functions by considering the corresponding operators in the space $W_p^{\ell_0+1}(\Omega)$ for $p > n$.

Let the functions $F: \bar{\Omega} \times R^{M(2m)} \to R^1$, $G_j: \bar{\Omega} \times R^{M(m_j)} \to R^1$, $j = 1, \dots, m$, belong to the spaces C^{ℓ_0+2-2m}, $C^{\ell_0+2-m_j}$, respectively, and let the conditions (i), (ii) from Sec. I be satisfied for them with some δ, $0 < \delta < 1 - \frac{n}{p}$.

Let $\mathcal{D}$ be an arbitrary bounded domain in the space $W_p^{\ell_0+1}(\Omega)$ with a boundary $\partial\mathcal{D}$, let $f(x)$, $g_j(x)$ be functions from the spaces $W_p^{\ell_0-2m+1}(\Omega)$, $B_p^{\ell_0-m_j-\frac{1}{p}+1}(\partial\Omega)$, where $B_p^s(\partial\Omega)$ denotes a Besov space. Under these conditions the operator $\mathfrak{A}$ defined above can be considered as an operator from $W_p^{\ell_0+1}(\Omega)$ into $W_p^{\ell_0+1}(\Omega, \partial\Omega)$:

$$W_p^{\ell_0+1}(\Omega, \partial\Omega) = W_p^{\ell_0-2m+1}(\Omega) \times \prod_{j=1}^{m} B_p^{\ell_0+1-m_j-\frac{1}{p}}(\partial\Omega).$$

Introduce a nonlinear operator $\mathcal{A}_2 : W_p^{l_0+1}(\Omega) \to [W_p^{l_0+1}(\Omega, \partial\Omega)]^*$ by the identity

$$\langle \mathcal{A}_2 u, \varphi \rangle = \sum_{|\alpha| \le l_0+1-2m} \int_\Omega \psi_p \{ \mathcal{D}^\alpha [F(x,u,\dots,\mathcal{D}^{2m}u) - f] \} \mathcal{D}^\alpha [L(u) + M(u)] \varphi \, dx +$$

$$+ \sum_{j=1}^{m} \sum_{|\beta| = l_0 - m_j} \int_{\partial\Omega} \int_{\partial\Omega} \psi_p \{ \mathcal{D}_x^\beta [G_j(x,u,\dots,\mathcal{D}_x^{m_j} u(x)) - g_j(x)] -$$

(28)

$$- \mathcal{D}_y^\beta [G_j(y,\dots,\mathcal{D}_y^{m_j} u(y)) - g_j(y)] \} \cdot [\mathcal{D}_x^\beta B_j(u)\varphi(x) - \mathcal{D}_y^\beta B_j(u)\varphi(y)] \frac{d_x S \, d_y S}{|x-y|^{n+p-2}} +$$

$$+ \sum_{j=1}^{m} \int_{\partial\Omega} \psi_p \{ G_j(x,u,\dots,\mathcal{D}^{m_j} u(x)) - g_j(x) \} \cdot B_j(u) \varphi \, dS,$$

where $\psi_p(t) = t + |t|^{p-2} \cdot t.$

THEOREM 8. <u>The operator $\mathcal{A}_2 : W_p^{l_0+1}(\Omega) \to [W_p^{l_0+1}(\Omega)]^*$ defined by (28) is continuous, bounded and satisfies the condition $(S)_+$ for $p > n$. If the problem (8), (9) has no solutions belonging to $\partial\mathcal{D}$, then the degree $Deg(\mathcal{A}_2, \bar{\mathcal{D}}, 0)$ of the mapping $\mathcal{A}_2$ of the set $\bar{\mathcal{D}}$ with respect to zero of the space $[W_p^{l_0+1}(\Omega)]^*$ is defined.</u>

Proof proceeds analogously to that of Theorem 6.

Notice that, if the condition (ii) is satisfied, the solutions of the problem (8), (9) coincide with those of the operator equation $\mathcal{A}_2 u = 0$ in the same way as was the case in Sec.I.

REMARK 3. For the case of boundary Dirichlet problem it is possible to obtain more simple constructions of operators $\mathcal{A}$ .It is connected with substitution of an apriori L_2 - estimates. utilizing to prove Theorem 6 on certain coercive estimates for pairs of linear elliptic operators which the author establiched in [I4] . This scheme will be illustrated with examples in Sec. 3.

3. SOLVABILITY OF NONLINEAR BOUNDARY VALUE PROBLEMS

In the present section we apply the topological methods developed above for essentialy nonlinear equations: the Dirichlet problem for Monge-Ampere equation and for general nonlinear equation of the thin layer.

I. Let us study the problem of existence of a solution regular in a closed domain, of the Dirichlet problem

(29) $$u_{xx}u_{yy}-u_{xy}^2=\varphi(x,u,u,\frac{\partial u}{\partial x},\frac{\partial u}{\partial y}),\quad (x,y)\in\Omega,$$

(30) $$u=g(x,y),\qquad (x,y)\in\partial\Omega.$$

Here Ω is a circle of radius R and centre at the origin of the coordinate system. We assume $g(x,y)\in C^{4,\lambda}(\partial\Omega)$, $\varphi(x,y,u,p,q)$ is a positive function of a class $C^{2,\lambda}(G)$, $0<\lambda<1$, where $G=\{(x,y,u,p,q):(x,y)\in\bar\Omega,\ (u,p,q)\in R^3\}$, and g, φ satisfy the conditions

(i) $$\varphi(x,y,u,p,q)<\psi(x^2+y^2)\cdot f(p^2+q^2)$$

provided $$u\le m=\max_{(x,y)\in\partial\Omega} g(x,y);$$

(ii) $$\iint_\Omega \psi(x^2+y^2)\,dx\,dy<\int_{-\infty}^{+\infty}\int_{-\infty}^{+\infty}\inf_{(\xi-p)^2+(\eta-q)^2<M_K} f^{-1}(\xi^2+\eta^2)\,dp\,dq,$$

M_K being the lower winding of the curve determined by the condition (30).

The existence of a solution, regular in the closed domain $\bar\Omega$, of the problem (29), (30) is proved in [I8] by the Newton-Kantorovich method under the conditions (i), (ii) and the additional assumption

(3I) $$\frac{\partial}{\partial u}\varphi(x,y,u,p,q)\ge 0.$$

Making use of the apriori estimates of a solution of the problem (29), (30), given in [I8], as well as of the topological method developed in [I2], the author with A.E.Siskov proved in [I9] the classical solvability of the problem (29), (30), assuming only (i), (ii) but without the assumption (3I).

Let us denote by $H_4^+(\Omega)$ the family of functions belonging to $W_2^4(\Omega)$ and satisfying the conditions

$$u_{xx}u_{yy}-u_{xy}^2>0,\qquad u_{xx}>0,\qquad (x,y)\in\bar\Omega.$$

THEOREM 9. [I9] Let $g(x,y)\in C^{4,\lambda}(\partial\Omega)$, $\varphi(x,y,u,p,q)\in C^{2,\lambda}(G)$ and let the conditions (i), (ii) be fulfilled. Then the problem (29) (30) has at least one solution in $H_4^+(\Omega)$.

REMARK 4. It follows from the imbedding theorems in the Sobolev spaces and from [I7] that the solution whose existence is asserted in Theorem I3, belong to $C^{4,\lambda}(\bar\Omega)$.

We will show how to reduce the problem (29), (30) to a nonlinear operator equation. It follows from [I8] and [I7] that under the assumptions of Theorem 9, solutions of the problem

$$(32)\qquad u_{xx}u_{yy}-u_{xy}^2=\tau\varphi(x,y,u,\frac{\partial u}{\partial x},\frac{\partial u}{\partial y})+(1-\tau)\psi(x^2+y^2)f(|\nabla u|^2),$$

$$(33)\qquad u(x,y)=\tau g(x,y)\ ,\quad (x,y)\in\partial\Omega,$$

which belong to $H_4^+(\Omega)$, satisfy the apriori estimate $\|u\|_4\leqslant K$ with a certain positive constant K, provided $0\leqslant\tau\leqslant 1$. Here and in what follows $\|\cdot\|_\ell$ stands for the norm in $W_2^\ell(\Omega)$.

Let $h(x,y)\in W_2^4(\Omega)$ be a harmonic function in Ω satisfying the boundary value condition (30). In $X=W_2^4(\Omega)\cap\mathring{W}_2^1(\Omega)$ let us define a domain $\mathcal{D}$ by

$$\mathcal{D}=\{v=u-\tau h:\ u\in H_4^+(\Omega),\ \|u\|_4<K+1,\ 0\leqslant\tau\leqslant 1\}.$$

For $v\in\mathcal{D}$ we consider the parametric family of differential equations

$$(34)\qquad \begin{aligned}F_\tau(x,y,v,\frac{\partial v}{\partial x},\frac{\partial v}{\partial y},\frac{\partial^2 v}{\partial x^2},\frac{\partial^2 v}{\partial x\partial y},\frac{\partial^2 v}{\partial y^2})\equiv(v_{xx}+\tau h_{xx})(v_{yy}+\tau h_{yy})-\\ -(v_{xy}+\tau h_{xy})^2-\tau\varphi(x,y,v+\tau h,\frac{\partial v}{\partial x}+\tau\frac{\partial h}{\partial x},\frac{\partial v}{\partial y}+\tau\frac{\partial h}{\partial y})-\\ -(1-\tau)\psi(x^2+y^2)f(|\nabla(v+\tau h)|^2)=0.\end{aligned}$$

By $N_{\bar{\mathcal{D}}}$ we denote the set of solutions of the equation (34) which belong to $\bar{\mathcal{D}}$. Then $N_{\bar{\mathcal{D}}}\cap\partial\mathcal{D}=\emptyset$.

Let $L_\tau(v)$ be a linear differential operator constructed for the function F_τ from (34) in an analogous way as the operator $L(v)$ was constructed for the function $F(x,\xi)$ in 2, Sec. I. Consider the family of differential operators

$$\mathcal{L}=\{L_\tau(v):\ v\in N_{\bar{\mathcal{D}}},\ \tau\in[0,1]\}.$$

According to [I4] we can construct a linear elliptic operator M of the second order and an inner product $[\ ,\ ]$ in $W_2^2(\Omega)$ so that

$$[L_\tau(v)u,Mu]\geqslant c_1\|u\|_4^2-c_2\|u\|_0^2$$

and

$$(Mu,u)\geqslant c_1\|u\|_1^2.$$

Analogously to we now define the family of nonlinear operators $A_\tau:\bar{\mathcal{D}}\to X^*$

$$(35)\qquad \langle A_\tau v,\xi\rangle=[F_\tau(x,y,v,\frac{\partial v}{\partial x},\frac{\partial v}{\partial y},\frac{\partial^2 v}{\partial x^2},\frac{\partial^2 v}{\partial x\partial y},\frac{\partial^2 v}{\partial y^2}),M\xi],\ \xi\in X.$$

Analogously to Theorem 6 it is verified that $\mathcal{A}_\tau$ are bounded demicontinuous operators and that the family of operators $\mathcal{A}_\tau$ satisfies the condition $(\alpha_0^{(\tau)})$. Since the equation $\mathcal{A}_\tau u=0$ has no solution with $u \in \partial\mathcal{D}$, we have $Deg(\mathcal{A}_1,\bar{\mathcal{D}},0)=Deg(\mathcal{A}_0,\bar{\mathcal{D}},0)$ by Theorem I. For $\tau=0$ the equation (34) has a unique solution in $\bar{\mathcal{D}}$. This solution represents a non-degenerate critical point of the field $\mathcal{A}_0(v)$ (in the terminology of [8]). Hence $Deg(\mathcal{A}_0,\bar{\mathcal{D}},0)\neq 0$ and the solvability of the problem (29), (30) is a consequence of Theorem 5.

REMARK 5. Using a topological method developed in [I2], A.E.Siskov obtained apriori estimates and solvability for a more general boundary value problem, which results by adding a quasilinear second order elliptic operator to the left hand side of (29).

2. In this section we establish the existence of a classical solution of the Dirichlet problem for the general nonlinear elliptic equation in a narrow strip. By $\{\Omega_h,\ 0<h\leq 1\}$ we denote the family of domain in R^n with infinitely differentiable boundaries, such that

(a) $\Omega_{h_1}\subset\Omega_{h_2}$ for $h_1<h_2$;

(b) there are open coverings $\{\mathcal{U}_i : i=1,\dots,I\}$ of the set $\bar{\Omega}_1$ and diffeomorphisms $\varphi_i : \mathcal{U}_i\cap\bar{\Omega}_1\to R^n$ of the class C^∞ satisfying $\varphi_i(\mathcal{U}_i\cap\bar{\Omega}_h)=S_h=\{x\in R^n : |x'|<1,\ 0\leq x_n\leq h\}$, $x=(x',x_n)$, $x'=(x_1,\dots,x_{n-1})$.

Let $\{\psi_i(x)\}$, $i=1,\dots,I$ be a partition of unity subordinate to the covering.

For nonnegative integers m, ℓ, κ and an arbitrary $p>1$ we denote by $W_p^{2m,\ell,\kappa}(\Omega_h)$ the closure of the set of functions infinitely differentiable in Ω_h, with respect to the norm

$$\|u\|^p_{W_p^{2m,\ell,\kappa}(\Omega_h)}=\sum_{i=1}^{I}\|\psi_i(\varphi_i^{-1}(y))\,u(\varphi_i^{-1}(y))\|^p_{W_p^{2m,\ell,\kappa}(S_h)},$$

where

$$\|v(y)\|^p_{W_p^{2m,\ell,\kappa}(S_h)}=\sum_{|\alpha|\leq 2m}\int_{S_h}\Big\{\sum_{j=0}^{\kappa}|\mathcal{D}_n^j\mathcal{D}^\alpha v(y)|^p+\sum_{|\beta|\leq\ell}{}'|\mathcal{D}^\beta\mathcal{D}^\alpha v(y)|^p\Big\}dy.$$

Here we use the usual multiindex notation; $\sum'$ indicates the summation over all multiindices with the last coordinate zero.

The space $C^{s,r,\lambda}(\Omega_h)$ for nonnegative integers s, r and $\lambda \in [0,1]$ consists of functions defined in Ω_h which have the finite norm

$$\|u\|_{C^{s,r,\lambda}(\Omega_h)} = \max_i \|\psi_i(\varphi_i^{-1}(y))\, u(\varphi_i^{-1}(y))\|_{C^{s,r,\lambda}(S_h)},$$

where

$$\|v(y)\|_{C^{s,r,\lambda}(S_h)} = \sum_{|\beta|\le s} {\sum_{|\alpha|\le r}}' \|\mathcal{D}^\beta \mathcal{D}^\alpha v(y)\|_{C^{0,\lambda}(S_h)}$$

and $\|\cdot\|_{C^{0,\lambda}}$ is the current norm in the space of functions satisfying the Hölder condition with exponent λ.

Further, we give a coercive estimate for pairs of linear elliptic operators in S_h. This estimate differs from those of [I4] by its uniformity for $h \in (0,1]$ and, on the other hand, by a special choice of the function space whose elements are involved in the estimate.

The next lemma in which we keep the notation from the paper [I4] for $\mathcal{L}^{\ell,\lambda}_{2m}(A,B,\Omega)$ is of crucial importance for the estimates of solutions of the nonlinear problem as well as for the proof of the existence Theorem.

LEMMA 5. <u>There are constants</u> C_1, C_2 <u>depending only on</u> A, B, m, n, ℓ, λ, <u>such that for</u> $q > C_1$, $\rho > qC_1$ <u>an arbitrary operator</u> $L(x,\mathcal{D}) \in \mathcal{L}^{\ell,\lambda}_{2m}(A,B,\Omega)$ <u>and an arbitrary function</u> $u(x) \in W_2^{2m,\ell,1}(S_h) \cap \mathring{W}_2^m(S_h)$ <u>that vanishes for</u> $|x'|$ <u>close to one, the estimate</u>

$$\int_{S_h} \{L(x,\mathcal{D})\mathcal{D}_n u \cdot M(\mathcal{D})\mathcal{D}_n u + \rho {\sum_{|\alpha|\le \ell}}' L(x,\mathcal{D})\mathcal{D}^\alpha u \cdot M(\mathcal{D})\mathcal{D}^\alpha u\}\,dx \geq$$

$$\geq \frac{A}{2}\int_{S_h} |\mathcal{D}_n^{2m+1} u|^2 dx + \frac{\rho A}{2}\int_{S_h} {\sum_{|\alpha|\le \ell}}' |\mathcal{D}^\alpha \mathcal{D}_n^{2m} u|^2 dx + \tag{36}$$

$$+ C_2 \rho q {\sum_{|\alpha|\le \ell}}' \sum_{|\beta|=m} {\sum_{|\gamma|=m}}' \int_{S_h} |\mathcal{D}^{\alpha+\beta+\gamma} u|^2 dx - C(\rho,q)\int_{S_h} \sum_{|\alpha|\le 2m} |\mathcal{D}^\alpha u|^2 dx,$$

<u>holds, where</u> $M(\mathcal{D}) = \mathcal{D}_n^{2m} + q[\mathcal{D}_1^2 + \dots + \mathcal{D}_{n-1}^2]^m$ <u>and</u> $C(\rho,q)$ <u>is a constant depending only on</u> A, B, m, n, λ, ρ, q.

In what follows we fix the number ℓ, $\ell \geq n+1$. We will consider the solvability in $W_2^{2m,\ell,1}(\Omega_h)$ of the nonlinear Diri -

chlet problem

(37) $$F(x, u, \dots, \mathcal{D}^{2m} u) = f(x), \quad x \in \Omega_h,$$

(38) $$\mathcal{D}^\alpha u = 0 \quad x \in \partial\Omega_h, \quad |\alpha| \le m-1.$$

The function $F(x, \xi)$ is assumed to be defined on $\bar{\Omega}_1 \times R^{M(2m)}$, to have continuous derivatives up to the order ℓ and to satisfy the uniform ellipticity condition

(39) $$\sum_{|\alpha|=2m} F_\alpha(x,\xi)\eta^\alpha \ge \nu|\eta|^{2m}$$

for $\eta \in R^n$ with a positive constant ν.

We shall assume that $F(x,0)=0$, the function $f(x)$ belongs to the space $W_2^{0,\ell,1}(\Omega_1) \cap C^{1,\lambda}(\bar{\Omega}_1)$ and

$$\|f\|_{W_2^{0,\ell,1}(\Omega_1)} + \|f\|_{C^{1,\lambda}(\bar{\Omega}_1)} \le R, \quad \|F(x,\xi)\|_{C^\ell(\bar{\Omega}_1 \times B_t)} \le g(t),$$

where $B_t = \{\xi \in R^{M(2m)} : |\xi| \le t\}$ and $g(t)$ is a nondecreasing positive function.

We include the problem (37), (38) in the parametric family of problems of the same type

(40) $$t F(x, u, \dots, \mathcal{D}^{2m} u) + (1-t) L_0 u = t f(x), \quad x \in \Omega_h$$

(4I) $$\mathcal{D}^\alpha u = 0, \quad x \in \partial\Omega_h, \quad |\alpha| \le m-1,$$

where $L_0(\mathcal{D}) = \sum_{|\alpha|=2m} a_\alpha \mathcal{D}^\alpha$ is a fixed elliptic operator with a constant of ellipticity ν.

THEOREM IO. <u>Let $u(x)$ be an arbitrary solution of the problem (40), (4I) satisfying</u>

(42) $$\|u(x)\|_{C^{2m-1,2,\lambda}(\Omega_h)} \le 1.$$

<u>Then there is a constant N depending on m, n, ℓ, ν, R, meas Ω_1 and on the functions $g(t)$, φ_i, ψ_i, such that</u>

(43) $$\|u(x)\|_{W_2^{2m,\ell,1}(\Omega_h)} \le N.$$

Proof of the existence theorem is based on the application of The methods from § I, with the choice of $X = W_2^{2m,\ell,1}(\Omega_h) \cap \mathring{W}_2^m(\Omega_h)$, $\mathcal{D} = \{u \in X : \|u\|_{W_2^{2m,\ell,1}(\Omega_h)} \le N+1\}$, where the number N is defined in Theorem IO. For $x \in U_i \cap \bar{\Omega}_h$ we pass from the variable x to the new variable y by the substitution $x = \varphi_i^{-1}(y)$

and denote by F_i , L_i the differential operators resulting from F , L_0 by the above change of variables. Consider the family of linear operators of the type

$$L_t(y, \mathcal{D}_y) = t \sum_{|\alpha| \le 2m} F_{i,\alpha}(y, u_i, \dots, \mathcal{D}_y^{2m} u_i) \mathcal{D}_y^{\alpha} + (1-t) L_i(y, \mathcal{D}_y).$$

for $u(x) \in \mathcal{D}$, $u_i(y) = u(\varphi_i^{-1}(y))$.

It is easily verified that the set of such operators is contained in the family $\mathcal{Z}_{2m}^{1,\lambda}(A, B, S_h)$ for certain values of A , B and hence according to Lemma 4 the set considered can be associated with an operator $M(\mathcal{D})$ and constants ρ , q such that (36) holds. In what follows, the numbers ρ , q are assumed to be chosen in the just mentioned way.

We define a family of nonlinear operators $A_t : \mathcal{D} \to X^*$ by the identity

(44)
$$\langle A_t u, v \rangle = \sum_{i=1}^{I} \int_{S_h} \psi_i^2(y) \{ \mathcal{D}_n^1 [t F_i(y, u_i, \dots, \mathcal{D}_y^{2m} u_i) + (1-t) L_i u_i - t f_i(y)] \cdot \mathcal{D}_n^1 M v_i + \rho \sum_{|\alpha| \le \ell}{}' \mathcal{D}^{\alpha} [t F_i(y, u_i, \dots, \mathcal{D}_y^{2m} u_i) + (1-t) L_i u_i - t f_i(y)] \mathcal{D}^{\alpha} M v_i \} dy.$$

LEMMA 5. <u>The family of operators</u> A_t <u>defined by (44) satisfies the condition</u> $(\alpha_0^{(t)})$.

The assertion follows from Lemma 4 and is verified analogously as the corresponding assertions of § 2.

We chose a number $h_0 > 0$ so that

(45)
$$K \cdot h_0^{\frac{1}{2} - \lambda} \cdot (N+1) \le 1$$

where the number N is defined from (43), and K is a certain constant defined by an enclosure $W_2^{2m,\ell,0}(\Omega_h)$ into $C^{2m-1,2\lambda}(\Omega_h)$.

THEOREM II. <u>Let</u> $0 < h \le h_0$ <u>, where</u> h_0 <u>is defined by the inequality (45). Then the problem (37), (38) has at least one solution, which belongs to</u> $W_2^{2m,\ell,1}(\Omega_h)$.

Proof. By virtue of Theorem 5, in order to prove Theorem II it suffices to verify that

(a) the equation $A_t u = 0$ has no solution $u \in \partial\mathcal{D}$;

(b) $Deg(A_0, \bar{\mathcal{D}}, 0) \ne 0$.

We shall verify (a) by contradiction. Let $u \in \partial\mathcal{D}$ and $A_t u = 0$. Then u is a solution of the problem (40), (41). Since $\|u\|_{W_2^{2m,\ell,1}(\Omega_h)} = N+1$, the function $u(x)$ in virtue of the inequality (45) satisfies the inequality (42) and consequent -

ly, by Theorem IO, the inequality (43) as well. The last inequality contradicts the condition $u \in \partial \mathcal{D}$.

The validity of (b) follows, for instance, from Theorem 3. Thus the assertions (a), (b) and hence Theorem II are proved.

REFERENCES

[I] J. LERAY, J. SCHAUDER: Topology and functional equations. Ann. Sci.Ecole Norm. Sup. (3) 5I (I934), 45-78.

[2] F.E.BROWDER: Topological methods for non-linear elliptic equations of arbitrary order. Pacif. J. Math. I7 (I966), I7-3I.

[3] O.A.LADYZENSKAYA, N.N.URAL'CEVA: Linear and quasilinear equations of elliptic type. Second edition, revised. (Russian), Izdat. "Nauka", Moscow, I967, 576 pp.

[4] F.E.BROWDER: Problemes non-lineaires. Les Presses de L'Universite de Montreal, I966, I53pp.

[5] Ju.A.DUBINSKII: Quasilinear elliptic and parabolic equations of arbitrary order. (Russian),Uspehi Mat. Nauk 23 (I968), No I (I39), 45-90.

[6] R.I.KACUROVSKII: Nonlinear monotone operators in Banach spaces. (Russian), Uspehi Mat. Nauk 23 (I968), No 2 (I40), I2I-I68

[7] J.Jr.EELLS: A setting for global analysis. Bull. Amer. Math. Soc. 72 (I966), 75I-807.

[8] I.V.SKRYPNIK: The solvability of nonlinear equations with monotone operators, (Ukrainian), Dopovidi Akad. Nauk Ukrain. RSR, Jer. A (I970), 32-35, 93.

[9] F.E.BROWDER, W.V.PETRYSHYN: Approximation methode and the generalized topological degree for non-linear mappings in Banach spaces. J. Funct. Anal. 3 (I969), 2I7-245.

[IO] I.V.SKRYPNIK: Solvability and properties of solutions of nonlinear elliptic equations. Contemporary problems in mathematics. (Russian), Itogi Nauki i Techniki, Moscow,9, I3I-254

[II] I.V.SKRYPNIK: Nonlinear higher order elliptic equations. (Russian), Izdat. "Naukova Dumka", Kiev, I973, 2I9 pp.

[I2] I.V.SKRYPNIK: Topological characteristics of nonlinear elliptic problems. Boundary value problems for partial differential equations. (Russian), Izdat. "Naukova Dumka", Kiev, I978, II8-I25.

[I3] I.V.SKRYPNIK: Topological characteristics of general nonlinear elliptic operators. (Russian), Dokl. Akad. Nauk SSSR 293

(I978), No 3.

[I4] I.V.SKRYPNIK: Coercive inequalities for pairs of linear elliptic operators. (Russian), Dokl. Akad. Nauk SSSR 293 (I978) No 2

[I5] J.G.BORISOVICH, V.G.ZVIAGIN, J.I.SAPRONOV: Nonlinear Fredholm's mappings and Leray-Schauder's theory. Uspechi Mat. Nauk, V. 32, No 4 (I977), 3-54.

[I6] F.E.BROWDER: Nonlinear elliptic boundary value problems and the generalized topological degree. Bull.Amer. Math. Soc. 76 (I970), 999-I005.

[I7] S.AGMON, A.DOUGLIS, L.NIRENBERG: Estimates near the boundary for solutions of elliptic partial differential equations sa - tisfying general boundary conditions I. Comm Pure Appl. Math. I2 (I959), 623-727.

[I8] I.N.BAKEL'MAN: Geometric methods of solution of elliptic equa - tions. (Russian), Izdat. "Nauka", Moscow, I965.

[I9] I.V.SKRYPNIK, A.E.SISKOV: On the solvability of the Dirichlet problem for the Monge-Ampere equations.(Russian), Dokl. Akad.Nauk USSR, ser. A, No 3 (I978), 2I6-2I9.

THE CONTACT GEOMETRY AND LINEAR DIFFERENTIAL EQUATIONS

B.Yu. Sternin and V.E. Shatalov
Moscow Institute of Civil
Aviation Engineers
Pulkovskaya ulitsa 6a
Moscow 125838, USSR

The paper aims mainly at an attempt to construct, proceeding from a unified point of view, a theory of asymptotic expansions (by smoothness) of differential equations solutions in different situations.

The role of symplectic geometry and, in particular, of Lagrangian manifolds in constructing the asymptotic solutions of differential equations containing a large (or small) parameter is well known. In constructing the theory of asymptotic solutions by smoothness we also use the concept of a Lagrangian manifold in symplectic space (see [1]- [4] and the references listed therein). But in this case there exists a supplementary structure in the space under consideration, that is, an action multiplicative group of positive numbers $\mathbb{R}_+$ on fibers of cotangent bundle T_0^*M without zero section to the manifold M. The Lagrangian manifolds under consideration are found to be $\mathbb{R}_+$ -invariant. It gives an idea that the exposition of the theory will be more natural if we use a <u>homogeneous</u> space $T_0^*M/\mathbb{R}_+$ with a <u>contact</u> but not a symplectic structure. Much less trivial is the complex - analytical case; the direct extension of the Fourier integral operators theory to the complex-analytical case is obviously not possible. It is connected with the fact that the kernel of Fourier integral operator is traditionally constructed as a distribution of the form

$$(1) \qquad K(x,y) = \int_{\mathbb{R}_n} e^{i\Phi(x,\tau)} a(x,\tau)\,d\tau$$

where $\Phi(x,\tau)$ is a homogeneous function of degree 1 and $a(x,\tau)$ is a homogeneous function of a certain degree k. Obviously, the construction of Fourier integral operators theory in complex-analytical case in such a form seems very difficult. Our task is to integrate real and complex analytical Fourier integral operators theories on the basis of contact geometry of phase space.

1. Homogeneous symplectic and contact geometry

We shall begin with the definition of a contact structure ([5], [6]). Let S be a smooth 2n-1 -dimensional manifold. We now consider the bundle

$$(2) \qquad T_o^*S/\mathbb{R}_+ \rightarrow S$$

As $T_o^*S \rightarrow T_o^*S/\mathbb{R}_+$ is a locally trivial bundle, any section $\alpha^*: S \rightarrow T_o^*S/\mathbb{R}_+$ of the bundle (2) is locally covered by a section $\alpha : S \rightarrow T_o^*S$ of a cotangent bundle, that is by a differential form of degree 1. At the same time $\alpha_1 = f \cdot \alpha_2$ (f being a positive smooth function) if α^* is locally covered by α_1 and α_2.

Definition 1. A section α^* of bundle (2) is called oriented contact structure if it is non-degenerate, i.e. for any 1-form α locally covering α^* the form $d\alpha\big|_{\ker\alpha}$ is non-degenerate.

Remark 1. To define the (non-oriented) contact structure, one has to substitute the group $\mathbb{R}_*$ for the group $\mathbb{R}_+$ in (2). Supposing S to be a complex analytic manifold and substituting $\mathbb{C}_*$ for $\mathbb{R}_+$ in (2) one can define the complex analytic contact structure.

Further we shall deal only with the $\mathbb{R}_+$ case; other cases are quite similar.

Let T be an (even-dimentional) smooth manifold with a closed non-degenerate 2-form ω on it (symplectic structure). Suppose $\mathbb{R}_+$ acts freely on T, and the set of orbits $S = T/\mathbb{R}_+$ admits the structure of a smooth manifold for which the canonical projection $T \rightarrow S$ is a smooth map. We denote the action $\lambda \cdot t (\lambda \in \mathbb{R}_+, t \in T)$ by $F_t(\lambda) = F_\lambda(t)$. Thus

$$F_t : \mathbb{R}_+ \rightarrow T; \quad F_\lambda : T \rightarrow T$$

Suppose the symplectic structure ω on T to be in accord with the action of $\mathbb{R}_+$, i.e.

$$(3) \qquad F_\lambda^*(\omega) = \lambda\omega.$$

We shall define the contact structure on the space $S = T/\mathbb{R}_+$. Let $X_t = F_{t*}(\frac{d}{d\lambda}\big|_1)$ be a radial vector in $t \in T$. Suppose $s \in S$, Y being a tangent vector in s . For t projecting on s and for tangent vector Y' in t projecting on Y we set

$$(4) \qquad \alpha_t(Y) = \omega(X_t, Y').$$

It should be noted that $\tilde{\alpha}_t$ is a correctly defined linear form on $T_s S$ linearly depending on $t \in \pi^{-1}(s)$ which leads to $\alpha^*_t = \mathrm{cls}\{\tilde{\alpha}_t\}$ to depend only on $s \in S$, i.e. the section α^* of bundle (2) is correctly defined. The non-degenerativeness of ω has not been used in determining α^*.

Proposition 1. An oriented contact structure is given by α^* if and only if a symplectic structure is given by ω.

Proposition 2. Suppose (T_1, ω_1) and (T_2, ω_2) are symplectic spaces with $\mathbb{R}_+$ action subject to the condition (3). Let $T_1/\mathbb{R}_+ = T_2/\mathbb{R}_+ = S$ and let contact structures on S defined by ω_1 and ω_2 coincide. Then T_1 is symplectically diffeomorphic to T_2, the corresponding diffeomorphism commuting with the $\mathbb{R}_+$ action.

Example 1. Suppose $T = T^*_o M$, $S = S^*M = T^*_o M/\mathbb{R}_+$, $\omega = dp \wedge dx$, $F_\lambda[(x,p)] = (x, \lambda p)$. Then X_t is $p\frac{\partial}{\partial p}$, $\tilde{\alpha}_t$ equals $p\frac{\partial}{\partial p} \lrcorner\, dp \wedge dx = pdx$. The vector tangent to $\{p = 1\}$ can be taken as the covering vector Y' in the chart $\{p_1 = 1\}$ on S^*M. Hence, α^* is defined by the restriction of pdx on $\{p_1 = 1\}$, i.e. $\alpha^* = dx^1 + \sum_{i=2}^{n} p_i\, dx^i$.

It should be remarked that the product $T_1 x T_2$ of symplectic manifolds is a symplectic manifold itself (the structure form $\omega_1 - \omega_2$, for example). At the same time the product $S_1 x\, S_2$ of contact manifolds does not admit any contact structure (because of its even-dimensionality). The definition of the contact product of two contact spaces can be given in the following way. Let $T_1 \to S_1$, $T_2 \to S_2$ be symplectic coverings of S_1 and S_2. The contact product is

$$(5) \qquad S_1 \underset{c}{x} S_2 = {}^{T_1 \,x\, T_2}/\, \mathbb{R}_+$$

with a corresponding complex structure. The mappings (projections and a diagonal mapping)

$$\pi^*_i : S_1 \underset{c}{x} S_2 \to S_i,\ i = 1, 2; \quad \Delta^* : S \to S \underset{c}{x} S$$

are defined in a canonical way.

Example 2. Suppose $T_1 = T_2 = T^*_o M$, $S_1 = S_2 = T^*_o M/\mathbb{R}_+$. Then $S_1 \underset{c}{x} S_2 = T^*_o M \,x\, T^*_o M/\mathbb{R}_+$ ($= S^*_o(MxM)$ by definition) which does not coincide with $S^*(M \,x\, M)$. The complement $\bar{S}(MxM)$ of $S^*_o(MxM)$ in $S^*(MxM)$ is a submanifold in $S^*(MxM)$ of codimension n.

We shall finally discuss the notions of symplectic and contact diffeomorphisms, vector fields, and corresponding Hamiltonians.

<u>Definition 2</u>. The diffeomorphism $G:T_1 \to T_2$ such as $G^*(\omega_2) = \omega_1$ and $F_\lambda \circ G = G \circ F_\lambda$ is called a <u>homogeneous</u> symplectic diffeomorphism.

<u>Definition 3</u>. The diffeomorphism $g:S_1 \to S_2$ such as $g^*(\alpha_2^*) = \alpha_1^*$ is called a <u>contact</u> diffeomorphism.

<u>Affirmation 1</u>. Any contact diffeomorphism is given by a <u>homogeneous</u> symplectic diffeomorphism, and vice versa.

Suppose g be a contact diffeomorphism. By virtue of Affirmation 1 a unique homogeneous symplectic diffeomorphism G is defined by g. Let i_G be the imbedding of the space T to the product T x T due to the diffeomorphism G. Then the diagram

$$\begin{array}{ccc} T & \xrightarrow{\;i_G\;} & T \times T \\ \downarrow & & \downarrow {\scriptstyle \pi(T \times T)} \\ S & \xrightarrow[\;i_g^*\;]{} & S \times_c S \end{array}$$

uniquely defines the imbedding i_g^*.

<u>Definition 4</u>. The vector field X' on a homogeneous symplectic space T such as its local one-parameter group $\{G_\lambda\}$ consists of homogeneous symplectic diffeomorphisms is called a <u>Hamilton</u> field.

<u>Definition 5</u>. The vector field X on a contact space S such as its local one-parameter group consists of contact diffeomorphisms is called a contact vector field.

Due to Affirmation 1 there exists a one-to-one correspondence between homogeneous Hamilton vector fields on a homogeneous symplectic space T, on one hand, and contact vector fields on the corresponding contact space S, on the other hand.

We recall that a function H, subject to the condition $-dH = X' \lrcorner\, \omega$, is called a Hamiltonian corresponding to the Hamilton vector field X' on the symplecting space T. The Hamiltonian is locally defined for any Hamilton vector field X' up to an additive constant. But a unique homogeneous Hamiltonian H of degree 1 corresponding to a vector field X' is globally defined if $\mathbb{R}_+$ acts on T and the field X' is homogeneous, i.e. $F_{\lambda *}(X') = X'$. The Hamiltonian can be defined as follows

$$H = \omega(X_t, X')$$

In this case we denote X' by V(H), and the corresponding contact vector field by X_H. The fact that any contact vector field X on the contact space S is determined uniquely by a homogeneous Hamiltonian H of order 1 on the space T follows from Affirmation 1.

If an (oriented) contact structure α^* on S is determined by a global 1-form α the situation becomes less complicated. In this case $\tilde{\alpha}_t = \chi(t)\cdot\alpha$ as $\tilde{\alpha}_t \in \mathrm{cls}[\alpha]$ where $\chi(t)$ is a positively homogeneous function on T of order 1. We can define the imbedding of the space S into the space T assuming $\chi|_S = 1$. Then the contact Hamiltonian can be defined as

(6) $$h = H|_S$$

its dependence on the contact vector field X_h is given by the formula

$$h = X_h \lrcorner \alpha .$$

If $T = T_o^*M$ and a Riemannian metrics is fixed on M we can put $\chi(x,p) = |p|$. In this case the imbedding of $S = S^*M$ into $T = T_o^*M$ is realised by a bundle of unit spheres on the space T_o^*M. Below we shall deal only with Hamiltonians of order 1; we shall consider the Hamiltonian $|p|^{1-r}$ H instead of any homogeneous Hamiltonian H of order r.

Let us give a definition to a contact analogue of the notion of the Lagrangian manifold.

Definition 6. A submanifold $l \subset S$ is called Legendre manifold if $\alpha^*|_l = 0$.

The next affirmation is evident.

Affirmation 2. Any Lagrangian manifold $L \subset T$ invariant with respect to $\mathbb{R}_+$ action determines a Legendre manifold $l \subset S$, and vice versa.

Thus, the oriented contact structure theory is identical with equivariant theory of symplectic structures. Similarly, the $\mathbb{R}_*$ -equivariant theory of symplectic structures gives the notion of a (non-oriented) contact structure, the $\mathbb{C}_*$-equivariant theory of complex-analytic symplectic structures giving a complex-analytic contact structure.

2. Fourier integral operators theory

We shall deal with the distributions of the following type (cf [13]):

$$u(x) = \frac{\Gamma(k/2)}{(2\pi)^{n/2}} \int (\Phi(x,\tau) + i\,0)^{-k/2} a(x,\tau)d\tau \ ;$$

$$0 < \arg \Phi(x,\tau) + i\varepsilon < \pi,$$

where $\Phi(x,\tau)$, $a(x,\tau) \in C^\infty$, supp $a(x,\tau)$ is compact, $\mathrm{grad}_{x,\tau}\Phi(x,\tau) \neq 0$, $\tau = (\tau_1, \ldots, \tau_m)$.

By $I_s(\Phi)$ we denote a space of distributions representable by integral (7) within the smooth functions where $k = m - 2s$.

The representation (7) is unambiguous. In fact, let us consider three transformations of the pair (Φ, a), $u(x)$ being invariant of them.

1°. The substitution of pair (Φ, a) by pair $(\Phi \cdot f, a \cdot f^{k/2})$ does not change the distribution $u(x)$ for a smooth strictly positive function $f(x,\tau)$.

2°. Let $\tau = \tau(x, \tau')$ be a diffeomorphism in a neighbourhood of the support of $a(x,\tau)$. Then the substitution of pair (Φ, a) by pair $\Phi(x, \tau(x,\tau'))$, $a(x, \tau(x,\tau')) \left| \frac{D\tau(x,\tau')}{D\tau'} \right|)$ does not change the distribution $u(x)$.

3°. The substitution of the pair (Φ, a) by pair $(\Phi(x,\tau) + \eta^2/2, a(x,\tau))$ or by pair $(\Phi(x,\tau) - \eta^2/2, i \cdot a(x,\tau))$ does not change the distribution $u(x)$.

Using transformations $1^o - 3^o$, we get the following

<u>Affirmation 3</u>. The spaces $I_S(\Phi)$ are connected with each other in the following way:

1) $I_S(\Phi') = I_S(\Phi)$ if $\Phi'(x,\tau) = \Phi(x,\tau) \cdot f(x,\tau)$, $f(x,\tau) > 0$;

2) $I_S(\Phi') = I_S(\Phi)$ if $\Phi'(x,\tau) = \Phi(x, \tau(x,\tau'))$, $\tau = \tau(x,\tau')$ being a diffeomorphism;

3) $I_S(\Phi') \supset I_S(\Phi)$ if $\Phi'(x, \tau, \eta) = \Phi(x,\tau) \pm \eta^2/2$.

But the investigation of elements of the space $I_S(\Phi)$ themselves is too delicate a problem. We shall deal with asymptotics of these elements "by smoothness", i.e. elements of factor-space

(8) $I_S(\Phi) / I_{S+N}(\Phi) = I^{S,N}(\Phi)$

An element of the latter space will be named asymptotics of order N. Let us consider the case when $N = 1$.

Let $\mathcal{J}(\Phi)$ be an ideal in ring $C^\infty([x,\tau])$ generated by functions $\Phi, \Phi_{\tau_1}, \ldots, \Phi_{\tau_m}$:

$$\mathcal{J}(\Phi) = \{\Phi, \Phi_{\tau_1}, \ldots, \Phi_{\tau_m}\}$$

We shall use

<u>Lemma 1.</u> Suppose $u(x) \in I_S(\Phi)$ and $a(x,\tau) \in \mathcal{J}(\Phi)$ in the represent-

ation (7). Then $u(x) \in I_{s+1}(\Phi)$. Inversely, if $u(x)$ representable as in (7) for $k = m - 2s$ belongs to $I_{s+1}(\Phi)$ then $a(x,\tau) \in \mathcal{J}(\Phi)$.

Corollary 1. $I_s(\Phi)/I_{s+1}(\Phi) = I_s(\Phi')/I_{s+1}(\Phi')$, i.e. $I^{s,1}(\Phi) = I^{s,1}(\Phi')$ if functions Φ and Φ' are connected with each other by any of the transformations 1°-3° (see Affirmation 3).

Definition 7. The function Φ_1 is called equivalent to the function $\Phi_2 (\Phi_1 \sim \Phi_2)$ if Φ_1 and Φ_2 can be transformed into the same function by a sequence of transformations 1°-3°.

Corollary 1 shows that $I^{s,1}(\Phi_1) = I^{s,1}(\Phi_2)$ if $\Phi_1 \sim \Phi_2$. Hence, it is natural to define the space $I_s(\operatorname{cls}\Phi) = \bigcup_{\Phi' \in \operatorname{cls}\Phi} I_s(\Phi')$. Further we shall not distinguish between $I_s(\Phi)$ and $I_s(\operatorname{cls}\Phi)$.

Now let us introduce a Legendre manifold in the space S^*M determined by Φ. For this we demand that $\Phi(x,\tau)$ be subjected to

Condition 1. The forms $d\Phi, d\Phi_{\tau_1}, \ldots, d\Phi_{\tau_m}$ are linearly independent in (x_o, τ_o).*

Let us determine the set C_Φ by formula

$$C_\Phi = \{(x,\tau) \mid \Phi(x,\tau) = 0, \ \Phi_\tau(x,\tau) = 0\}.$$

The set C_Φ is a smooth manifold due to the Condition 1. We shall determine the mapping

$$\alpha : [x,\tau] \rightarrow T^*([x])$$

$$\alpha(x,\tau) = (x, p) = (x, \Phi_x(x,\tau))$$

and notate $\alpha(C_\Phi) = \tilde{\mathcal{L}}(\Phi)$.

Affirmation 4. The mapping $\alpha : C_{(\Phi)} \rightarrow T_o^*([x])$ is an embedding, the manifold $\tilde{\mathcal{L}}(\Phi)$ being transversal to $\mathbb{R}_+$-orbits.

Corollary 2. The manifold $\tilde{\mathcal{L}}(\Phi)$ determines a manifold $l(\Phi)$ in space $S^*([x])$.

Affirmation 5. $l(\Phi)$ is a Legendre manifold. The above constructed Legendre manifold $l(\Phi)$ determined by Φ is a classifying object for spaces $I_s(\Phi)$ in the following sense:

Proposition 3. Φ_1 is equivalent to Φ_2 iff $l(\Phi_1) = l(\Phi_2)$. In particular, the spaces $I_s(\Phi_1)$ and $I_s(\Phi_2)$ coincide when $l(\Phi_1) = l(\Phi_2)$.

---

* Further on we shall deal with the local theory, i.e. with germs of functions; in particular, the support of $a(x,\tau)$ is to be sufficiently small.

Proof. (Sketch) One can see each function $\Phi(x,\tau)$ under Condition 1 be equivalent to $\Phi'(x, p_I)$ having the form

$$\Phi(x, p_I) = \pm (x^1 - S(x^I, p_{\bar I}) - x^{\bar I} p_{\bar I})$$

modulo numeration of coordinates (here I and $\bar I$ are supplementary subsets of $\{2, 3, \ldots, n\}$). Now it is enough to note that the function $S(x^I, p_{\bar I})$ is determined by a Legendre manifold uniquely.

We call any function $\Phi(x,\tau)$ under Condition 1 a determining function for $l(\Phi)$.

Further we shall give a more detailed consideration to the coincidence of asymptotics of functions U(x), having the form (7). Due to Lemma 1 an element of the space $I^{s,1}(\Phi)$ determined by integral (7) depends only on the restriction of the function $a(x,\tau)$ at C_Φ.

However, the restrictions of function $a(x,\tau)$ at C_Φ will differ, generally speaking, for different representations of one and the same function u(x) in $I^{s,1}(\Phi)$.

Therefore, we shall modify the definition of integral (7).

Let us introduce a function

$$F(\Phi,\mu) = \frac{\alpha^*\mu \wedge d\Phi \wedge d\Phi_{\tau_1} \wedge \ldots \wedge d\Phi_{\tau_m}}{v(x) \wedge d\tau}.$$

Here μ is a n - 1-form homogeneous of degree $-2s - 1$ on a homogeneous Lagrangian manifold L, containing $\tilde{\mathcal{L}}(\Phi)$; v(x) is a non-degenerated n-form on the space $[x]$; the form $\alpha^*\mu$ is freely continued from $C(\Phi)$ to its neighbourhood.

When M is a Riemannian manifold we can put $\mu = |p|^{-2s-1}|_L \pi^*\tilde\mu$ where $\mu \in \Lambda^{n-1}(l)$ is a measure on l, and $\pi : T^*_0 M \to S^*M$ is a canonical projection.

Hence

$$F(\Phi,\mu) = \frac{\alpha^*\mu \wedge d\Phi \wedge d\Phi_{\tau_1} \wedge \ldots \wedge d\Phi_{\tau_m}}{|\Phi_x|^{2s+1} v(x) \wedge d\tau}$$

Let us consider a modification of representation (7)

$$(9) \qquad \frac{\Gamma(k/2)}{(2\pi)^{m/2}} \int (\Phi(x,\tau) + i0)^{-\frac{k}{2}} a(x,\tau) \sqrt{F(\Phi,\mu)}\, d\tau$$

where a branch of argument of $F(\Phi, \mu)$ is fixed.

Proposition 4. If $\Phi_1 \sim \Phi_2$ and the restrictions of functions $a_1(x,\tau^{(1)})$

and $a_2(x,\tau^{(2)})$ on $l(\Phi)$ coincide, then the corresponding functions $u_1(x)$ and $u_2(x)$ determined by formulae of type (9) are connected by the relation

$$u_1(x) = e^{i\pi C_{12}}\, u_2(x) \pmod{I_{s+1}(\Phi)}$$

C_{12} being a whole number found with the help of formula

$$C_{12} = \frac{1}{2\pi}\Big[\arg F(\Phi_1,\mu) - \arg F(\Phi_2,\mu) + \frac{1}{2}\operatorname{ind}_-\frac{\partial^2\Phi_2}{\partial\tau^{(2)}\partial\tau^{(2)}} - \operatorname{ind}_-\frac{\partial^2\Phi_1}{\partial\tau^{(1)}\partial\tau^{(1)}}\Big].$$

Here by ind_- we denote a negative index of inertia of a corresponding symmetric matrix.

Numbers C_{12} determine a whole-number class $\operatorname{ind}[l,\mu] \in H^1(l,\mathbb{Z})$ of one-dimensional cohomologies of the Legendre manifold $l(\Phi)$.

Definition 8. We shall call a pair (l,μ) quantized if $\operatorname{ind}[l,\mu] = 0$.

Theorem 1. Let (l,μ) be a quantized pair, $\varphi \in C^\infty(l)$. Then there exists a choice of arguments $\arg F(\Phi,\mu)$ such as the formula

$$(10)\quad u(x) = \sum_\alpha \frac{\Gamma(k_\alpha/2)}{(2\pi)^{m_\alpha/2}} \int \big(\Phi_\alpha(x,\tau^{(\alpha)}) + i0\big)^{-k_\alpha/2} \varphi(x,\tau^{(\alpha)}) \times$$

$$\times\, e_\alpha(x,\tau^{(\alpha)}) \sqrt{F(\Phi_\alpha,\mu)}\, d\tau^{(\alpha)}$$

determines an element of module $I^{s,1}(l)$ independent of the choice of determining functions Φ_α on the neighbourhoods U_α and of the partition of unity subordinate to the covering $\{U_\alpha\}$. In (10) the functions φ, e_α are freely continued from l to a neighbourhood of C_Φ. We denote the element $u(x)$ constructed above by $K^s_{(l,\mu)}(\varphi)$.

The definition (10) of function $K^s_{(l,\mu)}(\varphi)$ makes a further generalization possible. If φ is a homogeneous function on L of degree r, and μ is a homogeneous measure on L of degree $-2r-2s-1$, then Theorem 1 is valid. In this case we again use the notation $K^s_{(l,\mu)}(\varphi)$.

Remark 2. If $u(x) \in I_s(l)$, then due to Lemma 1 it determines a function φ_0 on l such as the difference $u(x) - K^s_{(l,\mu)}(\varphi_0) \in I_{s+1}(l)$.

Applying the above operation to this difference and iterating the process we get a representation of $u(x)$ in the space $I^{s,N}(l)$ for any N:

$$u(x) = \sum_{k=0}^{N-1} K^{s+k}_{(l,\mu)}(\varphi_k)$$

where φ_k are functions on 1.

Applying the above construction for the product M x M and modifying formula (10) as follows

$$K(x,y) = \sum_{\alpha} \frac{\Gamma(k_\alpha/2)}{(2\pi)^{m_\alpha/2}} \int \left(\Phi_\alpha(x,y,\tau^{(\alpha)}) + i0\right)^{-\frac{k_\alpha}{2}} \times$$

$$\times \varphi(x,y,\tau^{(\alpha)})\, e_\alpha(x,y,\tau^{(\alpha)}) \sqrt{F(\Phi_\alpha,\mu)}\, d\tau^{(\alpha)}\, v(y), \qquad (11)$$

where

$$F(\Phi,\mu) = \frac{\alpha^*\mu \wedge d\Phi \wedge d\Phi_{\tau_1} \wedge \ldots \wedge d\Phi_{\tau_m}}{v(x)\wedge v(y) \wedge d\tau_1 \wedge \ldots \wedge d\tau_m}$$

we get the kernel K(x,y) being a function in the variables x and an n-form with respect to y.

Definition 9. The operator

$$f \mapsto \hat{\Phi}^{(r)}_{(1,\mu,\Phi)}(f) = \frac{i^n}{(2\pi)^{\frac{n+1}{2}}} \int K(x,y) f(y)$$

where K(x,y) is given by formula (11), $s = -r - \frac{n+1}{2}$ is called the Fourier integral operator on Legendre manifold $1 \subset S^*(M \times M)$ with measure μ, amplitude function φ. The number r is called the order of Fourier integral operator $\hat{\Phi}^{(r)}_{(1,\mu,\Phi)}$.

By $Op_r(1)$ we denote the set of Fourier integral operators of order r on 1.

Definition 10. Fourier integral operator with a kernel from $I_{-r-\frac{n+1}{2}}(1)$ is called a pseudodifferential operator (p.d.o.) if

1) the Legendre manifold 1 is determined by a mapping $\Delta^*: S^*M \to S^*M \underset{C}{\times} S^*M \subset S^*(M \times M)$ (a diagonal mapping described in Section 1);

2) the measure μ homogeneous of degree n on a Lagrangian manifold L is given by formula $\mu = X_t \lrcorner\, \omega^n = p\frac{\partial}{\partial p} \lrcorner\, (dp \wedge dx)^n$;

3) the amplitude function φ is a homogeneous function P(x,p) of degree r on T_0^*M; P(x,p) is called a symbol of p.d.o.

The number r is called the order of pseudodifferential operator $P(x, -\partial/\partial x)$. The local representation of the kernel K(x,y) of $P(x - \partial/\partial x)$ in the domain $p_1 \neq 0$ has the following form:

$$K(x,y) = \frac{i^n \Gamma(r+n)}{(2\pi)^n} \int \frac{P(x,1,p^*)dp^*}{[(x^1-y^1) + p^*(x'-y') + i0]^{r+n}}$$

If we interpret S^*M as a unit cosphere bundle (with a given Riemannian metrics), then the kernel $K(x,y)$ of p.d.o. with a symbol $P(x,p)$ can be given in the form

$$K(x,y) = \frac{i^n \Gamma(r+n)}{(2\pi)^n} \int \frac{P(x,p)\,\Omega_{(p)}}{[\langle p, x-y\rangle + i0]^{r+n}}$$

where $\Omega_{(p)} = \sum (-1)^k_{p_k}\, dp_1 \wedge \ldots \wedge \widehat{dp_k} \wedge \ldots \wedge dp_n$

is a volume element on a unit sphere, with integrating over fibers of cosphere bundle.

<u>Remark 3</u>. In case $P(x,p)$ is a homogeneous polynom in variable p, the described construction gives a <u>differential</u> operator $P(x, -\partial/\partial x)$.

<u>Remark 4</u>. Applying Remark 2 one can get a representation

$$\hat{p} = \sum_{k=0}^{N-1} \hat{\Phi}^{(r-k)}_{(1,\mu,P_{r-k})}$$

for any p.d.o modulo operators of order $r-N$.

3. <u>Composition with p.d.o formulae</u>

In this section two theorems for the calculation of composition between a p.d.o and a Fourier integral operator will be established.

Let $H(x, D_x)$ be a p.d.o of order r, $D_x = -\partial/\partial x$.

<u>Theorem 2</u>. The comparison

$$H(x, D_x)\circ \hat{\Phi}^{(r)}_{(1,\mu,\varphi)} \equiv \hat{\Phi}^{(r+r')}_{(1,\mu',H_r(x,p)|_L\varphi)} \pmod{Op_{r+r'-1}(1)}$$

is valid, where $\mu' = |p|^{2r}\mu$; a Riemannian metrics is fixed on M and S^*M is embedded in T^*_oM for $|p| = 1$; $H_{r'}(x,p)$ is the principal symbol of $H(x, D_x)$.

<u>Proof</u> (sketch) will be given for the case of differential operators. Obviously, it is sufficient to calculate the composition for local Fourier integral operator with kernel

$$F(x,y) = \frac{\Gamma(k/2)}{(2\pi)^{m/2}} \int \left(\Phi(x,y,\tau) + i0\right)^{-k/2} \varphi(x,y,\tau) \times \tag{12}$$

$$\times \sqrt{F(\Phi,\mu)}\ v(y)\,d\tau$$

(The coefficient $i^n/(2\pi)^{\frac{n+1}{2}}$ is omitted.)

The composition has a kernel

$$\tilde{K}(x,y) = \frac{\Gamma(k/2+r')}{(2\pi)^{m/2}} \int \left(\Phi(x,y,\tau)+i0\right)^{-\frac{k}{2}-r'} H_{r'}(x,\Phi_x(x,y,\tau))\times$$

$$\times\varphi(x,y,\tau)\sqrt{F(\Phi,\mu)}\ v(y)\,d\tau = \frac{\Gamma(k/2+r')}{(2\pi)^{m/2}} \int \left(\Phi(x,y,\tau)+i0\right)^{-\frac{k}{2}-r'}\times$$

$$\times H_{r'}\left(x, \frac{\Phi_x(x,y,\tau)}{|\Phi_x(x,y,\tau)|}\right)\varphi(x,y,\tau)\sqrt{F(\Phi,\mu)}\,d\tau\ v(y)$$

modulo operators from $Op_{r+r'-1}$.

The restriction of function $H_{r'}\left(x, \frac{\Phi_x(x,y,\tau)}{|\Phi_x(x,y,\tau)|}\right)\varphi(x,y,\tau)$ at Legendre manifold l coincides with $H_{r'}(x,p)\big|_l\,\varphi$.

Theorem 3. Let $H_{r'}(x,p)\big|_l = 0$, $\mu = |p|^{2r+n}\pi^*\tilde{\mu}$, $\tilde{\mu}$ being a measure invariant with respect to the contact vector field X_h determined by the contact Hamiltonian $h = H_r(x,p)_{S^*M}$. In this way the following comparison is valid

$$H(x,D_x)\circ\hat{\Phi}^{(r)}_{(l,\mu,\varphi)} \equiv \hat{\Phi}^{(r+r'-1)}_{(l,\mu',\hat{\mathcal{P}})} \pmod{Op_{s-r+2}(l)},$$

where $\mu' = |p|^{2(r+r'-1)}\pi^*\tilde{\mu}$, and the operator $\hat{\mathcal{P}}$ is given by the formula

$$\hat{\mathcal{P}}\varphi = \left\{X_h - H_1^{sub} - \left(\frac{n-1}{2} + 2r\right)|p|^{-1}\,V(H_1)\,|p|\right\}\Big|_l\,\varphi,$$

where

$$H_1(x,p) = |p|^{1-r'}H_{r'}(x,p);$$

$$H_1^{sub} = |p|^{1-r'}H_{r'-1}(x,p) - \frac{1}{2}\ \frac{H_1(x,p)}{\partial x\,\partial p} + \frac{1}{2}V(H_1)v(x)$$

Proof (sketch). Applying the operator $H(x,D_x)$ to the operator $\hat{\Phi}^{(r)}$

with kernel (12) we get the operator with kernel

$$\tilde{K}(x,y) = \frac{\Gamma(\frac{k}{2}+r')}{(2\pi)^{m/2}} \int (\Phi(x,y,\tau)+i0)^{-\frac{k}{2}-r'} H_1(x,\Phi_x(x,y,\tau))\varphi(x,y,\tau)\times$$

$$\times\sqrt{F(\Phi,\mu)}\,d\tau\, v(y) + \frac{\Gamma(\frac{k}{2}+r'-1)}{(2\pi)^{m/2}} \int (\Phi(x,y,\tau)+i0)^{-\frac{k}{2}-r'+1} \Big\{ H_{r'p}(x,$$

$$\Phi_x(x,y,\tau))\frac{\partial}{\partial x} + \frac{1}{2} H_{r'pp}(x,\Phi_x(x,y,\tau))\Phi_{xx}(x,y,\tau) + H_{r'-1}(x,$$

$$\Phi_x(x,y,\tau))\Big\}\varphi(x,y,\tau)\sqrt{F(\Phi,\mu)}\,d\tau\, v(y)$$

modulo operators from $Op_{r+r'-2}(1)$.

The condition $H_r(x,p)\big|_1 = 0$ is equivalent to the equality

$$H_1(x,\Phi_x(x,y,\tau))\big|_{C_\Phi} = 0$$

Due to Condition 1 with the aid of Hadamard lemma we get the functions $B_i(x,y,\tau)$, $i = 0, 1, \ldots, m$, such as

$$(13)\quad H_1(x,\Phi_x(x,y,\ \ )) + \sum_{i=1}^{m} B_i(x,y,\tau)\Phi_{\tau_i}(x,y,\tau) +$$

$$+ B_o(x,y,\tau)\,\Phi(x,y,\tau) = 0$$

The construction of Legendre manifold can be described as follows. Let $J_1([x, y,\tau])$ be a space of 1-jets of C^∞-functions on the space $[x,y,\tau]$; the coordinates in $J_1([x, y,\tau])$ we denote by $(x,y,\tau, p, q, \eta, \xi)$. Any function $\Phi(x,y,\tau)$ determines a submanifold $j_1(\Phi)$ in $J_1([x, y,\tau])$ by

$$(14)\quad p = \Phi_x(x,y,\tau),\ q = -\Phi_y(x,y,\tau),\ \eta = \Phi_\tau(x,y,\tau),\ \xi = \Phi(x,y,\tau).$$

Condition 1 provides the transversality of intersection of $j_1(\Phi)$ with the plane $\{\eta = 0,\ \xi = 0\}$. The intersection is C_Φ. The plane $\{\eta = 0,\ \xi = 0\}$ is projected on S^*M by α^*. Due to Affirmation 5, $\alpha^*(C_\Phi) = l(\Phi)$ is a Legendre manifold. The mapping α^* can be decomposed as $\alpha^* = \pi\circ\alpha$. We shall use a contact Hamiltonian

$$\mathcal{H}(x,y,\tau, p, q, \eta, \xi) = H_1(x,p) + \sum_{i=1}^{m} B_i(x,y,\tau)\eta_i +$$

$$+ B_o(x,y,\tau)\xi$$

on the space $\mathcal{J}_1([x, y, \tau])$. Due to (13) and (14) we get $\mathcal{H}|_{j_1(\Phi)} = 0$. Hence $X_{\mathcal{H}}$ is tangent to $j_1(\Phi)$.

Affirmation 6. The field $X_{\mathcal{H}}$ is tangent to C_Φ. The image of the field $X_{\mathcal{H}}$ on Legendre manifold $l(\Phi)$ is a contact vector field X_h. Now we shall transform the expression for the function K(x,y) as follows

$$\tilde{K}(x,y) = \frac{\Gamma(k/2+\tau'-1)}{(2\pi)^{m/2}} \int (\Phi(x,y,\tau)+i0)^{-\frac{k}{2}-\tau'+1} \left[X_{\mathcal{H}} + |\Phi_x(x,y,\tau)|^{-\tau'+1} \times\right.$$

$$\times H_{\tau' pp}(x, \Phi_x(x,y,\tau))\Phi_{xx}(x,y,\tau) + \sum_{i=1}^{m} \frac{\partial B_i(x,y,\tau)}{\partial \tau_i} + \left(\frac{k}{2}+\tau'-1\right)B_0(x,y,\tau) +$$

$$+(1-\tau)|\Phi_x(x,y,\tau)|^{-\tau'+1} H_{\tau' p_i}(x, \Phi_x(x,y,\tau))\Phi_{x^i}(x,y,\tau)\Phi_{x^i x^j}(x,y,\tau) +$$

$$\left. + |\Phi_x(x,y,\tau)|^{-\tau'+1} H_{\tau'-1}(x, \Phi_x(x,y,\tau))\right] \sqrt{F(\Phi,\mu)}\, \varphi(x,y,\tau)\, d\tau\, v(y).$$

Commutating $X_{\mathcal{H}}$ and $F(\Phi, \mu)$ we get the affirmation of the theorem, using the Sobolev lemma.

4. Continuity and composition of Fourier integral operators

We shall further formulate continuity and composition theorems for Fourier integral operators. We shall consider, for simplicity, the case when the corresponding Legendre manifolds are determined by contact diffeomorphisms.

Definition 11. A Fourier integral operator $\hat{\Phi}^{(r)}_{(g,\varphi)}$ will be called associated with a constant diffeomorphism g if

1) the Legendre manifold l is determined by an embedding

$$i^*_g : S^*M \to S^*M \underset{c}{\times} S^*M \subset S^*(M \times M)$$

2) the measure μ has a form $\mu = X_t \lrcorner\, \omega^n = p\frac{\partial}{\partial p} \lrcorner\, (dp \wedge dx)^n$.

The following theorems hold.

Theorem 4. The operator

$$\hat{\Phi}^{(r)}_{(g,\varphi)} : H^s(M) \to H^{s-r}(M)$$

is continuous for any s, $H^s(M)$ being Sobolev spaces.

Theorem 5.

$$\hat{\Phi}^{(r_2)}_{(g_2,\varphi_2)} \circ \hat{\Phi}^{(r_1)}_{(g_1,\varphi_1)} = \hat{\Phi}^{(r_1+r_2)}_{(g_2\circ g_1,\ \varphi_1+g_1^*\varphi_2)}$$

holds up to operators of order $r_1 + r_2 - 1$.

The proofs of the theorems use a priori estimations for an auxiliary Cauchy problem (see [4], for example).

5. Regularization of equations of principal type

Definitions of equations of principal type are given below (see [7,8], for example).

Definition 12. The equation*

(15) $$Hu = f$$

is called

1) (microlocally) an equation of principal type in $\alpha \in S^*M$ if its contact vector field does not vanish in α ;

2) (locally) an equation of principal type in $x \in M$ if all the trajectories of a contact vector field X_H lying in the set

(16) $$\text{char } H = \{\alpha \in S^*M \mid H(\alpha) = 0\}$$

leave the fibre of bundle S^*M above x in finite time;

3) (semi-globally) an equation of principal type if for any pair of compacts K_1, $K_2 \subset M$ there exists a number $T(K_1, K_2)$ such that all the trajectories of the field X_H lying in char H and beginning at t=0 above K_1 lie above the complement of K for $|t| \geq T(K_1, K_2)$.

Let us give the corresponding definitions for the solvability of equation (15).

Definition 13. The equation (15) is called

1) microlocally solvable in $\alpha \in S^*M$ if for any $f \in H^s(M)$ there exists a function $u \in H^s(M)$ such that the wave front** $WF(Hu - f)$ of the difference $Hu - f$ does not intersect a neighbourhood of α and $\|u\|_s \leq C \|f\|_s$;

* We recall that the order of H is equal to unity and its principal symbol is real.

** For the definition of the wave front see, for example, [3].

2) locally solvable in $x \in M$ if for any $f \in H^s(M)$ there exists a function $u \in H^s(M)$ such that $Hu - f = 0$ in a neighbourhood of x and $\|u\|_s \leq C \|f\|_s$;

3) semi-globally solvable if for any compact K the kernel N(K) of operator H^* adjoint to $H : H^s_{loc}(K) \to H^s_{loc}(K)$ is independent of s , finite-dimensional and for any $f \in H^s_{loc}(K)$ orthogonal to N(K) there exists a function $u \in H^s_{loc}(K)$ such that $Hu = f$ in a neighbourhood of K with $\|u\|_s \leq C \|f\|_s$.

Here K is a closure of a domain K; $H_{loc}(K)$ is a space of functions determined in a neighbourhood (dependent on the function) of K and belonging in the neighbourhood to the space H_{loc}. We identify the functions coinciding in K.

Theorem 6. If the equation

$$Hu = f$$

is microlocally (locally, semi-globally) of principal type, then it is microlocally (correspondingly, locally or semi-globally) solvable.

Proof (sketch). We consider a modification of operators introduced in Section 2. By l_o we denote a set in $S^*(M \times M)$ given by $l_o = \Delta^* \cap \text{char } H$. Let l be a phase flow of this set along a contact vector field X_H (here H is a raising of H to $T^*_o M \times T^*_o M$ with the help of a projection onto the first factor). Suppose $X_H(\alpha) \neq 0$. Then l is obviously a submanifold in $S^*(M \times M)$, transversal to X_H in a neighbourhood of $\tilde{\alpha}$.

We introduce the coordinates $(x^1, \ldots, x^n, p_2^*, \ldots, p_n^*) = (x, p^*)$ in a neighbourhood of α on S^*M with the help of equations $p_i^* = p_i/p_1$, where $(x^1, \ldots, x^n, p_1, \ldots, p_n)$ is a canonical coordinate system on T^*M determined by a coordinate system $(x^1, \ldots, x^n)$ on M (to be brief, we consider the case $p_1 > 0$). The coordinate expression for the field X_H is

$$X_H = -\left(\sum_{i=2}^{n} p_i^* H_{p_i^*}(x,1,p^*) - H(x,1,p^*)\right)\frac{\partial}{\partial x^1} + \sum_{i=2}^{n} H_{p_i^*}(x,1,p^*) \times \frac{\partial}{\partial x^i} - \sum_{i=2}^{n}\left(H_{x^i}(x,1,p^*) - p_i^* H_{x^1}(x,1,p^*)\right)\frac{\partial}{\partial p_i^*}.$$

The inequality $X_H \neq 0$ yields either $H_{p^*_{i_o}}(x, 1, p^*) \neq 0$ or

$(H_{x^{i_o}} - p^*_{i_o} H_{x^1})(x, 1, p^*) \neq 0$ for a certain $i_o \in \{2, \ldots, n\}$.

The first case is called regular, the second one - singular. We shall use a coordinate system $(x^1, \ldots, x^n, y^1, \ldots, y^n, p_1^*, \ldots,$

$p_n^*, q_2^*, \ldots, q_n^*)$ on $S^*(M \times M)$, where (x,p) are the coordinates on the first factor T^*M, (y, q) are the coordinates on the second factor and

$$p_i^* = \frac{p_i}{q_1}, \quad q_i^* = \frac{q_i}{q_1}$$

(to be brief, we consider the case $q_1 > 0$). The function

$$(17) \quad \Phi^{i_o} = \Big[y^1 + \sum_{\substack{i=2 \\ i \neq i_o}}^{n} y^i q_i^* \Big]\Big|_1 - \Big[y^1 + \sum_{\substack{i=2 \\ i \neq i_o}}^{n} y^i q_i^* \Big] =$$

$$= \Phi^{i_o}(x, y, q_2^*, \ldots, \widehat{q_{i_o}^*}, \ldots, q_n^*)$$

is a determining function for the manifold 1 in the regular case, $(x^1, \ldots, x^n, y^{i_o}, q_2^*, \ldots, \widehat{q_{i_o}^*}, \ldots q_n^*)$ being the coordinates on 1.

In the singular case the coordinates on 1 are $(p_2^*, \ldots, p_n^*, u, q_{i_o}^*, y^2, \ldots, \widehat{y^{i_o}}, \ldots y^n)$, where $u = y^1 + q_{i_o}^* y^{i_o}$, and the determining function is

$$(18) \quad \Phi_{i_o}(x,y, p_2^*, \ldots, p_n^*, u, q_{i_o}^*) = -\Big[x^1 + \sum_{i=2}^{n} p_i^* x^i\Big]\Big|_1 +$$

$$+ (u - y^1 - q_{i_o}^* y^{i_o}) + \Big[x^1 + \sum_{i=2}^{n} p_i^* x^i\Big]$$

The set 1_o divides the set 1 into two parts 1_+ and 1_- according to the sign of the parameter t of the shift along the vector field X_H. Supposing that the inequality $x^{i_o} > y^{i_o}$ holds on 1_+ in the regular case and $p_{i_o}^* > q_{i_o}^*$ holds on 1_+ in the singular case we introduce operators $\hat{\Phi}^{i_o}_{(1_+, \mu, \varphi)}$ and $\hat{\Phi}_{i_o}(1_+, \mu, \varphi)$ as Fourier integral operators with kernels

$$K^{i_o}(x,y) = \frac{i^n \Gamma(n-\frac{1}{2})}{(2\pi)^n} \Theta(x^{i_o} - y^{i_o}) \times$$

$$(19)\quad \times\int \frac{\varphi(x^1,\dots,x^n,y^{i_0},q_2^*,\dots,\widehat{q_{i_0}^*},\dots,q_n^*)}{[\Phi^{i_0}(x,y,q_2^*,\dots,\widehat{q_{i_0}^*},\dots,q_n^*)+i0]^{n-\frac{1}{2}}}\times$$

$$\times\sqrt{F(\Phi^{i_0},\mu)}\, dq_2^*\dots\widehat{dq_{i_0}^*}\dots dq_n^*\,,$$

$$K_{i_0}(x,y)=\frac{i^n\Gamma(n+\frac{3}{2})}{(2\pi)^n}\int\frac{\varphi(p_2^*,\dots,p_n^*,u,q_{i_0}^*,y^2,\dots,\widehat{y^{i_0}},\dots,y^n)}{[\Phi_{i_0}(x,y,p_2^*,\dots,p_n^*,u,q_{i_0}^*)+i0]^{n+\frac{3}{2}}}\times$$

$$(20)\quad \times\sqrt{F(\Phi_{i_0},\mu)}\,\theta(p_{i_0}^*-q_{i_0}^*)\,dp_2^*\dots dp_n^*\,du\,dq_{i_0}^*$$

<u>Remark 5</u>. The definition of the operator $\hat{\Phi}_{i_0(1_+,\mu,\varphi)}$ in the singular case uses in principle <u>the Fourier-Radon transformation</u>. In fact, formula (20) can be rewritten in the form

$$K_{i_0}(x,y)=\frac{i^n\Gamma(n+\frac{3}{2})}{(2\pi)^n}\int\sqrt{F(\Phi_{i_0},\mu)}\Big|_{\tilde{u}=y^1+q_{i_0}^*y^{i_0}}\theta(p_{i_0}^*-q_{i_0}^*)\times$$

$$\times\frac{\varphi(p_2^*,\dots,p_n^*,u,q_{i_0}^*,y^2,\dots,\widehat{y^{i_0}},\dots,y^n)\,dp_2^*\dots dp_n^*\,du\,dq_{i_0}^*}{[S(p_2^*,\dots,p_n^*,u,q_{i_0}^*,y^2,\dots,\widehat{y^{i_0}},\dots,y^n)+(u-\tilde{u})+[x^1+\sum_{i=2}^n p_i^*x^i]+i0]^{n+\frac{3}{2}}}$$

using the notation

$$S(p_2^*,\dots,p_n^*,u,q_{i_0}^*,y^2,\dots,y^{i_0},\dots,y^n)=$$
$$=-\Big[x^1+\sum_{i=2}^n p_i^*x^i\Big]\Big|_1\,.$$

In the last formula the Fourier-Radon transformation from the variables $(u, q_{i_0}^*)$ to the variables (y^1, y^2) can be easily seen.

The following formulae

$$(21)\quad H(x,-\frac{\partial}{\partial x})\circ\hat{\Phi}^{i_0}_{(1_+,\mu,\varphi)}=\Phi^{i_0}_{(1_+,\mu,\varphi)}+S^{i_0}(x,-\frac{\partial}{\partial x})$$

$$(22)\quad H(x, -\tfrac{\partial}{\partial x})\circ\hat{\Phi}_{i_o}(l_+, \mu, \varphi) = \hat{\Phi}_{i_o}(l_+, \mu, \varphi) + S_{i_o}(x - \tfrac{\partial}{\partial x})$$

are valid up to the operators of order -1 (using the notations of Theorem 3). Here $S^{i_o}(x, -\frac{\partial}{\partial x})$, $S_{i_o}(x, -\frac{\partial}{\partial x})$ are p.d.o. with symbols

$$(23)\quad S^{i_o}(x, p) = \Delta p_i H(x, p)\left[\sqrt{F}\cdot\varphi|_{l_o}\right]$$

$$(24)\quad S_{i_o}(x, p) = \left\{\Delta_{x^{i_o}} H(x, p) - p_{i_o}\,\Delta_{x^1} H(x, p)\right\}\left[\sqrt{F}\cdot\varphi|_{l_o}\right],$$

where Δp_{i_o}, $\Delta_{x^{i_o}}$, Δ_{x^1} are difference derivatives on char H:

$$\Delta_{p_{i_o}} H(x,p) = \frac{H(x,p)}{p_{i_o} - p_{i_o}|_{\mathrm{char}\,H}}, \quad \Delta_{x^{i_o}} H(x,p) = \frac{H(x,p)}{x^{i_o} - x^{i_o}|_{\mathrm{char}\,H}},$$

$$\Delta_{x^1} H(x,p) = \frac{H(x,p)}{x^1 - x^1|_{\mathrm{char}\,H}}.$$

With the help of formulae (21), (22) we construct a microlocal regularizator $\hat{R}$ of equation (15), i.e. the operator R such that $H(x, -\frac{\partial}{\partial x})\circ\hat{R}$ is a p.d.o. with a symbol equal to unity in a neighbourhood of α up to the operators of arbitrarily low order. The operator $\hat{R}$ is constructed in the form $\hat{\Phi}^{i_o}(l_+, \mu, \varphi) + T(x, -\frac{\partial}{\partial x})$ in the regular case and $\hat{\Phi}_{i_o}(l_+, \mu, \varphi) + T(x, -\frac{\partial}{\partial x})$ in the singular case, $T(x, -\frac{\partial}{\partial x})$ being a certain p.d.o. Supposing $u = \hat{R}f$ we get the first approximation of the theorem. To prove the second part of the theorem, we construct a local regularizator $\hat{R}$, i.e. such an operator that the composition $H(x, -\frac{\partial}{\partial x})\circ\hat{R}$ is a p.d.o. with a symbol equal to unity above a neighbourhood u of x up to the operators of arbitrarily low order. Such an operator is a sum of operators of the form $\hat{\Phi}^{i_o}(l_+, \mu, \varphi)$, $\hat{\Phi}_{i_o}(l_+, \mu, \varphi)$, of Fourier integral operators on l with amplitudes, having supports in l_+ and of a p.d.o.

In this case the operator $\hat{R}^*$ adjoint to $\hat{R}$ is the left local regularizator for the operator $\hat{H}^*$ adjoint to $\hat{H} = H(x, -\frac{\partial}{\partial x})$. Now let v be a solution of the equation

$$(25)\quad \hat{H}^* v = 0$$

with a support compact in u. The inclusion $v \in C^\infty(M)$ and the fact that the space N(U) of the solutions of equation (25) with a support in a neighbourhood U of x is finite-dimensional follows from the formula

$$0 = \hat{R}^* \circ \hat{H}^* v = v + \hat{Q} v$$

We suppose U to have a compact closure.

If U is sufficiently small the equation (25) has only a trivial solution with a support in U. This fact follows from the inclusion $N(U') \subset N(U)$ for $U' \subset U$ and from the non-existence of C^∞-function with the support in a point. Further the proof is based on the methods of functional analysis (see [7]).

The proof of the third part of the theorem is analogous to that of the second part. As l is everywhere a regular immersion we can construct a <u>semi-global</u> <u>regularizator</u> $\hat{R} : H^s_{comp}(M) \to H^s_{loc}(M)$ for which the relation

$$\hat{H} \circ \hat{R} = 1 + \hat{Q}$$

is valid, where $\hat{Q} : H^s_{comp}(M) \to H^s_{loc}(M)$ is continuous for every $s' > s$. But in this case the space N(K) of the solutions of equation (25) staying finite-dimensional is, generally speaking, not a zero space for any compact $K \subset M$. Thus, the fulfilment of a finite number of orthogonality conditions to the right part of equation (15) is necessary and sufficient for the solvability of this equation in the interior part K of any compact K.

Theorem 6 fully investigates the principal type equations from the point of view of their solvability. We shall note that the methods described in this paper (after some modification) can be applied to the investigation of the solvability of equations with contact (<u>not Hamilton</u> !) stationary points. The solvability of the equations in this case is determined by the principal and subprincipal symbols of H; the solution can be singular even if the right part is infinitely smooth. The assertion about the infinite smoothness of elements of the space N(K) is not valid for such equations (see [8]).

Finally it may be noted that the geometry of contact manifolds in a <u>complex</u> situation can be applied to the investigation of the solvability of differential equations on <u>complex</u> -analytic manifolds. In particular, such considerations permitted us to investigate "in large" the <u>characteristical</u> Cauchy problem in complex-analytical case. The solution of the last problem demanded a further (and non-trivial) generalization of the theory connected with a study of Legendre manifolds with singularities (see [9], [10], [11]).

References

1. V.P. Maslov, Teoriya vozmushchenii i asimptoticheskie metody (Theory of Perturbations and Asymptotic Methods), Moscow State University, Moscow, 1965 (in Russian)(French translation: Theorie des perturbations et methodes asymptotiques, Dunod, Ganthier-Villars, Paris, 1972).
2. V.P. Maslov, Operatornye metody (Operator Methods), Nauka, Moscow, 1973 (in Russian). MR 56 = 3647.
3. L. Hormander, Fourier integral operators. I, Acta Math. 127(1971), 79-183. MR 52 = 9299.
4. V.E. Nazaikinskii, V.G. Oshmyan, B.Yu. Sternin, and V.E. Shatalov, Fourier integral operators and the canonical operator, Uspehi Mat. Nauk 36:2(1981), 81-140.
= Russian Math. Surveys 36:2(1981), 93-161.
5. V.I. Arnold, Matematicheskie metody klassicheskoi mekhaniki (Mathematical Methods of Classical Mechanics), Nauka, Moscow, 1974 (in Russian). MR 57 = 14032.
6. V.V. Lychagin, Local classification of first-order non-linear partial differential equations, Uspehi Mat. Nauk 30:1(1975), 101-171. MR 54 = 8691.
= Russian Math. Surveys 30:1(1975), 105-175.
7. J.J. Duistrermaat and L. Hormander, Fourier integral operators. II, Acta Math. 128 (1972), 183-269. MR 52 = 9300.
8. B.Yu. Sternin, On regularization of the equations of subprincipal type, Uspehi Mat. Nauk 37:2(1982), 235-236.
= Russian Math. Surveys 37:2(1982).
9. V.E. Shatalov, Global asymptotic expansions in characteristic Cauchy problem for complex-analytical functions, Uspehi Mat. Nauk 35:4(1980), 181-182.
= Russian Math. Surveys 35:4(1980).
10. B.Yu. Sternin and V.E. Shatalov, Legendre uniformization of analytical functions with ramification, Mat. Sb. 113(1980), 263-284.
11. B.Yu. Sternin and V.E. Shatalov, Characteristic Cauchy problem on a complex-analytic manifold, in : Uravneniye na mnogoobraziyakh (Equations on Manifolds), Voronezh State University, Voronezh, 1982, 83-104 (in Russian). **See this volume.**
12. B.Yu. Sternin and V.E. Shatalov, On a method of solution of equations with single characteristics, Mat. Sb. 116(1981), 29-71.
13. M.V. Fedoryuk, Singularities of the kernels of Fourier integral operators and the asymptotic behaviour of the solution of the mixed problem, Uspehi Mat. Nauk 32:6(1977),67-115. MR 57 = 13580.
= Russian Math. Surveys 32:6(1977), 67-120.

CLASSICAL AND NON-CLASSICAL DYNAMICS WITH CONSTRAINTS

A.M.Vershik

Leningrad State University,
Leningrad, 198904, USSR

This article is purposed to give a detailed and, to a large extent, self-contained account of results, and to raise a number of questions on dynamics with constraints on the tangent bundle of a smooth manifold. The classical problems of this kind are the problems of non-holonomic mechanics, non-classical ones - the problems of optimal control and economical dynamics. The investigation of these topics from the standpoint of global analysis was started fairly recently (see /1, 2, 3/). Such a treatment needs a detailed study of geometry of the tangent bundle, connections and other notions necessary for general Lagrangian mechanics and, particularly, for the theory of non-holonomic problems. The present article continues the investigations of geometry of the tangent bundle and dynamics on it. For standard facts from the geometry of manifolds and the Riemannian geometry see /13, 18, 19/.

0.1. Lagrangian dynamics

The Lagrangian formalism is based on a procedure that allows one to construct invariantly a special vector field given an arbitrary C^2-smooth function on the tangent bundle (this function is called Lagrangian). This construction, formally, makes use of only two canonical objects present in the tangent bundle of any manifold: the principal tensor and the fundamental vertical field. These structures are defined invariantly in /1, 2, 3/.

Lagrangian mechanics studies the structure of trajectories of special vector fields as dependent on the Lagrangian - integrability, stability, integrals, etc.

All the entities and facts of classical mechanics (forces, virtual displacements, variational principles) can be completely immersed in the terms of the tangent bundle geometry. In addition to that there appears a number of geometrical notions insufficiently used till now, but, probably, important for mechanics - the notion of connection, in the first place*. The translation of Lagrangian mechanics into the language of geometry was initiated in the works /1, 2, 3/, but it is not entirely completed by now. The d'Alembert principle - the most general local principle of mechanics, which is also valid for dynamics with constraints - was formulated invariantly in the paper /2/.

It should be noticed that Hamiltonian mechanics in invariant form (symplectic dynamics) gained much more expansion than Lagrangian one. This is quite clear, because the symplectic structure is a universal object of analysis and geometry (see /4/). Yet, from the point of view of mechanics, symplectic dynamics is more scanty in entities, it has no equivalents for certain mechanical patterns (force, constraint, etc.). Besides that, the presence of two parallel formalisms (Lagrangian and Hamiltonian) in the quantum theory demonstrates that they can not completely replace each other for the classical theory too. In this article we primarily use the Lagrangian formalism, passing to symplectic geometry only in connection with the reduction and examples (see Section 3).

0.2. Dynamics with constraints

At least three branches - non-holonomic mechanics, optimal control and economical dynamics - lead to necessity of considering the following generalization of Lagrangian dynamics: given a submanifold (e.g. a subbundle) or a distribution (a field of tangent subspaces) in the tangent bundle TQ of a manifold Q, one should construct a dynamics so as to the trajectories could not leave the given submanifold, or the vector field would belong to the given distribution. In the non-classical case this submanifold has a boundary or, even, corners.

Practically speaking, these problems were almost nowhere considered from the standpoints of global analysis and coordinatefree differential geometry, save the study of non-holonomic dynamics in /2/. The article /2/ was initially aimed to comprehend invariant non-holonomic mechanics and to revise Lagrangian mechanics in conformity to this comprehension. The notion of a field of cones or polytops in

---

* It is useful to recall here the following fundamental result due to Levi-Civita (and later to Synge et al.) which connects mechanics with geometry: a mechanical system with quadratic Lagrangian moves by inertia along the geodesics of the corresponding Riemannian manifold. Levi-Civita also defined the Riemannian connection. In /1, 2/ (see Section 2) this theorem was expanded for the non-holonomic case, i.e. for the non-Riemannian connections.

the tangent bundle (see /4, 5/) is important in optimal control and economical dynamics. Certainly, this notion implicitly appears in a number of works on optimal control and geometry.

There exist two different constructions of dynamics with constraints, each of these leading to reasonable mathematical problems. A choice between the two possibilities lies outside mathematics.

Let a submanifold of the tangent bundle be given either as the set of zeros or as the set of non-positivity of a system of functions and, besides that, we are given a Lagrangian or, more generally, an objective function. Then one can:

1) consider the conditional variational problem of minimizing a certain functional (action, time, etc.) provided the trajectories belong to the given submanifold, and obtain finally the Euler-Lagrange equation, i.e. a vector field, with the help of an appropriate variant of the Lagrange method,

2) consider a certain (see below) projection of the vector field of the unconditional problem (on the whole tangent bundle) in every point of the given submanifold onto the tangent space of this submanifold, and again obtain finally a vector field.

The both vector fields are tangent to the submanifold of constraints, are special (by the mode of construction) and, by the same token, determine a dynamics with constraints. But, generally speaking, these fields do not coincide. In the first case the constraints are as if included in the Lagrangian, in the second case only the reactions of constraints take effect.

It is the second construction that should be used for description of the movement of mechanical systems. In a number of special cases (holonomic constraints, Chaplygin's case - a special case of linear constraints, etc.), the both constructions lead to the same vector field*. The second construction, as it was demonstrated in /2/, corresponds to the general d'Alembert principle, the latter in this situation being equivalent to no variational principle. Of cource, the principle is a postulate verified by practice, and by no means a theorem.

An implicit abusing of the two approaches favoured confusions in the foundations of non-holonomic mechanics until recently. From the

---

* Of a considerable number of works on non-holonomic mechanics, mention here /7, 8, 9, 10/ where one can find further references. The author knows no work using an invariant approach and written later than /2/ (see /23/).

first works of classics, these confusions are present, in a form, in almost all mathematical textbooks of variational calculus. For example, in mathematics the conditional Lagrange problems with non-integrable conditions on derivatives are called non-holonomic. This usage could suggest to the reader that the problems of non-holonomic mechanics are the conditional variational problems with non-integrable constraints. As it was said above, this isn't true*.

The first construction is employed in optimal control and other applications of dynamics with constraints, this fact being quite natural. However, using the second construction is also possible.

Many technical difficulties appearing in the treatment of dynamics with constraints are due to the absence or insufficient using of the adequate geometrical apparatus. We mean here, at the first place, the coordinate-free theory of distributions and connections. Making use of these notions instead of cloudy "quasiccordinates", "commutation relations" and sophisticated procedures of exclusion simplifies the account and turns the whole dynamics into the source of new clear mathematical problems whose solving will promote appearing of new effective applications.

0.3. Contents and the main facts

In this article we give a detailed account of the theory of the problems with constraints on the tangent bundle. In Section 1 we briefly, with certain innovations and simplifications, recall the results from /1, 2/ - we give an invariant derivation of the Euler-Lagrange equations for an arbitrary Lagrangian and an invariant treating of the equations with constraints.

In Section 2 we analyse the systems with a quadratic Lagrangian (the Newton equations), and the systems with linear constraints. For this purpose we define a new geometric object - the reduced connection on a subbundle of the tangent bundle of a Riemannian manifold. To the author's mind, this object corresponds to the old coordinate notion of "non-holonomic manifold" appearing in a number of geometrical works written in the 1930s-40s (V.V.Vagner, J.Schouten, G.Vranceanu). The geodesic flow for this connection is the main object to study. The most interesting is the case of an absolutely non-holonomic constraint: in this case the manifold is endowed with

---

* Of all the textbooks known to the author, the difference between the "problems with non-holonomic constraints" and non-holonomic mechanics is underlined in /24/ only.

a new (non-Riemannian) metric, and the Hopf-Rinow theorem in the usual formulation is false for this metric. Further in Section 2 we consider the general problem with linear constraints and quadratic Lagrangian, we give a complete proof of the following theorem: inertial motion of a system with quadratic Lagrangian and linear constraints occurs along the geodesics of the projectional connection in the subbundle assigned to the constraints. In a less precise form this theorem is given in /1, 2/. Section 3 is devoted to considering the main example - the group problem with constraints. We formulate the problems about rolling without sliding as group ones and then consider the inertial motion. The main statement of this Section is the reduction theorem analogous to that of symplectic dynamics. We reduce the problem to dynamics on the Lie coalgebra. It is interesting to find out the cases of its integrability.

In Section 4 we briefly consider stating non-classical problems for the field of indicatrices. The peculiarity of these problems is that the constraints (indicatrix) are given by inequalities and determine manifolds with boundary or with corners. The problems of optimal control (stating in contingencies - differential inclusions) and of economical dynamics where appear the fields of convex sets in bundles - analogues of distributions of subspaces in the classical case - can be reduced to this form (see /5, 6, 21/). Note that it is necessary here to use extremal stating of problem, because the constraints work in active form (through control) and not through reactions. One should underline that these problems are very close to the problem of infinite-dimensional convex programming. For the symplest case it was remarked in /5, 11/. We formulate a set of unsolved questions.

1. Geometry of the tangent bundle and Lagrangian mechanics

1.1. General definitions

Let Q be a smooth (always C^∞-smooth) connected manifold without boundary, TQ - the tangent bundle, $\pi : TQ \to Q$ - the canonical projection, $T^2Q = T(TQ)$ - the second tangent bundle, $d\pi :$ $T^2Q \longrightarrow TQ$. The tangent spaces at the points $q \in Q$ and $(q, v) \in TQ$ are T_q and $T_{q,v}$ respectively. A vector field X on TQ is called special if $d\pi X_{q,v} = v$. "Special vector field" is an equivalent to the terms "virtual displacement" in mechanics and "second order equation" in analysis. The vertical tangent vectors, i.e. the vectors tangent to the fiber $\widetilde{T}_q \subset TQ$ over q of the vector bundle TQ, form the subspace $\widetilde{T}_{q,v} \subset T_{q,v}$. Evidently, $\widetilde{T}_{q,v}$ can be identified with $\widetilde{T}_q$ and, by the same token, with T_q. Thus, we have the following canonical monomorphism

$$\gamma_{q,v}: T_q \longrightarrow T_{q,v}$$

and a 1-1-tensor $\tau_{q,v} = \gamma_{q,v} d\pi_{q,v}$ in TQ.

The tensor field $\tau = \{\tau_{q,v}\}$ is called the principal tensor field on TQ. This 1-1-field, as a map of vector fields, annihilates all the vertical vector fields (because they are annihilated by $d\pi$) and only them. The range of τ is also vertical vector fields.

The coordinate form of τ is:

$$\tau\left(a\frac{\partial}{\partial q} + b\frac{\partial}{\partial v}\right) = a\frac{\partial}{\partial v}$$

The dual tensor field τ^* acts on forms:

$$\tau^*(a\,dq + bdv) = bdq$$

the range and the kernel of τ^* are horisontal 1-forms on TQ (i.e. the forms that annihilate the vertical fields).

The vertical field on TQ with coordinates $\Phi_{q,v} = \gamma_{q,v} v = v_i\frac{\partial}{\partial v_i}$ will be called fundamental. A field X is special if and only if $\tau X = \Phi$. Indeed, in local coordinates $X_{q,v} = v_i\frac{\partial}{\partial q_i} + \ldots$, and $(\tau X)_{q,v} = v_i\frac{\partial}{\partial v_i} = \Phi$.

These two notions - τ and Φ - assist to formulate all of Lagrangian mechanics in invariant way*. Produce the basic definitions.

Let L be the Lagrangian - a smooth function on TQ, $\tau^*(dL)$ (i.e. $\frac{\partial L}{\partial v_i} v_i$) - the horisontal 1-form which is called the impulse field of the Lagrangian. In mechanical treating horisontal 1-forms denote forces, and their integrals along integral curves of a special field - the work done by force. The Lagrangian 2-form is $\Omega_L = d(\tau^* dL)$ (cf. /3/ Ch. 11 § 1), in coordinate notation

$$\frac{\partial^2 L}{\partial v_i \partial v_j} dq_i \wedge dv_j + \frac{\partial^2 L}{\partial v_i \partial q_j} dq_i \wedge dq_j$$

---

* The tensor τ was defined by the author and L.D.Faddeev in 1968. A report on invariant construction was made in Leningrad Mathematical Society in May 1970 (see Uspekhi Mat. Nauk, 1973, v.28, N 4, p.230). Later the author came to know about the book /3/, publi-

The image of this 2-form under the Legendre transformation (provided the Hessian $\frac{\partial^2 L}{\partial v_i \partial v_j}$ is nondegenerate) is the canonical 2-form $dp \wedge dq$ on the cotangent bundle. Energy (the Hamiltonian) is $H_L = dL(\Phi) - L$, and the Lagrangian force on the virtual displacement (special field) X is $\Omega_L(X, \cdot) - dH_L$. This is a horisontal 1-form, its value on a vector field Y (i.e. $\Omega_L(X, Y) - dH_L(Y)$) can be treated as the work of the Lagrangian force along X on Y.

The d'Alembert principle (the principle of virtual displacements) in our terms is formulated as follows: on the vector field that determines the real trajectories of motion the Lagrangian force equals the exterior force (or zero, if the exterior force equal zero), i.e.

$$\Omega_L(X, \cdot) = dH_L + \omega, \qquad \Omega_L(X, \cdot) = dH_L \tag{1}$$

respectively. Here ω is the exterior force. The Euler-Lagrange equations in coordinate language are:

$$\frac{d}{dt}\frac{\partial L}{\partial \dot{q}_i} = \frac{\partial L}{\partial q_i} \tag{2}$$

If the 2-form Ω_L is nondegenerate, then one can easily find the vector field itself (i.e. solve the equations with respect to second derivatives) in Hamiltonian form:

$$X_L = \Pi_L(dH_L) \tag{3}$$

where Π_L is the map of 1-forms into vector fields (a bivector) defined by the formula $\Omega_L(\Pi_L(\omega), Y) = \omega(Y)$. The fact that the field X_L is special follows from local formulae (also see /2/). The condition of the nondegeneracy of Ω_L is equivalent to $\det\left(\frac{\partial^2 L}{\partial v_i \partial v_j}\right) \neq 0$, the latter condition being invariant because the fibers in TQ are linear. The equation (3) is the motion equations written in Hamiltonian form. The matrix $\Gamma_L = \left(\frac{\partial^2 L}{\partial v_i \partial v_j}\right)$ determines a quadratic form on the fiber and is connected with Π_L by the relati-

---

shed in French in 1969 and translated into Russian in 1973, where were considered the field termed as "vertical endomorphism" and the invariant form of the Euler equations (without constraints). The work /2/ has been accomplished in 1971 and was sent to press in early 1973.

on $\Gamma_L^{-1} = \Pi_L \tau^* = -\tau \Pi_L$ (see /2/)*.

The case when L is a positive definite quadratic form on the fibers of TQ, i.e. a Riemannian metric on Q, is of special interest. In this case the formula (1) can be given the form of the Newton equations making use of the Riemannian connection on Q (see Section 3). In addition remark that the Lagrangian can be considered as a closed (rather than exact, as generally) 1-form on TQ. All the arguments are valid for this case, but the Hamiltonian is defined only locally. This theory is studied in recent years /12/.

1.2. Constraints

Suppose we consider certain dynamical systems on a smooth manifold X. By constraint one should mean a submanifold S in TX, and by system on X with constraint S - a dynamical system for which the velocities at every point $x \in X$ belong to S. According to this, the most general notion of constraint in mechanics (and in the theory of the second order equations in general) is defined as a submanifold in T^2Q. Nevertheless, the necessity of so general a definition hardly appeared wheresoever. With a loss of generality, one can consider the submanifold to be a subbundle in T^2Q, i.e. the constraint to be a distribution in TQ. It is just this definition that is accepted in /1, 2/. It is more convenient to consider the codistributions, i.e. the subbundles of $T^*(TQ)$. Then the mentioned above distribution can be regarded as the annihilator of a codistribution. After these preliminaries proceed to the following definition.

Definition 1. Constraint on the phase space TQ is a codistribution Θ on TQ, dynamical system concordant with the constraint Θ is a special vector field on TQ that is annihilated by Θ at every point. (Local form: $\Theta = \mathrm{Lin}\ \{ a_{ik}dq_k + b_{ik}dv_k,\ i = 1, \dots m \}$).

We does not require here the codistribution to be integrable, but the most current example is that very case: Θ is given as the linear span of differentials of a set of functions on TQ, these functions (their zeros or level sets) determine a constraint in the usual sense of this word. Such a constraint can be called functional. In this case all vector fields and forms are considered only on the level sets of these functions rather than on the whole TQ without particular reservations. A classical example of this kind: linear

---

* Remark that in /2/ p.132, 7-th line from below one should read "factor-space by horisontal forms" instead of "horisontal forms". The tensor τ in the present paper has the same meaning as τ^* in /2/.

constraints - functions on TQ linear with respect to velocities - $\theta = \varphi_i(q, v) = \sum_k a_{ik}(q)v_k$, $i = 1, \ldots, m$, then

$\theta = \mathrm{Lin}\ \sum_k a_{ik}dv_k + \sum_{k,j} \frac{\partial a_{ik}}{\partial q_j} dq_j$. The corresponding subset of TQ is the subbundle assigned to the codistribution $\mathrm{Lin}\{ \sum_k a_{ik}dv_k\}$.

The principal tensor field permits one to define the notion of constraint reactions - by that we mean the horisontal codistribution $\tau^*\theta$ where θ is the constraint. A 1-form belonging to $\tau^*\theta$ is called constraint reaction force. A constraint is called admissible if $\dim \tau^*\theta = \dim \theta$, i.e. if the codistribution θ has no horisontal covectors at any point (recall that the kernel of τ^* is horisontal 1-forms). A constraint is called ideal if it annihilates the fundamental vector field Φ .

Statement 1. 1) If a constraint is admissible, then there exist special vector fields concordant with this constraint. 2) If a constraint is ideal, then 1-forms - constraint reactions vanish on cycles lifted on TQ from Q (i.e. "do no work").

Proof. 1) Since the special fields are exactly those satisfying the condition $\tau X = \Phi$, we have to state solvability of the linear system

$$\tau X = \Phi \ , \quad \theta(X) = 0$$

Let

$$X = \sum v_i \frac{\partial}{\partial q_i} + \sum f_i \frac{\partial}{\partial v_i}$$

$$\theta = \{ \sum_i (\theta^1_{ki}dq_i + \theta^2_{ki}dv_i)\}_{k=1}^m$$

$$\tau^*\theta = \{ \sum_i \theta^2_{ki}dq_i \}_{k=1}^m$$

Just as $\dim \tau^*\theta = \dim \theta$, so $rg(\{\theta^1_k\}, \{\theta^2_k\}) = rg\{\theta^2_k\}$, but $\theta(X) = \sum_i (\theta^1_{ki}v_i + \theta^2_{ki} f_i) = 0$, hence the system

$\sum_i \theta^2_{ki} f_i = \sum_i \theta^1_{ki}v_i$ $(k = 1, \ldots, m)$ is solvable (with respect to $\{f_i\}$). It is easy to see that the number of linearly independent special vector fields in this case is not less than $\dim Q - \dim \theta$.

2) Let ξ be a cycle in Q, i.e. $\xi : S^1 \longrightarrow Q$, $\tilde{\xi}$ - its lifting in TQ. Then

$$\int_{\tilde{\xi}} \tau^*\theta = \int_{S^1} \langle \tau^*\theta, \dot{\xi}\rangle = \int_{S^1} \langle \theta, \tau\dot{\xi}\rangle = \int_{S^1} \langle \theta, \Phi\rangle = 0$$

Let a distribution Θ be given as the set of zeros of a system of functions φ_i on TQ, $\Theta = \mathrm{Lin}\{d\varphi_i\}$.

Statement 2. If the functions φ_i are homogeneous on v with degree one, then the constraint Θ is ideal.

Proof. It follows from the Euler theorem that $\sum_j v_j \frac{\partial \varphi_i}{\partial v_j} = \lambda \varphi_i$, and $\sum_j v_j \frac{\partial \varphi_i}{\partial v_j} = 0$ on the set of zeros of the functions φ_i, i.e. $\langle d\varphi_i, \Phi \rangle = 0$.

Now proceed to derivation of the equations of dynamics with constraints (cf. /2/). Let $\alpha = \{\alpha_i\}_{i=1}^m$ be a constraint distribution on TQ, L - the Lagrangian, H_L and Ω_L - the corresponding Hamiltonian and the Lagrangian 2-form (see above). As before, we proceed from the d'Alembert principle:

$$\Omega_L(X, \cdot) = dH_L + \omega \qquad (1')$$

where ω is the constraint reaction force that causes the desired vector field X to be concordant with the constraints

$$\alpha(X) = 0 \qquad (4)$$

Theorem 1. If the Lagrangian L is nondegenerate, and the Hessian $\left(\frac{\partial^2 L}{\partial v_i \partial v_j}\right)$ is positive defined for all (q, v), then for every admissible constraint α there exists a special vector field X concordant with the constraints (4) and satisfying the d'Alembert principle (1), where ω is a certain constraint reaction force.

Proof. We intend to solve the systems (1'), (4) with respect to X and ω. Since Ω_L is nondegenerate, there exist a 2-0-antisymmetric field Π_L of maps from 1-forms to vector fields such that $\Omega_L(\Pi_L(\rho), Y) = \rho(Y)$. Then we have $X = \Pi_L(dH_L + \omega)$, and by (4) $\langle \alpha, \Pi_L(dH_L) + \Pi_L(\omega) \rangle = 0$ or $\langle \alpha, \Pi_L(dH_L) \rangle = -\langle \alpha, \Pi_L(\omega) \rangle$, $X = X_L + \Pi_L(\omega)$, where $X_L = \Pi_L(dH_L)$.

By the definition of constraint reaction we intend to search for ω in the form $\tau^* \rho$, $\rho \in \alpha$, i.e.

$$\langle \alpha, X_L \rangle = -\langle \alpha, \Pi_L(\tau^* \rho) \rangle \qquad (5)$$

Since the constraint α is admissible, τ^* preserves the dimension of α, but as the Hessian and, hence, Π_L are positive defined, the restriction of Π_L on $\tau^* \alpha$ is also positive defined, thus, nondegenerate. The determinant of the system (5) with respect to ρ is non-zero at every point:

$$\rho = \sum_{j=1}^{m} \lambda_j \alpha_j , \quad \sum_j \lambda_j \langle \alpha_j(\Pi_L \tau), \alpha_i \rangle = -\alpha_i(X_L) \quad (6)$$

Here λ_j are the desired coefficients of expansion - the Lagrange multipliers. The theorem is proven.

Remarks. 1. The desired vector field is the sum of a special vector field for the constraint-free problem and an additional vector field $\Pi_L(\omega)$, the latter field being vertical, since ω is a horisontal 1-form. Thus, one can obtain the field for the problem with constraints from the constraint-free problem field by means of a special projection. This projection consists in adding a certain vertical field (the horisontal component being unchanged), whereupon the field becomes tangent to the constraint. This projection operator depends on the Lagrangian (see Section 2).

2. If the Hessian, even if nondegenerate, was not positive defined, its restriction on a certain codistribution, possibly, would have not the maximal rank, i.e. the system (6) would have the zero determinant for a certain constraint. In the same manner, if the constraint was not admissible, the rank of the system could lessen for a certain Lagrangian. In this sense the both conditions are necessary for existence of a solution.

3. One can allow violation of these conditions on lower dimensional manifolds in TQ (and this is inevitable for certain problems). In this case the desired vector field can have singular points too.

4. The coordinate notation of the equations (6) is

$$\frac{d}{dt}\frac{\partial L}{\partial \dot{q}_i} - \frac{\partial L}{\partial q_i} = \sum_{j=1}^{m} \lambda_j \alpha_j^i \qquad i = 1, \ldots, m \qquad (7')$$

Here one finds the values λ_j from the conditions of concordance between the solution and the constraints:

$$\sum_{i=1}^{n} \alpha_j^i \, dq_i + \tilde{\alpha}_j^i \, d\dot{q}_i = 0 \qquad j = 1, \ldots, m \qquad (7'')$$

where $\alpha_j = (\alpha_j^1, \ldots, \alpha_j^n, \tilde{\alpha}_j^1, \ldots, \tilde{\alpha}_j^n)$ are the coordinates of the constraint forms in variables $q_1, \ldots, q_n, v_1, \ldots, v_n$, and $d\dot{q}_i$ in (7'') are taken from (7'). In the case of functional constraints (exact forms α) the equations (7'') change into functional relations

$$\varphi_j(q, \dot{q}) = 0 \qquad j = 1, \ldots, m$$

i.e. into common constraint equations. In particular, for linear homogeneous functional constraints they change into the conditions

$$\sum_{i=1}^{n} a_j^i(q)\dot{q}_i = 0 \qquad j = 1, \ldots, m$$

and so on.

Underline once more that (7') and (7'') are not, generally speaking, the Euler-Lagrange equations for a conditional variational problem: the Lagrange multipliers enter the right-hand side of the Euler equation rather than the Lagrangian.

Thus, universally adopted motion equations of non-holonomic Lagrangian mechanics are derived from the general d'Alembert principle by means of invariant structures in the tangent bundle.

2. Riemannian metric and reduced connection

2.1. Various classical formulations for Riemannian manifolds

If $2L = g$ is a Riemannian metric on Q, i.e. a qudratic symmetric positive defined form on T_q and by the same token a function on TQ that is quadratic on fibers, then the vector field of the dynamical system with the Lagrangian L admits many descriptions. Usually, one proves their equivalence in textbooks of Riemannian geometry, but, unfortunately, mostly in implicit form. Cite the most important formulations, several simplifications having been fixed priviously.

Let $L = H_L = \frac{1}{2} g$, then T^*Q can be identified with TQ by means of the metric*, and Ω_L is the canonical 2-form on TQ. The manifold TQ is also Riemannian with the metric G defined on T^2Q by the formula $T_{q,v} = T_{q,v}^{vert} \oplus T_{q,v}^{hor}$ (as Eucledean spaces). Here T^{vert} and T^{hor} are the vertical and horisontal subspaces of $T_{q,v}$ (they exist by the existence of the Riemannian connection). Each of these subspaces is canonically isomorphic to T_q ($d\pi : T_{q,v}^{hor} \sim T_q$, $T_{q,v}^{vert} = \widetilde{T}_{q,v}$ - see Section 1), hence is endowed with the form g_q. The blockwise form of the tensors τ and τ^* (with respect to the decomposition $T = T^{vert} \oplus T^{hor}$) is: $\tau = \begin{pmatrix} 0 & 1 \\ 0 & 0 \end{pmatrix}$, $\tau^* = \begin{pmatrix} 0 & 0 \\ 1 & 0 \end{pmatrix}$. Horisontal forms (fields) are connected with vertical forms (fields) by fixed isometry, because every vector from T_q has the horisontal lifting in $T_{q,v}$, and $T_{q,v}^{vert} = T_q$. Finally, the bivector Π_L in matrix form is

---

* One can identify TQ and T^*Q in the case of arbitrary nondegenerate Lagrangian too (the Legendre transformation), but it is important that in our case this identification is linear on fibers, hence it preserves linearity of constraints.

the isometry $\begin{pmatrix} 0 & 1 \\ -1 & 0 \end{pmatrix}$. All these facts follow from elemetary geometry of Riemannian spaces.

The vector field X_L corresponding to the motion with the Lagrangian $L = \frac{1}{2} g$ can be described in the following manners.

A. X_L is a Hamiltonian field in TQ: $X_L = \Pi_L(dg)$, i.e. $\Omega_L(X_L, \cdot) = dg(\cdot)$.

The field X_L is horisontal with respect to the Riemannian connection because it determines the geodesic flow. Hence its integral curves permit another description.

B. Integral curves of the field X_L are the curves $x(\cdot)$ in Q satisfying the Newton equation $\nabla_{\dot{x}} \dot{x} = 0$ (in coordinate notation $\ddot{x}^k = \Gamma_{ij}^{k} \dot{x}^i \dot{x}^j$) lifted to TQ.

C. The field X_L is the Euler equation field for the variational problem of minimizing length (the principle of least action).

The equivalence of the formulations A and B is an important fact that is proven usually by comparing formulas or by the equivalence with C. However, A and B are local principles (d'Alembert's and Gauss's), and their nature is other than that of C. In the succesion of lessening generality they are placed in the following manner: A, C, B*.

2.2. Linear constraints and reduced connection

Now proceed to the specialization of the Section 1 equations for the case of the Riemannian metric and linear constraints. In mechanics one considers generally the linear constraints. That accounts for linearity of constraints in the majority of non-holonomic applied problems (rolling, etc.). Moreover, the possibility of realizing non-linear constraints in mechanical problems has been an open problem for a great while (see /9/).

Let Q be a Riemannian manifold with metric g , β - a codistribution on Q, $\beta^{\perp}$ - the distribution annihilating β . From here on we make no distinction between distributions and codistributions since 1-forms are identified with vector fields. However, we determine

---

* The principle B seems to have more generality than formulated here. We think that the objects of the type of connection and covariant derivative exist for a more general class of Lagrangians than Riemannian metrics.

constraint by its annihilator, hence we denote the distribution of admissible vectors as $\beta^{\perp}$. The set of pairs $T^{\beta}Q = \{(q, v) : \beta_q(v) = 0\}$ is a subbundle of TQ. The submanifold $T^{\beta}Q$ determines constraint and is described by the equation

$$\varphi_i(q, v) \equiv \langle a_i(q), v\rangle = 0 \qquad i = 1, \ldots, m$$

where $a_i(q)$ are 1-forms forming a basis in β. The forms that correspond to our constraint are defined on $T^{\beta}Q$ only and are determined by the set of differentials of the functions φ_i. We does not need these forms because of the following evident statement (true for arbitrary manifolds).

Statement 3. Reactions of the linear constraints defined by a codistribution β on Q form the codistribution $(d\pi)^*\beta$ on TQ (see Section 1.2). The linear constraints are admissible and ideal.

Proof. For i = 1, ..., m

$$\tau^*(d\varphi_i) = \tau^*\{\partial_q a_i dq + a_i dv\} = \{a_i dq\} = (d\pi)^* a_i$$

Ideality follows from homogeneity on v, admissibility is evident.

Now define connections in $T^{\beta}Q$. By the definition, any connection in $T^{\beta}Q$ is an adjoint connection in the principal fiber bundle of partial frames $B^{\beta}Q$, where the fiber over $q \in Q$ is the space of all frames in the subspace $\beta^{\perp}_q$, and the structure group is $GL(\dim \beta^{\perp})$. Since $T^{\beta}Q$ is a vector bundle, connection can be determined by a covariant derivative for the fields concordant with $\beta^{\perp}$.

At first prove the following general lemma (without using the metric).

Lemma. Let X be an arbitrary manifold with a linear connection determined by a covariant differentiation ∇. Suppose we are given a subbundle $T^{\beta}X$ in TX, and for every fiber R_x (as a subspace of T_x) we have a projection $F_x : T_x \longrightarrow R_x$ depending on x smoothly. Then $\widetilde{\nabla}_x Y \overset{\text{def}}{=\!=} F\nabla_x Y$ determines a connection in the subbundle.

Proof. Verify that the axioms of connection are satisfied for $\widetilde{\nabla}$ (see /13/). Of the four relations, only the following one is of interest:

$$\widetilde{\nabla}_x(\lambda Y) = F\nabla_x(\lambda Y) = F(\nabla_x Y + (X\lambda)Y) =$$

$$= \widetilde{\nabla}_x Y + X\lambda \cdot FY = \widetilde{\nabla}_x Y + (X\lambda)Y$$

since $FY = Y$ for $Y \in T^{\beta}X$. Hence $\widetilde{\nabla}$ is a connection in $T^{\beta}X$.

Let Q be a Riemannian manifold and β, $\beta^{\perp}$ - distributions in Q.

Define for any point q the ortogonal projection $F_q: T_q \longrightarrow \beta_q^{\perp}$ by means of the Riemannian metric on Q, and then the connection $\widetilde{\nabla} = \nabla^{\beta}$ by the lemma. This connection (on $T^{\beta}Q$) we call reduced one.

Theorem-Definition 2. Coordinate notation (in local coordinates) of the reduced connection ∇^{β} in the subbundle $T^{\beta}Q$ is:

$$\nabla^{\beta}_{X_i} X_j = \sum_k \Gamma^k_{ij} X_k, \qquad i,j,k = 1, \ldots, m \qquad (8)$$

where X_i, X_j, X_k are coordinate fields from $\beta^{\perp}$, Γ^k_{ij} the Christoffel symbol of the Riemannian connection.

Proof. Since F is the ortogonal projection onto $\beta^{\perp}$, the formula (8) follows from the formula of covariant differential.

Remarks. 1. If the distribution $\beta^{\perp}$ is involutary, then ∇^{β} is the induced Riemannian connection on the fibers of the bundle determined by $\beta^{\perp}$.

2. The reduced connection can be extended to a connection (non-Riemannian) in TQ, but hardly there exists any canonical extension.

3. $\nabla^{\beta}_X Y - \nabla^{\beta}_Y X - [X,Y] = F([X,Y]) - [X,Y]$, and this expression is non-zero if $\beta^{\perp}$ is non-involutory. On the other hand, the symbol Γ^k_{ij} is symmetrical. This means that one can not define torsion by the given formula. It well may be true that the torsion for ∇^{β} can not be defined at all.

4. The curvature form is rather sophisticated and is studied insufficiently. Of the most interest is to find the holonomy groups for ∇^{β}.

2.3. Inertial motion with quadratic Lagrangian and linear constraints

Now apply the results of Sections 1 and 2.1, 2.2 to the initial problem.

Theorem 3. Let Q be the position space, $L = \frac{1}{2} g$, where g is a Riemannian metric on Q. Let β be a distribution on Q, $T^{\beta}Q = \{(q, v) : \beta_q(v) = 0\}$ - the assigned subbundle of TQ. Then the motion equations of the dynamical system with the Lagrangian L and linear constraints β are equations of the geodesics for the reduced connection ∇^{β}.

Proof. By theorem 1 the desired field in $T^{\beta}Q$ exists, because the constraint is admissible, and is given by the formula $X = X_L + \Pi_L(\omega)$

where X_L is the vector field for the system without constraints (geodesic field), and ω is a 1-form from the codistribution of the constraint reactions. This means that ω is a covector (hence, vector) field on Q lying in β . Thus $\Pi_L(\omega)$ (see above) is a vertical field, it is chosen so that the field X be concordant with the constraint, i.e. the vertical projection of X lie in the image of $\beta^{\perp}$. On the other hand, the connection ∇^{β} can be presented as

$\nabla^{\beta} = F\nabla = \nabla + S$, where S is a 1-2-tensor (the difference between any two covariant differentiations is a 1-2-tensor), i.e. $S = F^{\perp}\nabla$. Hence, any horisontal vector of the connection ∇^{β} differs from a horisontal vector of the connection ∇ in a vertical vector lying in β ($F^{\perp}$ is a projection on β) and is concordant with the constraint. The decomposition into vertical and horisontal components is unique, hence the special horisontal field of the connection ∇^{β} at the points $(q, v) \in T^{\beta}Q$ and the field X coincide.

Corollary. The motion equation for our problem can be presented as

$$\nabla^{\beta}_{\dot{x}} \dot{X} = 0 \tag{9}$$

or

$$\ddot{X}^k = -\sum_{ij} \Gamma^k_{ij} \dot{X}^i \dot{X}^j \qquad i,j,k = 1, \ldots, m;\ m = \dim \beta^{\perp} \tag{10}$$

where the coordinates are chosen so that $\dot{X}^1, \ldots, \dot{X}^m$ form a basis in the distribution $\beta^{\perp}$. If there are exterior forces, potential, etc., then one makes a change in (9) in the usual way:

$$\nabla^{\beta}_{\dot{x}} \dot{X} = \omega_1$$

where ω_1 is the (vector) field of exterior forces (the gradient for the potential case).

If the distribution $\beta^{\perp}$ is involutary, then (9) is the equation of geodesics in the fiber of the bundle, and if the distribution is geodesic (i.e. its fiber is completely geodesic) then the solution of (9) coincide with the solution of the constraintfree problem.

Remark. As we have said already, for the dynamical problem one usually uses the field X_L (in our terms - the field of geodesics) only. Connections were not generally used even for constraint-free problems. As one can see from above, its role is rather essential.

2.4. Absolute non-involutority and the Hopf-Rinow theorem

Of the most interest is the case when the involutary hall of the distribution $\beta^{\perp}$ generates the whole bundle.

Recall that by the Frobenius theorem this means that the brackets of the vector fields concordant with $\beta^{\perp}$ generate the whole Lie algebra of vector fields on Q.

Problem. To describe the distributions of dimension m lying in general position and involutary properties of these distributions*.

Statement 4. The connection ∇^{β} on a complete Riemannian manifold is complete (cf. /13/ Ch. 3 § 6).

Corollary. If the distribution $\beta^{\perp}$ is absolutely non-involutary, then for any two points q_1, $q_2 \in Q$ there exists a continuous curve connecting these points and consisting of pieces of the geodesics of the connection ∇^{β}. In other words, the geodesic flow is transitive.

At the same time not every two points can be immediately connected by a ∇^{β}-geodesic. The following interesting problem appears here. Define on a Riemannian manifold Q with an absolutely non-involutary distribution $\beta^{\perp}$ the following new metric:

$$r_{\beta}(q_1, q_2) = \inf\{ \ell(\tau) : \tau(0) = q_1, \ \tau(1) = q_2, \dot{\tau} \in \beta^{\perp}\}$$

where τ is a smooth curve, $\ell(\cdot)$ - the length with respect to the initial Riemannian metric. This infimum is not assumed on the geodesic connection ∇^{β}, hence the classical Hopf-Rinow theorem isn't valid for this case.

Problem. To describe intrinsically the r_{β}-type metrics on Riemannian manifolds**.

An example of this kind was considered by Gershkovich.

3. Group mechanics with constraints

3.1. Problem of rolling and its group model

The most popular example of a non-holonomic problem is the following one: two bodies (e.g. a ball and a plane) move (by inertia or in a certain field) so that the linear velocities of the both bodies

---

* One can suppose that for $m>1$ the distribution generates the whole T_q for the points $q \in Q$ lying in general position, and the dimension of the involutary hull reduces at singular points.

** Compare with the "space of geodesics" defined by Busemann (see /14/).

at the point of contact coincide (no sliding). The reader interested in the traditional technique of derivation of the constraint equations is referred to special literature, and here we describe (apparently for the first time) the group model of this problem in a reasonably general formulation. The group example is a central one here, as in the case of ordinary dynamics.

Let G_1 and G_2 be two arbitrary connected Lie groups determining position of each body in mobil coordinate system, G_1° and G_2° - their stationary subgroups (of motions provided the point of contact fixed), and let we have an isomorphism s: $\mathfrak{g}_1/\mathfrak{g}_1^{\circ} \simeq \mathfrak{g}_2/\mathfrak{g}_2^{\circ}$ where $\mathfrak{g}_i$ are the corresponding Lie algebras (coincidence of velocities at the point of contact). Now consider $Q = G_1 \times G_2$ as the position space and single out in $(TQ)_e = \mathfrak{g}_1 + \mathfrak{g}_2$ the subspace of those pairs (a_1, a_2) for which $st_1(a_1) = t_2(a_2)$ where $t_i : \mathfrak{g}_i \longrightarrow \mathfrak{g}_i/\mathfrak{g}_i^{\circ}$, $i = 1, 2$ - the canonical projection. This is a linear subspace of $\mathfrak{g}_1 + \mathfrak{g}_2$, transferring this subspace to all points of $G = G_1 \times G_2$ by means of left translations we obtain a distribution (linear elements with coinciding velocities at the points of contact). This is the principal model. Now one can consider a Lagrangian (left-invariant for inertial motions) and apply the developed methods to form the motiom equations. For a more general scheme one doesn't need to specialize the group as the direct product of two (or more) groups. The most natural form of our scheme is the following one. Let G be an arbitrary connected Lie group, $\mathfrak{m}$ - a subspace in the Lie algebra $\mathfrak{g}$ (mostly - a complement to a subalgebra). The distribution $\beta^{\perp}$ is the distribution of the left translates of $\mathfrak{m}$ on TG. Such a model will be called a group one.

Examples. 1. $G = SO(3)$, $\mathfrak{g} = so(3)$. This is the rotation of a solid body with a fixed point and zero linear velocity at another point.

2. $G = SO(3) \times R^2$, $\mathfrak{g} = so(3) + R^2$,

$\mathfrak{m} = \{(a, h) : \pi a = h\} \subset \mathfrak{g}$,

where $\pi : so(3) \longrightarrow so(3)/so(2) \simeq R^2$. This is the rolling of a ball on a plane.

3.2. Reduction

Reduce the proposed group model to a system on the Lie algebra for the case of a left-invariant Lagrangian. Remark that the problems with constraints are not symplectic, i.e. the appearing fields do not preserve, in general, a 2-form, hence the question about integrals and reduction for these systems should be considered separately. Reduction of the order of these systems is less than that of symplectic systems. The following statement refers to a general system with constraints.

Statement 5 (see /2/). If the constraint α is ideal, then the energy H_L is a motion integral for system with the Lagrangian L.

Proof. $\frac{d}{dt} H_L = X(dH_L) = \Omega_L(\Pi_L(dH_L), X) =$

$$= \Omega_L(\Pi_L(dH_L), \Pi_L(dH_L)) + \Omega_L(\Pi_L(dH_L), \Pi_L(\omega)) =$$

$$= \omega(\Pi_L(dH_L)) = \omega(X_L) = \tau^* \rho(X_L) = \rho(\tau X_L) = \rho(\Phi) = 0.$$

Theorem 4. Let the position space be a Lie group G, and the constraints (linear) and the Lagrangian (quadratic) of the system are left-invariant. Then the phase flow of the system is the skew product with the base being the flow whose vector field is the projection of the Euler field (without constraints) onto the subspace of constraints in the Lie coalgebra, and the fiber - a conditionally periodic flow on the group.

Proof. Identify TG and T^*G by means of the Lagrangian*. The field X that describes the motion of the system commutes with the left-invariant fields A on G. Indeed, according to the premisses we have

$$[X, A] = [X_L, A] + [X, \Pi_L(\omega)] = 0$$

Hence the orbit partition of the natural (Hamiltonian) action of the group G in T^*G is invariant under the flow of the field X. Thus the base of the skew product is $T^{\mathfrak{f}}G/G \cong \mathfrak{M} \subset \mathfrak{g}^*$ (here $\mathfrak{g}^*$ is the Lie coalgebra, and $\mathfrak{M}$ is the subspace determining the constraint) and the fiber is the group G. The flow on the fiber is determined by the motion on orbits, i.e. on the group, and commutes with the left shifts, hence it reduces to the flow on left classes with respect to a maximal torus, the latter flow being conditionally periodic on each of these classes. The trajectories are determined by their initial vectors at a certain point, e.g. at the unit element of the group.

Proceed now to the base. Remind that for the constraint-free systems one can define the moment map $dH : T^*G \longrightarrow \mathfrak{g}$ - see, e.g. /4/. On the Lie coalgebra we have the Euler equation: $\dot{a} = \{dH(a), a\}$ where $a \in \mathfrak{g}^*$, $\{\cdot,\cdot\}$ is the Poisson bracket. The Euler equation determines the motion in base for the constraint-free system. According

---

* or, by means of the Killing form for semi-simple groups.

to theorem 3, in our case the vector field defined on the subspace of constraints $\mathfrak{M}$ is the ortogonal projection of the Euler field onto $\mathfrak{M} \subset \mathfrak{g} = \mathfrak{g}^*$. Thus, the equation on $\mathfrak{M}$ assumes the form

$$\dot{a} = P\{dH(a), a\}$$

where P is the ortogonal (in the sence of Lagrangian) projection from $\mathfrak{g}$ onto $\mathfrak{M}$. The theorem is proven.

Remarks. 1. In fact we assert that the projection of the initial field onto constraints commutes with the group factorization. But, to distinct from the generalised Noether theorem, here one can not assert that the elements from the center of the enveloping algebra are integrals of motion*. Hence the obtained system in $\mathfrak{M}$ does not preserve, generally speaking, the coadjoint action orbits.

2. If the Lagrangian is not supposed to be quadratic, the reduction is also possible, but in this case the Legendre transformation on the fiber of TG is non-linear, the factor-space T^*G/G has no linear structure, and the reduced system is rather sophisticated.

Example. Consider the problem of inertial rolling of two n - dimensional hard bodies. In this case we have (in the notations of Section 3.1.):

$$G_1 = G_2 = SO(n), \quad G_1{}^0 = G_2{}^0 = SO(n-1), \quad G = G_1 \times G_2,$$

$$\mathfrak{g} = \mathfrak{g}_1 + \mathfrak{g}_2 = so(n) + so(n), \quad \mathfrak{g}_1{}^0 = \mathfrak{g}_2{}^0 = so(n-1),$$

$$\mathfrak{M} = \operatorname{diag} \mathfrak{g}^0 + (\mathfrak{g}_1{}^0 + \mathfrak{g}_2{}^0),$$

where $\operatorname{diag} \mathfrak{g}^0 = \{(a, a) : a \in \mathfrak{g}^0\} \subset \mathfrak{g}_1 + \mathfrak{g}_2$. Hence, the elements of $\mathfrak{M}$ has the form $\{(a+b, a+c),\ a \in so(n-1);\ b, c \in so(n-1)\}$. Let P be the projection onto $\mathfrak{M}$**. Then $(i, j = 1, \dots, n-1)$

---

* Reduction of Hamiltonian systems with respect to regular actions of Lie groups (generalizations of the Noether theorem) was considered and rediscovered by a great many of researchers. For Lie groups the author defined it as early as 1968 (the reports were made at seminars in Leningrad and Moscow Universities and at a conference in Tsakhkadzor in 1969). The well-known 2-form discovered by A.A.Kirillov in 1962 (see /4/) naturally appears from the canonical form in T^*G in process of the reduction. A list of literature on this topic (by no means complete) is given in the book /4/ and in the survey /20/, the latter being devoted to the symmetries of more complicated nature.

** The ortogonal projection P does not depend on the inertia tensors in our case.

$$P(x_{ij}^1, x_{ij}^2) = \begin{pmatrix} x_{ij}^1 + x_{ij}^2 & x_{ij}^1 \\ -x_{ij}^1 & 0 \end{pmatrix} + \begin{pmatrix} x_{ij}^1 + x_{ij}^2 & x_{ij}^2 \\ -x_{ij}^2 & 0 \end{pmatrix}$$

Let I_1, I_2 be the inertia tensors of the bodies. Then the motion equations in $\mathfrak{M}$ has the form

$$\dot{a}^{(1)} = P(a^{(1)^2} I_1 - I_1 a^{(1)^2}),$$

$$\dot{a}^{(2)} = P(a^{(2)^2} I_2 - I_2 a^{(2)^2}),$$

$$(a^{(1)}, a^{(2)}) \in \mathfrak{M}.$$

The projection P connects the both equations.

The qualitative character of motion in this system is unknown to the author.

In the same manner one can take another stationary subgroup (e.g. SO(n - k) that corresponds to the Stifel manifold; however, it is unknown whether the constraint-free problem is integrable or not).

4. Non-classical problems with constraints

The following problem arises in optimal control and economical dynamics.

Let Q be a manifold, TQ - its tangent bundle, and we are given a field of submanifolds (with boundary or with corners) in fibers of TQ. We have to minimize a certain objective functional of the boundary value problem, the tangent vectors to admissible curves lying in the given field:

$$\inf \{ F(q(\cdot)) : \dot{q} \in B(q) \subset T_q, \quad q(a) = \bar{q}, \quad q(b) = \bar{\bar{q}} \}.$$

The problem in this stating is called "problem in contingencies" (differential inclusions). One passes from the traditional stating in the following way: B(q) is the set of right-hand sides of differential equations when controls run over the admissible region. Usually B(q) are convex solid, particularly, polyhedral sets (cones or polytops). In connection with this it is important to consider the theory of fields of such sets as a generalization of the theory of distributions (involutarity, singularities, etc.). For another ground (connected with other problems) see /6, 21/.

It would be interesting to apply for these problems the following principle adopted in mechanics that was under consideration in previous Sections: one can construct the Euler equation field for the constraint-free problem and then project this field onto the constraint manifold as we have done it above. The field in interior points does not change under this projection, but jumps (switches) can appear on the boundary. Constraints of this kind are not ideal in general.

Problem. To find out problems of optimal control for which the desired field is obtained by the cited principle.

Mostly the field of sets $B(\cdot) \subset TQ$ is defined by means of restrictions. For example, let $B(\cdot)$ be the field of polyhedral sets $B(q) = \{ v : A(q)v \leqslant b(q) \}$ where $b(\cdot)$ is a map from Q to R^m and $A(\cdot) \in \mathrm{Hom}(R^n, R^m)$. In this case we have the following general problem:

$$\inf \{ \Phi (q(\cdot)) : \quad A(q)\dot{q} \leqslant b(q), \ q \in Q \}$$

Thus, this is a field of problems of linear (if Φ is linear) or convex programming in the tangent bundle TQ. Similar problems were considered in the linear theory of optimal control and also (for $Q = R^1$ or $Q = [a, b] \subset R^1$) in the theory of continuous ecenomical models.

Problem. To connect Pontriagin's optimality principle for the described problem with the duality theorem of convex programming*.

Hypotetically these two statements are equivalent, one can easily verify this fact for the case when $A(\cdot)$, $b(\cdot)$ are constant on $Q = R^n$.

A problem close to the classical ones from Sections 2 and 3 consists in finding a minimum in the following situation. Let Q be a Riemannian (or Finsler) manifold, $B(\cdot)$ - a field of convex sets (polytops, spheres, etc.) in TQ. It is convenient to suppose that $B(q)$ is central-symmetrical for all q. One should find the shortest (in the sense of metric) curves that are admissible for the restrictions $\dot{q} \in B(q)$. The obtained new metric (non-Riemannian) seems to be similar to the metric from Section 2.4. Manifolds with these metrics, probably, posess unusual properties.

Problem. To study metrics on smooth manifolds appearing in these problems (cf. Section 2.4.)**.

Raise one more problem. In non-classical dynamics (as well as in arbitrary dynamics) the most interesting example of a problem with constraints is the group one. Let a Lie group act transitively on a

---

* The author didn't happen to come across this stating of the problem, but, possibly, similar questions were stated and solved. There is a considerable number of works that refer to the discussed question indirectly. We confine ourselves to the reference to the monograph /11/.

** This problem is profoundly studied by V.Ja. Gershkovich /21/.

manifold M, and the field of restrictions is invariant under this action. If the objective functional is invariant too, then one can reduce the problem (in the same manner as in Section 3) to a single tangent space, e.g. to Lie algebra if the manifold is a group. The difficulty appearing here is that the reduction permits one to study only the Cauchy problem rather than the boundary value problem (the same as in Section 3). This circumstance isn't substantial for classical mechanics without constraints, because usually (e.g., for Riemannian complete manifolds) the set of solutions of the Cauchy problem includes solutions of all boundary value problems (the Hopf-Rinow theorem). Here this isn't the case and one should search for a solution in considering not only the result of reduction but the whole skew product.

Problem. To describe the reduction for group problems of optimal control and the structure of the skew product.

REFERENCES

1. Vershik A.M., Faddeev L.D. Differential geometry and Lagrangian mechanics with constraints. - Dokl. Akad. Nauk SSSR 202:3 (1972), 555-557 = Soviet Physics Doklady 17:1 (1972), 34-36.

2. Vershik A.M., Faddeev L.D. Lagrangian mechanics in invariant form. - In: Problems of Theoretical Physics. vol.2, Leningrad, 1975 = Selecta Math. Sov. 1:4 (1981), 339-350.

3. Godbillon C. Géométrie différentielle et méchanique analytique, Hermann, Paris, 1969.

4. Arnold V.I. Mathematical Methods of Classical Mechanics, Springer-Verlag, New York, 1980.

5. Vershik A.M. Several remarks on infinite-dimensional problems of linear programming. - Usp. Mat. Nauk 25:5 (1970), 117-124.

6. Vershik A.M., Chernjakov A.G. Fields of convex polytops and Pareto-Smale optimum. - In: Optimization, Novosibirsk, 1982. (To be publish in Selecta Mathematica Sovietica).

7. Chaplygin S.A. Studies on Dynamics of Non-Holonomic Systems, Moscow, 1949 (in Russian).

8. Dobronravov V.V. Foundations of Mechanics of Non-Holonomic Systems, Moscow, 1970 (in Russian).

9. Neimark I.I., Fufaev N.A. Dynamics of Non-Holonomic Systems, Moscow, 1967 (in Russian).

10. Gohman A.V. Differential-Geometric Foundations of the Classical Dynamics of Systems, Saratov, 1969 (in Russian).

11. Ter-Krikorov A.M. Optimal Control and Mathematical Economics, Moscow, 1977 (in Russian).

12. Novikov S.P. Variational methods and periodic solutions of the Kirchhoff type equations. - Funct. Anal. Appl. 15:4 (1981).

13. Kobayashi S., Nomizu K. Foundations of Differential Geometry, vol. 1, Interscience, New York, 1963.

14. Busemann H. The Geometry of Geodesics, Academic Press, New York, 1955.

15. Manakov S.V. Remarks on integrating the Euler equations of dynamics of n-dimensional rigid body. - Funct. Anal. Appl. 10:4 (1976)

16. Mistchenko A.S., Fomenko A.T. Generalized Liouville's method of integrating Hamiltonian systems. - Funct. Anal. Appl. 12:2 (1978).

17. Alekseev V.M., Tihomirov V.M., Fomin S.V. Optimal Control, Moscow, 1981 (in Russian).

18. Bishop R.L., Crittenden R.J. Geometry of Manofolds, Academic Press, New York, 1964.

19. Stenberg S. Lectures on differental geometry. N.J., 1964.

20. Perelomov A. Integrable systems of classical mechanic and Lie algebras. Systems with constraints. ITEP - 116, preprint, Moscow, 1983 (in Russian).

21. Gershkovich V. Twoside estimations of a metric which generated by absolutely nonholonomic distribution on a Riemanian manifold. Soviet Doklady, 1984.

22. Vershik A., Chernjakov A. Critical points of fields of convex polytopes and the Pareto-Smale optimum with respect to a convex cone. Soviet Math. Dokl. Vol. 26 (1982), No.2.

23. Gliklikh Yu. Riemanian parallel translation in nonlinear mechanics. See this volume.

24. Smirnov V. A course of higher mathematics. Vol. IV, p.1, Moscow, 1974 (in Russian).

Vol. 952: Combinatorial Mathematics IX. Proceedings, 1981. Edited by E. Billington, S. Oates-Williams, and A.P. Street. XI, 443 pages. 1982.

Vol. 953: Iterative Solution of Nonlinear Systems of Equations. Proceedings, 1982. Edited by R. Ansorge, Th. Meis, and W. Törnig. VII, 202 pages. 1982.

Vol. 954: S.G. Pandit, S.G. Deo, Differential Systems Involving Impulses. VII, 102 pages. 1982.

Vol. 955: G. Gierz, Bundles of Topological Vector Spaces and Their Duality. IV, 296 pages. 1982.

Vol. 956: Group Actions and Vector Fields. Proceedings, 1981. Edited by J.B. Carrell. V, 144 pages. 1982.

Vol. 957: Differential Equations. Proceedings, 1981. Edited by D.G. de Figueiredo. VIII, 301 pages. 1982.

Vol. 958: F.R. Beyl, J. Tappe, Group Extensions, Representations, and the Schur Multiplicator. IV, 278 pages. 1982.

Vol. 959: Géométrie Algébrique Réelle et Formes Quadratiques, Proceedings, 1981. Edité par J.-L. Colliot-Thélène, M. Coste, L. Mahé, et M.-F. Roy. X, 458 pages. 1982.

Vol. 960: Multigrid Methods. Proceedings, 1981. Edited by W. Hackbusch and U. Trottenberg. VII, 652 pages. 1982.

Vol. 961: Algebraic Geometry. Proceedings, 1981. Edited by J.M. Aroca, R. Buchweitz, M. Giusti, and M. Merle. X, 500 pages. 1982.

Vol. 962: Category Theory. Proceedings, 1981. Edited by K.H. Kamps D. Pumplün, and W. Tholen, XV, 322 pages. 1982.

Vol. 963: R. Nottrot, Optimal Processes on Manifolds. VI, 124 pages. 1982.

Vol. 964: Ordinary and Partial Differential Equations. Proceedings, 1982. Edited by W.N. Everitt and B.D. Sleeman. XVIII, 726 pages. 1982.

Vol. 965: Topics in Numerical Analysis. Proceedings, 1981. Edited by P.R. Turner. IX, 202 pages. 1982.

Vol. 966: Algebraic K-Theory. Proceedings, 1980, Part I. Edited by R.K. Dennis. VIII, 407 pages. 1982.

Vol. 967: Algebraic K-Theory. Proceedings, 1980. Part II. VIII, 409 pages. 1982.

Vol. 968: Numerical Integration of Differential Equations and Large Linear Systems. Proceedings, 1980. Edited by J. Hinze. VI, 412 pages. 1982.

Vol. 969: Combinatorial Theory. Proceedings, 1982. Edited by D. Jungnickel and K. Vedder. V, 326 pages. 1982.

Vol. 970: Twistor Geometry and Non-Linear Systems. Proceedings, 1980. Edited by H.-D. Doebner and T.D. Palev. V, 216 pages. 1982.

Vol. 971: Kleinian Groups and Related Topics. Proceedings, 1981. Edited by D.M. Gallo and R.M. Porter. V, 117 pages. 1983.

Vol. 972: Nonlinear Filtering and Stochastic Control. Proceedings, 1981. Edited by S.K. Mitter and A. Moro. VIII, 297 pages. 1983.

Vol. 973: Matrix Pencils. Proceedings, 1982. Edited by B. Kågström and A. Ruhe. XI, 293 pages. 1983.

Vol. 974: A. Draux, Polynômes Orthogonaux Formels – Applications. VI, 625 pages. 1983.

Vol. 975: Radical Banach Algebras and Automatic Continuity. Proceedings, 1981. Edited by J.M. Bachar, W.G. Bade, P.C. Curtis Jr., H.G. Dales and M.P. Thomas. VIII, 470 pages. 1983.

Vol. 976: X. Fernique, P.W. Millar, D.W. Stroock, M. Weber, Ecole d'Eté de Probabilités de Saint-Flour XI – 1981. Edited by P.L. Hennequin. XI, 465 pages. 1983.

Vol. 977: T. Parthasarathy, On Global Univalence Theorems. VIII, 106 pages. 1983.

Vol. 978: J. Ławrynowicz, J. Krzyż, Quasiconformal Mappings in the Plane. VI, 177 pages. 1983.

Vol. 979: Mathematical Theories of Optimization. Proceedings, 1981. Edited by J.P. Cecconi and T. Zolezzi. V, 268 pages. 1983.

Vol. 980: L. Breen. Fonctions thêta et théorème du cube. XIII, 115 pages. 1983.

Vol. 981: Value Distribution Theory. Proceedings, 1981. Edited by I. Laine and S. Rickman. VIII, 245 pages. 1983.

Vol. 982: Stability Problems for Stochastic Models. Proceedings, 1982. Edited by V.V. Kalashnikov and V.M. Zolotarev. XVII, 295 pages. 1983.

Vol. 983: Nonstandard Analysis-Recent Developments. Edited by A.E. Hurd. V, 213 pages. 1983.

Vol. 984: A. Bove, J.E. Lewis, C. Parenti, Propagation of Singularities for Fuchsian Operators. IV, 161 pages. 1983.

Vol. 985: Asymptotic Analysis II. Edited by F. Verhulst. VI, 497 pages. 1983.

Vol. 986: Séminaire de Probabilités XVII 1981/82. Proceedings. Edited by J. Azéma and M. Yor. V, 512 pages. 1983.

Vol. 987: C.J. Bushnell, A. Fröhlich, Gauss Sums and p-adic Division Algebras. XI, 187 pages. 1983.

Vol. 988: J. Schwermer, Kohomologie arithmetisch definierter Gruppen und Eisensteinreihen. III, 170 pages. 1983.

Vol. 989: A.B. Mingarelli, Volterra-Stieltjes Integral Equations and Generalized Ordinary Differential Expressions. XIV, 318 pages. 1983.

Vol. 990: Probability in Banach Spaces IV. Proceedings, 1982. Edited by A. Beck and K. Jacobs. V, 234 pages. 1983.

Vol. 991: Banach Space Theory and its Applications. Proceedings, 1981. Edited by A. Pietsch, N. Popa and I. Singer. X, 302 pages. 1983.

Vol. 992: Harmonic Analysis, Proceedings, 1982. Edited by G. Mauceri, F. Ricci and G. Weiss. X, 449 pages. 1983.

Vol. 993: R.D. Bourgin, Geometric Aspects of Convex Sets with the Radon-Nikodým Property. XII, 474 pages. 1983.

Vol. 994: J.-L. Journé, Calderón-Zygmund Operators, Pseudo-Differential Operators and the Cauchy Integral of Calderón. VI, 129 pages. 1983.

Vol. 995: Banach Spaces, Harmonic Analysis, and Probability Theory. Proceedings, 1980–1981. Edited by R.C. Blei and S.J. Sidney. V, 173 pages. 1983.

Vol. 996: Invariant Theory. Proceedings, 1982. Edited by F. Gherardelli. V, 159 pages. 1983.

Vol. 997: Algebraic Geometry – Open Problems. Edited by C. Ciliberto, F. Ghione and F. Orecchia. VIII, 411 pages. 1983.

Vol. 998: Recent Developments in the Algebraic, Analytical, and Topological Theory of Semigroups. Proceedings, 1981. Edited by K.H. Hofmann, H. Jürgensen and H. J. Weinert. VI, 486 pages. 1983.

Vol. 999: C. Preston, Iterates of Maps on an Interval. VII, 205 pages. 1983.

Vol. 1000: H. Hopf, Differential Geometry in the Large, VII, 184 pages. 1983.

Vol. 1001: D.A. Hejhal, The Selberg Trace Formula for PSL(2, ℝ). Volume 2. VIII, 806 pages. 1983.

Vol. 1002: A. Edrei, E.B. Saff, R.S. Varga, Zeros of Sections of Power Series. VIII, 115 pages. 1983.

Vol. 1003: J. Schmets, Spaces of Vector-Valued Continuous Functions. VI, 117 pages. 1983.

Vol. 1004: Universal Algebra and Lattice Theory. Proceedings, 1982. Edited by R.S. Freese and O.C. Garcia. VI, 308 pages. 1983.

Vol. 1005: Numerical Methods. Proceedings, 1982. Edited by V. Pereyra and A. Reinoza. V, 296 pages. 1983.

Vol. 1006: Abelian Group Theory. Proceedings, 1982/83. Edited by R. Göbel, L. Lady and A. Mader. XVI, 771 pages. 1983.

Vol. 1007: Geometric Dynamics. Proceedings, 1981. Edited by J. Palis Jr. IX, 827 pages. 1983.

Vol. 1008: Algebraic Geometry. Proceedings, 1981. Edited by J. Dolgachev. V, 138 pages. 1983.

Vol. 1009: T.A. Chapman, Controlled Simple Homotopy Theory and Applications. III, 94 pages. 1983.

Vol. 1010: J.-E. Dies, Chaînes de Markov sur les permutations. IX, 226 pages. 1983.

Vol. 1011: J.M. Sigal. Scattering Theory for Many-Body Quantum Mechanical Systems. IV, 132 pages. 1983.

Vol. 1012: S. Kantorovitz, Spectral Theory of Banach Space Operators. V, 179 pages. 1983.

Vol. 1013: Complex Analysis - Fifth Romanian-Finnish Seminar. Part 1. Proceedings, 1981. Edited by C. Andreian Cazacu, N. Boboc, M. Jurchescu and I. Suciu. XX, 393 pages. 1983.

Vol. 1014: Complex Analysis - Fifth Romanian-Finnish Seminar. Part 2. Proceedings, 1981. Edited by C. Andreian Cazacu, N. Boboc, M. Jurchescu and I. Suciu. XX, 334 pages. 1983.

Vol. 1015: Equations différentielles et systèmes de Pfaff dans le champ complexe - II. Seminar. Edited by R. Gérard et J.P. Ramis. V, 411 pages. 1983.

Vol. 1016: Algebraic Geometry. Proceedings, 1982. Edited by M. Raynaud and T. Shioda. VIII, 528 pages. 1983.

Vol. 1017: Equadiff 82. Proceedings, 1982. Edited by H.W. Knobloch and K. Schmitt. XXIII, 666 pages. 1983.

Vol. 1018: Graph Theory, Łagów 1981. Proceedings, 1981. Edited by M. Borowiecki, J.W. Kennedy and M.M. Sysło. X, 289 pages. 1983.

Vol. 1019: Cabal Seminar 79-81. Proceedings, 1979-81. Edited by A.S. Kechris, D.A. Martin and Y.N. Moschovakis. V, 284 pages. 1983.

Vol. 1020: Non Commutative Harmonic Analysis and Lie Groups. Proceedings, 1982. Edited by J. Carmona and M. Vergne. V, 187 pages. 1983.

Vol. 1021: Probability Theory and Mathematical Statistics. Proceedings, 1982. Edited by K. Itô and J.V. Prokhorov. VIII, 747 pages. 1983.

Vol. 1022: G. Gentili, S. Salamon and J.-P. Vigué. Geometry Seminar "Luigi Bianchi", 1982. Edited by E. Vesentini. VI, 177 pages. 1983.

Vol. 1023: S. McAdam, Asymptotic Prime Divisors. IX, 118 pages. 1983.

Vol. 1024: Lie Group Representations I. Proceedings, 1982-1983. Edited by R. Herb, R. Lipsman and J. Rosenberg. IX, 369 pages. 1983.

Vol. 1025: D. Tanré, Homotopie Rationnelle: Modèles de Chen, Quillen, Sullivan. X, 211 pages. 1983.

Vol. 1026: W. Plesken, Group Rings of Finite Groups Over p-adic Integers. V, 151 pages. 1983.

Vol. 1027: M. Hasumi, Hardy Classes on Infinitely Connected Riemann Surfaces. XII, 280 pages. 1983.

Vol. 1028: Séminaire d'Analyse P. Lelong - P. Dolbeault - H. Skoda. Années 1981/1983. Edité par P. Lelong, P. Dolbeault et H. Skoda. VIII, 328 pages. 1983.

Vol. 1029: Séminaire d'Algèbre Paul Dubreil et Marie-Paule Malliavin. Proceedings, 1982. Edité par M.-P. Malliavin. V, 339 pages. 1983.

Vol. 1030: U. Christian, Selberg's Zeta-, L-, and Eisensteinseries. XII, 196 pages. 1983.

Vol. 1031: Dynamics and Processes. Proceedings, 1981. Edited by Ph. Blanchard and L. Streit. IX, 213 pages. 1983.

Vol. 1032: Ordinary Differential Equations and Operators. Proceedings, 1982. Edited by W.N. Everitt and R.T. Lewis. XV, 521 pages. 1983.

Vol. 1033: Measure Theory and its Applications. Proceedings, 1982. Edited by J.M. Belley, J. Dubois and P. Morales. XV, 317 pages. 1983.

Vol. 1034: J. Musielak, Orlicz Spaces and Modular Spaces. V, 222 pages. 1983.

Vol. 1035: The Mathematics and Physics of Disordered Media. Proceedings, 1983. Edited by B.D. Hughes and B.W. Ninham. VII, 432 pages. 1983.

Vol. 1036: Combinatorial Mathematics X. Proceedings, 1982. Edited by L.R.A. Casse. XI, 419 pages. 1983.

Vol. 1037: Non-linear Partial Differential Operators and Quantization Procedures. Proceedings, 1981. Edited by S.I. Andersson and H.-D. Doebner. VII, 334 pages. 1983.

Vol. 1038: F. Borceux, G. Van den Bossche, Algebra in a Locali Topos with Applications to Ring Theory. IX, 240 pages. 1983.

Vol. 1039: Analytic Functions, Błażejewko 1982. Proceedings. Edited by J. Ławrynowicz. X, 494 pages. 1983

Vol. 1040: A. Good, Local Analysis of Selberg's Trace Formula. I, 128 pages. 1983.

Vol. 1041: Lie Group Representations II. Proceedings 1982-198 Edited by R. Herb, S. Kudla, R. Lipsman and J. Rosenberg. I 340 pages. 1984.

Vol. 1042: A. Gut, K.D. Schmidt, Amarts and Set Function Processe III, 258 pages. 1983.

Vol. 1043: Linear and Complex Analysis Problem Book. Edited b V.P. Havin, S.V. Hruščëv and N.K. Nikol'skii. XVIII, 721 pages. 198

Vol. 1044: E. Gekeler, Discretization Methods for Stable Initial Valu Problems. VIII, 201 pages. 1984.

Vol. 1045: Differential Geometry. Proceedings, 1982. Edited b A.M. Naveira. VIII, 194 pages. 1984.

Vol. 1046: Algebraic K-Theory, Number Theory, Geometry and Analysis. Proceedings, 1982. Edited by A. Bak. IX, 464 pages. 1984.

Vol. 1047: Fluid Dynamics. Seminar, 1982. Edited by H. Beirão Veiga. VII, 193 pages. 1984.

Vol. 1048: Kinetic Theories and the Boltzmann Equation. Semin 1981. Edited by C. Cercignani. VII, 248 pages. 1984.

Vol. 1049: B. Iochum, Cônes autopolaires et algèbres de Jord VI, 247 pages. 1984.

Vol. 1050: A. Prestel, P. Roquette, Formally p-adic Fields. V, 167 pag 1984.

Vol. 1051: Algebraic Topology, Aarhus 1982. Proceedings. Edited I. Madsen and B. Oliver. X, 665 pages. 1984.

Vol. 1052: Number Theory. Seminar, 1982. Edited by D.V. Ch novsky, G.V. Chudnovsky, H. Cohn and M.B. Nathanson. V, 3 pages. 1984.

Vol. 1053: P. Hilton, Nilpotente Gruppen und nilpotente Räu V, 221 pages. 1984.

Vol. 1054: V. Thomée, Galerkin Finite Element Methods for Parab Problems. VII, 237 pages. 1984.

Vol. 1055: Quantum Probability and Applications to the Quan Theory of Irreversible Processes. Proceedings, 1982. Edited L. Accardi, A. Frigerio and V. Gorini. VI, 411 pages. 1984.

Vol. 1056: Algebraic Geometry. Bucharest 1982. Proceedings, 19 Edited by L. Bădescu and D. Popescu. VII, 380 pages. 1984.

Vol. 1057: Bifurcation Theory and Applications. Seminar, 1983. Ed by L. Salvadori. VII, 233 pages. 1984.

Vol. 1058: B. Aulbach, Continuous and Discrete Dynamics Manifolds of Equilibria. IX, 142 pages. 1984.

Vol. 1059: Séminaire de Probabilités XVIII, 1982/83. Proceedi Edité par J. Azéma et M. Yor. IV, 518 pages. 1984.

Vol. 1060: Topology. Proceedings, 1982. Edited by L.D. Faddeev A.A. Mal'cev. VI, 389 pages. 1984.

Vol. 1061: Séminaire de Théorie du Potentiel. Paris, No. 7. ceedings. Directeurs: M. Brelot, G. Choquet et J. Deny. Rédact F. Hirsch et G. Mokobodzki. IV, 281 pages. 1984.